de Gruyter Expositions in Mathematics 43

Modules over Discrete Valuation Domains

by

Piotr A. Krylov and Askar A. Tuganbaev

Walter de Gruyter · Berlin · New York

Authors

Piotr A. Krylov
Tomsk State University
pr. Lenina, 36
634050 Tomsk
Russian Federation

E-Mail: krylov@math.tsu.ru

Askar A. Tuganbaev
Russian State University of Trade
and Economics
Smolnaya st., 36
125993 Moscow
Russian Federation

E-Mail: tuganbaev@gmail.com

Mathematics Subject Classification 2000: First: 16-02, Second: 03C60, 03Exx, 13-XX, 16-XX, 20Kxx

Key words: Algebra, module theory, discrete valuation domain

∞ Printed on acid-free paper which falls within the guidelines
of the ANSI to ensure permanence and durability.

ISSN 0938-6572

ISBN 978-3-11-020053-9

Bibliographic information published by the Deutsche Nationalbibliothek

The Deutsche Nationalbibliothek lists this publication in the Deutsche Nationalbibliografie; detailed bibliographic data are available in the Internet at http://dnb.d-nb.de.

Typeset using the authors' Tex-files: Kay Dimler, Müncheberg.
Printing and binding: Hubert & Co. GmbH & Co. KG, Göttingen.
Cover design: Thomas Bonnie, Hamburg.

Contents

Introduction

There are sufficiently many books about modules over arbitrary rings (many of them are included in the bibliography). At the same time, books on modules over some specific rings are in short supply. Modules over discrete valuation domains certainly call for a special consideration, since these modules have specific properties and play an important role in various areas of algebra (especially of commutative algebra).

This book is the study of modules over discrete valuation domains. It is intended to be a first systematic account on modules over discrete valuation domains. In every part of mathematics, it is desirable to have many interesting open problems and a certain number of nice theorems. The theory presented in the book completely satisfies these conditions.

Discrete valuation domains form the class of such local domains which are very close to division rings. However, it is convenient for us to choose such a definition of a discrete valuation domain under which a division ring is not a discrete valuation domain. A discrete valuation domain is a principal ideal domain with unique (up to an invertible factor) prime element. In the theory of modules over discrete valuation domains, the role of prime elements is very important, and the nature of various constructions related to prime elements is clearly visible. Among discrete valuation domains, complete (in the p-adic topology) discrete valuation domains stand out. Typical examples of such domains are rings of p-adic integers and formal power series rings over division rings.

It is well known that all localizations of (commutative) Dedekind domains with respect to maximal ideals are discrete valuation domains. It follows from the general localization principle that it is sufficient to study many problems of the theory of modules over Dedekind domains in the case of modules over discrete valuation domains. For example, primary modules over a Dedekind domain coincide with primary modules over localizations of this ring with respect to maximal ideals.

It is necessary to emphasize close various interrelations between the theory of modules over discrete valuation domains and the theory of Abelian groups. These theories have many points of contact. This is a partial case of the principle of localization of problems, since Abelian groups coincide with modules over the ring of integers $\mathbb{Z}$ which is a commutative principal ideal domain. This implies that the theories have close ideas, methods, results, and lines of researches.

In many areas of the theory of Abelian groups, we deal with so-called *p-local* groups (i.e., modules over the localization $\mathbb{Z}_p$ of the ring $\mathbb{Z}$ with respect to the ideal generated by the prime integer p). The ring $\mathbb{Z}_p$ is a subring in the field $\mathbb{Q}$ of rational numbers; $\mathbb{Z}_p$ consists of rational numbers whose denominators are not divisible by p, and p-local groups coincide with Abelian groups G such that $G = qG$ for all prime integers $q \neq p$. With slight changes, many results on p-local groups and their proofs remain true for modules over arbitrary discrete valuation domains. In the theory of Abelian groups, modules over the ring $\widehat{\mathbb{Z}}_p$ of p-adic integers are very useful (such modules are called *p-adic modules*). The ring $\widehat{\mathbb{Z}}_p$ is the completion in the p-adic topology of the ring $\mathbb{Z}$ and the ring $\mathbb{Z}_p$. In addition, $\widehat{\mathbb{Z}}_p$ is a complete discrete valuation domain.

One of central positions in the theory of Abelian groups is occupied by p-groups, which are also called primary groups. It is appropriate to say that all modules over a fixed discrete valuation domain can be partitioned into three classes: primary modules, torsion-free modules, and mixed modules. Abelian p-groups coincide with primary $\mathbb{Z}_p$-modules as well as primary $\widehat{\mathbb{Z}}_p$-modules. We note that primary modules over discrete valuation domains are essentially presented in the literature by the theory of Abelian p-groups. With a suitable correction, main definitions, methods, and results of this theory can be transferred to primary modules. We almost are not involved in this process, since that the theory of Abelian p-groups is extensively presented in the books of Fuchs [93] and Griffith [126].

In the Kaplansky's book [166], several important theorems on modules over discrete valuation domains were included for the first time. There are three more familiar books which have appreciably affected formation of the theory of modules over discrete valuation domains. These books are the books of Baer [28] and Fuchs [92, 93]. Some topics related to the theory of endomorphism rings have their origin in the Baer's book, where the theory is developed for vector spaces (e.g., see the studies in Section 15 and Chapter 7). In the light of what has been said on the theory of Abelian groups, the reference to books of Fuchs is natural. Chapters 4, 7, and 8 of our book are related to the books of Krylov–Mikhalev–Tuganbaev [183] and Göbel–Trlifaj [109].

All main areas of the theory of modules over discrete valuation domains are presented in the book. The authors try to present main ideas, methods, and theorems which can form a basis of studies in the theory of modules over discrete valuation domains, as well as over some other rings. Some of the items presented in the book are also included in the papers [185] and [186].

Properties of vector spaces over division rings and their linear operators are assumed to be familiar; we use them without special remarks.

In comments at the end of chapters, we present some results not included in the main text and short historical remarks; we also outline other areas of studies and call attention to the literature for further examination of the field. Similar remarks are also presented in some sections. This will help to the reader to pass to the study of journal papers.

In the beginning of every chapter, we outline the content of the chapter. All sections contain exercises beginning with Section 2. Some exercises contain results from various papers. We present 34 open problems which seem to be interesting. The bibliography is quite complete, although it is possible that we did not consider some papers.

To work with the book, the reader needs to know basic results of the general theory of rings and modules. We also use certain some topological and category-theoretical ideas.

The authors assume that the book is useful to young researchers as well as experienced specialists. The book can be recommended to students and graduates studying algebra. We accept the Zermelo–Fraenkel axiomatic system from the set theory (including the choice axiom and the Zorn lemma which is equivalent to the choice axiom). The terms ‘class’ and ‘set’ are used in the ordinary set-theoretical sense. The end of the proof of some assertion is denoted by the symbol □.

The first chapter is auxiliary. In Chapter 2, we present foundations of the theory of modules over discrete valuation domains. Chapter 3 is devoted to some questions about endomorphism rings of divisible primary modules and complete torsion-free modules. In Chapter 4, we study the problem of existence of an isomorphism from an abstract ring onto the endomorphism ring of some module. In Chapter 5, torsion-free modules are studied. In Chapter 6, mixed modules are studied. In Chapter 7, we analyze the possibility of an isomorphism of two modules with isomorphic endomorphism rings. In Chapter 8, we consider several questions on transitive or fully transitive modules.

Chapter 1

Preliminaries

In Chapter 1, we consider the following topics:

- some definitions and notation (Section 1);
- endomorphisms and homomorphisms of modules (Section 2);
- discrete valuation domains (Section 3);
- primary notions of the theory of modules (Section 4).

The first two sections contain some necessary standard information about modules. Some notation and terms are presented. In Section 2, we also consider the endomorphism ring of the module which is one of important objects of the study in the book. The material of these sections is included in the book for convenience. In Section 3 discrete valuation domains are defined and their main properties are studied. In Section 4, we lay the foundation of the theory of modules over discrete valuation domains.

1 Some definitions and notation

We assume that the reader is familiar with basic notions of the theory of rings and modules such that a ring, a module, a subring, an ideal, a submodule, the factor ring, the factor module, a homomorphism, and other notions. In the text we permanently use various elementary results on rings and modules (such as isomorphism theorems), basic properties and several constructions of rings and modules (e.g., direct sums), and some standard methods of the work with these objects. It is impossible to list all these properties. In any case, for reading the book, it is sufficient to know the theory of rings and modules within one of the three following books: F. Anderson and K. Fuller "Rings and categories of modules" ([1]), I. Lambek "Rings and modules" ([198]), F. Kasch "Modules and rings" ([170]). The considered (quite simple) category properties, can be also found in the book of S. MacLane [217]. Sometimes we touch on several aspects of the theory of topological rings and modules (e.g., see the book [2] of V. Arnautov, S. Glavatsky, and A. Mikhalev [2]). The two-volume monograph of L. Fuchs [92, 93] is an acknowledged manual in the theory of Abelian groups.

We do present here some material related to terminology and notation. Other required definitions and results will be presented when needed.

All rings considered in the book are associative. By definition, all rings have identity elements (the only exclusions are Sections 17 and 18). The identity element is preserved under homomorphisms; it is contained in subrings and the action of the identity element on modules is the identity mapping (i.e., modules are unitary).

When we speak about an order in the set of right, left, or two-sided ideals (two-sided ideals are also called ideals), we always assume the order with respect to set-theoretical inclusion. Minimality and maximality of ideals are considered with respect to this order. Each of the three sets of ideals of the given ring forms a partially ordered set. In fact, we have three lattices. For two right (left or two-sided) ideals A and B of the ring R, the least upper bound is the intersection $A \cap B$ and the greatest lower bound is the sum $A + B$, where $A + B = \{a + b \mid a \in A,\ b \in B\}$.

If R and S are two rings and $\varphi\colon R \to S$ is a ring homomorphism, then $\mathrm{Im}(\varphi)$ and $\mathrm{Ker}(\varphi)$ denote the image and the kernel of the homomorphism φ, respectively. The ring of all $n \times n$ matrices over the ring S with $n > 1$ is denoted by S_n.

We specialize some details related to interrelations between decompositions of rings and idempotents. A subset $\{e_1, \ldots, e_n\}$ of the ring R is called a *complete orthogonal system of idempotents* if $e_i^2 = e_i$, $e_i e_j = 0$ for $i \neq j$, and $\sum_{i=1}^{n} e_i = 1$.

For a complete orthogonal system of idempotents, we have the Pierce decomposition of the ring into a direct sum of right ideals:

$$\{e_1, \ldots, e_n\} \to R = e_1 R \oplus \cdots \oplus e_n R.$$

In addition, the right ideal $e_i R$ is not decomposable into a direct sum of right ideals if and only if the idempotent e_i is *primitive*. This means that every relation $e_i = f + g$, where f and g are orthogonal idempotents, implies that either $e_i = f$ or $e_i = g$. If we additionally assume that the idempotents e_i are *central* (therefore, they are contained in the center of the ring R), then we obtain a decomposition of the ring R into a direct sum of two-sided ideals $e_i R$. In this case, the ideal $e_i R$ is a ring with identity element e_i and the direct sum $e_1 R \oplus \cdots \oplus e_n R$ can be identified with the product of the rings $e_1 R, \ldots, e_n R$. The *direct product* of some family of rings $R_1, \ldots, R_n$ is denoted by

$$R_1 \times \cdots \times R_n \quad \text{or} \quad \prod_{i=1}^{n} R_i$$

(we can also write $R_1 \oplus \cdots \oplus R_n$ or $\bigoplus_{i=1}^{n} R_i$).

For a ring R, an element r of R is called a *non-zero-divisor* if $sr \neq 0$ and $rs \neq 0$ for every $0 \neq s \in R$. Otherwise, r is called a *zero-divisor*. Let R be a subring of the ring S. The ring S is called the *right classical ring of fractions* of the ring R if the following conditions hold:

(1) All non-zero-divisors of the ring R are invertible in the ring S.

(2) All elements of the ring S have the form ab^{-1}, where $a, b \in R$ and b is a non-zero-divisor of the ring R.

We say that a ring R satisfies the *right Ore condition* if for any element $a \in R$ and each non-zero-divisor $b \in R$, there exist elements $a', b' \in R$ such that b' is a non-zero-divisor and $ab' = ba'$.

It is well known that the ring R has the right classical ring of fractions if and only if it satisfies the right Ore condition. We obtain that the right classical ring of fractions of a domain with the right Ore condition is a division ring. The left classical ring of fractions and the left Ore condition are defined similarly.

Unless otherwise stated, we usually consider left modules; it was mentioned above that the modules are unitary. Similar to ideals of rings, submodules of the given module form a partially ordered set with respect to inclusion, and we also have the lattice consisting of all submodules.

We use the same notation $\mathrm{Ker}(\varphi)$ and $\mathrm{Im}(\varphi)$ for the kernel and the image of the module homomorphism φ. If $\varphi\colon M \to N$ is a homomorphism of modules and A is a submodule of the module M, then $\varphi_{|A}$ is the *restriction* of φ to A. The restriction to A of the identity mapping of the module M is called the *embedding* from the submodule A in M.

The *direct sum* of modules A_i, $i \in I$, where I is some subscript set, is denoted by $\bigoplus_{i \in I} A_i$ or $A_1 \oplus \cdots \oplus A_n$ provided $I = \{1, 2, \ldots, n\}$. We assume that we have the direct sum $M = \bigoplus_{i \in I} A_i$. For every subscript $i \in I$, we have the coordinate embedding $\varkappa_i\colon A_i \to M$ and the coordinate projection $\pi_i\colon M \to A_i$ (details on direct sums are presented in Section 2). Setting $\varepsilon_i = \pi_i \varkappa_i$, we obtain an idempotent endomorphism of the module M, i.e., $\varepsilon_i^2 = \varepsilon_i$. Clearly, we can identify π_i and ε_i if it is convenient.

We assume that $M = A \oplus B$ and $x = a + b$, where $a \in A$ and $b \in B$. Then the elements a and b are the components of the element x. More generally, assume that either $M = \prod_{i \in I} A_i$ (the *direct product* of the family of modules A_i with $i \in I$) or $M = \bigoplus_{i \in I} A_i$. We write the element x of the module M either in the vector form $x = (\ldots, a_i, \ldots)$ or in the brief form $x = (a_i)$, where a_i is the *component* of the element x contained in the summand A_i (we also say the "coordinate").

If M is a module and n is a positive integer, then $\bigoplus_n M$ or M^n is the direct sum of n copies of the module M.

Let M be a left module over a ring R or a left R-module for brevity. For every subset X of the module M, we denote by RX the submodule of the module M generated by X. The submodule RX is the intersection of all submodules of the module M containing X. This is obvious that RX consists of all sums of the form $r_1x_1 + \cdots + r_nx_n$, where $r_i \in R$ and $x_i \in X$. If $RX = M$, then X is called a *generator system* of the module M. A module is said to be *finitely generated* if it has a finite generator system. We say that M is a *cyclic* module with generator x if $M = Rx$ for some $x \in M$.

By definition, the *sum* $\sum_{i \in I} A_i$ of the submodules A_i with $i \in I$ consists of the set of all sums of the form $a_{i_1} + \cdots + a_{i_n}$, where $a_{i_j} \in A_{i_j}$. This sum coincides with the submodule generated by the union of all submodules A_i.

If R and S are two rings, then an *R-S-bimodule* ${}_RM_S$ is an Abelian group M such that M is a left R-module and a right S-module and $(rx)s = r(xs)$ for all elements $r \in R$, $x \in M$, and $s \in S$. The ring R (more precisely, the additive group of R) can be naturally considered as a left R-module and a right R-module (these modules are also called regular modules). More precisely, we have an R-R-bimodule R.

We assume that R is a commutative ring. In this case, every left R-module M can be turned into a right R-module and conversely with the use of the relation $rx = xr$, where $r \in R$ and $x \in M$. We obtain the R-R-bimodule M.

In Section 2, we present several familiar properties of induced exact sequences of modules. Now we consider the following details. A short exact sequence of modules $0 \to A \xrightarrow{\varkappa} B \xrightarrow{\pi} C \to 0$ is said to be *split* if $B = \operatorname{Im}(\varkappa) \oplus C'$ for some module C' (we have $C' \cong C$). Every submodule A of the module M provides an exact sequence

$$0 \to A \xrightarrow{\varkappa} M \xrightarrow{\pi} M/A \to 0,$$

where $\varkappa$ is an embedding and π is the *canonical homomorphism* such that $x \to x + A$ for all $x \in M$.

Section 2 also contains main properties of the tensor product of modules which is often used in the book.

The theory of modules over discrete valuation domains and the theory of Abelian groups are congenial theories. Many sections of these theories are developed in similar ways. The reason is that discrete valuation domains and the ring of integers are close to each other, since Abelian groups and modules over the ring of integers are the same objects. Commutative discrete valuation domains and the ring of integers are contained in some special class of rings. They are examples of commutative principal ideal domains.

In some questions of the theory of modules over discrete valuation domains, categories and the category language are very useful. We give the definition of a category and consider some important related notions. An additional information about categories is considered in Sections 24 and 27. In Sections 24 and 29, we define the following three categories with a module origin: the category of quasi-homomorphisms, the Walker category, and the Warfield category. These categories will be essentially used later.

A class $\mathcal{E}$ of objects $A, B, C, \dots$ is called a *category* if for any two objects $A, B \in \mathcal{E}$, there is a set of morphisms $\mathrm{Hom}_{\mathcal{E}}(A, B)$ with the composition

$$\mathrm{Hom}_{\mathcal{E}}(A, B) \times \mathrm{Hom}_{\mathcal{E}}(B, C) \to \mathrm{Hom}_{\mathcal{E}}(A, C)$$

such that the following two assertions hold.

(1) The composition is associative.

(2) For every object $A \in \mathcal{E}$, there exists a morphism $1_A \in \mathrm{Hom}_{\mathcal{E}}(A, A)$ such that $1_A f = f$ and $g 1_A = g$ every time, when $f \in \mathrm{Hom}_{\mathcal{E}}(A, B)$ and $g \in \mathrm{Hom}_{\mathcal{E}}(B, A)$.

The morphism 1_A is called the *identity morphism* of the object A.

The category $\mathcal{E}$ is said to be *additive* if the following conditions (3) and (4) hold.

(3) For any two objects $A, B \in \mathcal{E}$, the set $\mathrm{Hom}_{\mathcal{E}}(A, B)$ is an Abelian group and the composition of morphisms is bilinear, i.e.,

$$g(f_1 + f_2) = gf_1 + gf_2 \quad \text{and} \quad (f_1 + f_2)h = f_1 h + f_2 h$$

for all

$$g \in \mathrm{Hom}_{\mathcal{E}}(C, A), \quad f_i \in \mathrm{Hom}_{\mathcal{E}}(A, B), \quad \text{and} \quad h \in \mathrm{Hom}_{\mathcal{E}}(B, D).$$

(4) There exist finite direct sums in $\mathcal{E}$. This means that for given objects

$$A_1, \dots, A_n \in \mathcal{E},$$

there exist an object $A \in \mathcal{E}$ and morphisms $e_i \in \mathrm{Hom}_{\mathcal{E}}(A_i, A)$ such that if $f_i \in \mathrm{Hom}_{\mathcal{E}}(A_i, B)$ ($i = 1, \dots, n$), then there exists the unique morphism $f \in \mathrm{Hom}_{\mathcal{E}}(A, B)$ such that $e_i f = f_i$ for all $i = 1, \dots, n$.

The object A is called the *direct sum* of objects $A_1, \dots, A_n$ and the morphisms $e_1, \dots, e_n$ are called *embeddings*. In this case, we write $A = A_1 \oplus \dots \oplus A_n$ and also say that there exists a *direct decomposition* of the object A.

In the additive category $\mathcal{E}$, the set $\mathrm{Hom}_{\mathcal{E}}(A, A)$ is a ring with identity element 1_A, which is called the *endomorphism ring* of the object A; it is denoted by $\mathrm{End}_{\mathcal{E}}(A)$.

For two categories $\mathcal{C}$ and $\mathcal{E}$, the category $\mathcal{C}$ is called a *subcategory* of $\mathcal{E}$ if $\mathcal{C}$ satisfies the following conditions:

(1) All objects of the category $\mathcal{C}$ are objects of the category $\mathcal{E}$.

(2) $\mathrm{Hom}_{\mathcal{C}}(A, B) \subseteq \mathrm{Hom}_{\mathcal{E}}(A, B)$ for any two objects $A, B \in \mathcal{C}$.

(3) The composition of any morphisms in $\mathcal{C}$ is induced by their composition in $\mathcal{E}$.

(4) All identity morphisms in $\mathcal{C}$ are identity morphisms in $\mathcal{E}$.

A subcategory $\mathcal{C}$ of the category $\mathcal{E}$ is said to be *full* if $\mathrm{Hom}_{\mathcal{C}}(A, B) = \mathrm{Hom}_{\mathcal{E}}(A, B)$ for any two objects $A, B \in \mathcal{C}$.

A morphism $f \in \mathrm{Hom}_{\mathcal{E}}(A, B)$ is called an *isomorphism* if there exists a morphism $g \in \mathrm{Hom}_{\mathcal{E}}(B, A)$ such that $fg = 1_A$ and $gf = 1_B$. In this case, we say that the objects A and B are isomorphic to each other in the category $\mathcal{E}$.

We assume that we have two categories $\mathcal{E}$ and $\mathcal{D}$. A *covariant* (resp., *contravariant*) functor $F\colon \mathcal{E} \to \mathcal{D}$ from the category $\mathcal{E}$ into the category $\mathcal{D}$ consists of the mapping $\mathcal{E} \to \mathcal{D}$, $A \to F(A)$, $A \in \mathcal{E}$, and mappings

$$\mathrm{Hom}_{\mathcal{E}}(A, B) \to \mathrm{Hom}_{\mathcal{D}}(F(A), F(B))$$

$$(\text{resp.,} \quad \mathrm{Hom}_{\mathcal{E}}(A, B) \to \mathrm{Hom}_{\mathcal{D}}(F(B), F(A))\,)$$

$(f \to F(f))$ such that they preserve the composition of morphisms and identity morphisms, i.e.,

$$F(fg) = F(f)F(g) \quad (\text{resp.,} \quad F(fg) = F(g)F(f)) \quad \text{and} \quad F(1_A) = 1_{F(A)}$$

for all objects and morphisms $A, f, g \in \mathcal{E}$.

The *identity functor* $1_{\mathcal{E}}$ of the category $\mathcal{E}$ defined by the relations

$$1_{\mathcal{E}}(A) = A \quad \text{and} \quad 1_{\mathcal{E}}(f) = f$$

for all $A, f \in \mathcal{E}$ is a covariant functor from the category $\mathcal{E}$ into $\mathcal{E}$.

Let F and G be two covariant functors from a category $\mathcal{E}$ into a category $\mathcal{D}$. A correspondence $\phi\colon F \to G$ associating a morphism $\phi_A\colon F(A) \to G(A)$ in $\mathcal{D}$ with every object $A \in \mathcal{E}$ is called a *natural transformation* $\phi\colon F \to G$ if for every morphism $f\colon A \to B$ in the category $\mathcal{E}$, we have the relation $F(f)\phi_B = \phi_A G(f)$ in the category $\mathcal{D}$. If ϕ_A is an isomorphism for every object $A \in \mathcal{E}$, then ϕ is called a *natural equivalence*. The morphism ϕ_A is called a *natural isomorphism* between $F(A)$ and $G(A)$.

One says that two categories $\mathcal{E}$ and $\mathcal{D}$ are equivalent if there exist two covariant functors

$$F\colon \mathcal{E} \to \mathcal{D} \quad \text{and} \quad G\colon \mathcal{D} \to \mathcal{E}$$

such that the functor FG (defined as the composition of the functors F and G from the left to the right) is equivalent to the identity functor $1_{\mathcal{E}}$ and the functor GF is equivalent to the identity functor $1_{\mathcal{D}}$. In this case, we say that the functors F and G define an *equivalence* of the categories $\mathcal{E}$ and $\mathcal{D}$.

In the theory of modules additive functors are usually used. A functor F is said to be *additive* if $F(f + g) = F(f) + F(g)$ for any two morphisms $f, g \in \mathcal{E}$ such that the morphism $f + g$ is defined.

2 Endomorphisms and homomorphisms of modules

We present some well known definitions and results related to endomorphism rings, group homomorphisms, and tensor products. In this section R is an arbitrary ring.

We write homomorphisms to the left side of arguments. To avoid the use of anti-isomorphic rings, we define the composition of homomorphisms as follows. Let $\alpha\colon M \to N$ and $\beta\colon N \to L$ be two homomorphisms of modules. Then the composition $\alpha\beta$ of α and β is the mapping $M \to L$ such that $(\alpha\beta)(a) = \beta(\alpha(a))$ for every $a \in M$. It is clear that the composition is a homomorphism. In some works, homomorphisms are written to the right side of the arguments, i.e., $(a)\alpha$ is used instead of $\alpha(a)$. Then for the composition $\alpha\beta$, we have

$$(a)(\alpha\beta) = ((a)\alpha)\beta, \qquad a \in M.$$

Let M and N be two R-modules. We denote by $\operatorname{Hom}_R(M, N)$ the set of all homomorphisms from the module M into the module N. (Sometimes we write "R-homomorphisms" for brevity.) The set $\operatorname{Hom}_R(M, N)$ is nonempty, since it contains the zero homomorphism $0\colon M \to N$, where $a \to 0$ for all $a \in M$. We can define the pointwise addition of homomorphisms, where

$$(\alpha + \beta)(a) = \alpha(a) + \beta(a)$$

for $\alpha, \beta \in \operatorname{Hom}_R(M, N)$ and $a \in M$. Then $\alpha + \beta$ is a homomorphism from M into N.

A homomorphism $M \to M$ is called an *endomorphism* of the module M. We set $\operatorname{End}_R(M) = \operatorname{Hom}_R(M, M)$. Endomorphisms can be multiplied, where the product $\alpha\beta$ coincides with the composition of the endomorphisms.

Endomorphisms of the module M which are bijections are called *automorphisms*. The identity mapping 1_M, where $1_M(a) = a$ for all $a \in M$, is an automorphism of the module M. Let $\operatorname{Aut}_R(M)$ be the set of all automorphisms of the module M. There exists the operation of multiplication of automorphisms in $\operatorname{Aut}_R(M)$.

Theorem 2.1. (a) *The set* $\operatorname{Hom}_R(M, N)$ *is an Abelian group with respect to addition of homomorphisms.*

(b) *The set* $\operatorname{End}_R(M)$ *is an associative ring with identity element.*

(c) *The set* $\operatorname{Aut}_R(M)$ *is a group with respect to multiplication of automorphisms. It coincides with the group of invertible elements of the ring* $\operatorname{End}_R(M)$.

Proof. (a) The commutativity and the associativity of addition homomorphisms are directly verified. The zero homomorphism is the zero element. For a homomorphism $\alpha\colon M \to N$, we define the homomorphism $-\alpha\colon M \to N$ by the relation $(-\alpha)(a) = -\alpha(a)$ for every $a \in M$. Then $\alpha + (-\alpha) = 0$, i.e., $-\alpha$ is the opposite element for α.

(b) By (a), $\mathrm{End}_R(M)$ is an Abelian group with respect to addition. The composition of mappings satisfies the associativity law. The identity element is the identity mapping 1_M. It is directly verified that two distributivity laws hold.

(c) Let γ be an automorphism of the module M. There exists the converse mapping γ^{-1}; in addition, γ^{-1} is an automorphism of the module M and $\gamma\gamma^{-1} = \gamma^{-1}\gamma = 1_M$. Considering (b), we obtain that $\mathrm{Aut}_R(M)$ is a group. It is clear that an endomorphism α of the module M is an automorphism if and only if α is an invertible element of the ring $\mathrm{End}_R(M)$. Consequently, the group $\mathrm{Aut}_R(M)$ coincides with the group of invertible elements of the ring $\mathrm{End}_R(M)$. □

The group $\mathrm{Hom}_R(M, N)$ is called the *group of homomorphisms* from the module M into the module N, the ring $\mathrm{End}_R(M)$ is called the *endomorphism ring* of the module M, and $\mathrm{Aut}_R(M)$ is called the *automorphism group* of the module M. In these three objects, the subscript R is often omitted. Homomorphism groups and endomorphism rings of right modules are defined similarly.

In some cases, the Abelian group $\mathrm{Hom}_R(M, N)$ can be considered as a module. Let S and T be additional two rings, M be an R-S-bimodule, and let N be an R-T-bimodule. If $\alpha \in \mathrm{Hom}_R(M, N)$, $s \in S$, and $t \in T$, then we set

$$(s\alpha)(a) = \alpha(as) \quad \text{and} \quad (\alpha t)(a) = \alpha(a)t$$

for every $a \in M$. Then $s\alpha, \alpha t \in \mathrm{Hom}_R(M, N)$. Using this method, we can turn the group $\mathrm{Hom}_R(M, N)$ into a left S-module and a right T-module. More precisely, we obtain an S-T-bimodule. For example, the group $\mathrm{Hom}_R(R, N)$ is a left R-module. For a commutative ring R, the group $\mathrm{Hom}_R(M, N)$ is an R-module for any two R-modules M and N (see Section 1 about bimodules).

We assume that we have R-S-bimodules M and M_i $(i \in I)$ and R-T-bimodules N and N_i $(i \in I)$. Then there are canonical S-T-bimodule isomorphisms

$$\mathrm{Hom}_R\Big(M, \prod_{i \in I} N_i\Big) \cong \prod_{i \in I} \mathrm{Hom}_R(M, N_i)$$

and

$$\mathrm{Hom}_R\Big(\bigoplus_{i \in I} M_i, N\Big) \cong \prod_{i \in I} \mathrm{Hom}_R(M_i, N).$$

Proposition 2.2. (a) *For every R-module M, we have an isomorphism $\mathrm{Hom}_R(R, M) \cong M$ of left R-modules.*

(b) *There is an isomorphism of rings $\mathrm{End}_R({}_RR) \cong R$.*

Proof. (a) We define a mapping $f\colon \operatorname{Hom}_R(R, M) \to M$ by the relation $f(\alpha) = \alpha(1)(a)$. We define a mapping $f\colon \operatorname{Hom}_R(R, M) \to M$ by the relation $f(\alpha) = \alpha(1)$ for every $\alpha\colon R \to M$. We have

$$f(r\alpha) = (r\alpha)(1) = \alpha(1 \cdot r) = r\alpha(1) = rf(\alpha).$$

Consequently, f is a homomorphism of left R-modules. Since $\alpha(r) = \alpha(r \cdot 1) = r\alpha(1)$ for every $r \in R$, we have that $\alpha = 0$ for $f(\alpha) = 0$. For $a \in M$, we set $\alpha(r) = ra$, $r \in R$. Then $\alpha \in \operatorname{Hom}_R(R, M)$ and $f(\alpha) = \alpha(1) = a$, i.e., f is an epimorphism. Consequently, f is an isomorphism. It is clear that the inverse mapping for f is the mapping $g\colon M \to \operatorname{Hom}_R(R, M)$, $g(a) = \alpha$, $a \in M$, where $\alpha(r) = ra$ for all $r \in R$.

(b) Let $f\colon \operatorname{End}_R({}_RR) \to R$ be the mapping from (a), i.e., $f(\alpha) = \alpha(1)$ for every $\alpha\colon R \to R$. Then f is an additive isomorphism. If $\alpha, \beta \in \operatorname{End}_R({}_RR)$, then

$$(\alpha\beta)(1) = \beta(\alpha(1)) = \beta(\alpha(1) \cdot 1) = \alpha(1)\beta(1).$$

Therefore $f(\alpha\beta) = f(\alpha)f(\beta)$. □

We obtain that every endomorphism of the module ${}_RR$ is the right multiplication of the ring R by some element of R.

The endomorphism ring $\operatorname{End}_R(M)$ has various relations to the initial ring R. We set $S = \operatorname{End}_R(M)$. First, the action of endomorphisms in S on the group M induces the structure of a right S-module on M (here it is more convenient to write endomorphisms to the right side of arguments). More precisely, as follows from the definition of an endomorphism, we have the R-S-bimodule ${}_RM_S$. Furthermore, let $T = \operatorname{End}_S(M)$ be the endomorphism ring of the right S-module M. We write endomorphisms of the right S-module M_S from the left side. For a fixed element $r \in R$, the mapping $h_r\colon a \to ra$, $a \in M$, is an endomorphism of the S-module M (i.e., an element of the ring T).

The correspondence $r \to h_r$ defines a ring homomorphism $R \to T$. If M is a faithful R-module (this means that the relation $rM = 0$ implies the relation $r = 0$), then we obtain an embedding $R \to T$. The ring T is called the *biendomorphism ring* of the module ${}_RA$.

Now we consider some elementary properties of the endomorphism ring. Let M be an R-module, $M = A \oplus B$ be a direct decomposition of M, and let π be the homomorphism $M \to A$ such that $\pi(a + b) = a$ for any two elements $a \in A$ and $b \in B$. The homomorphism π is called the *projection* from the module M onto the direct summand A with kernel B. We denote by $\varkappa$ the embedding $A \to M$. Then $\pi\varkappa \in \operatorname{End}(M)$ (we recall that the subscript R can be omitted) and $(\pi\varkappa)^2 = \pi\varkappa$. Therefore $\pi\varkappa$ is an idempotent of the ring $\operatorname{End}(M)$ (the idempotents of $\operatorname{End}(M)$ are called *idempotent endomorphisms* of the module M). We set

$\varepsilon = \pi\varkappa$ and identify π with ε. Consequently, we assume that the projection π is an endomorphism of the module M such that π acts on A as the identity mapping and π annihilates B. It is clear that $1-\varepsilon$ is also an idempotent of the ring $\operatorname{End}(M)$ which is orthogonal to ε. In addition, we have

$$A = \varepsilon M \quad \text{and} \quad B = (1-\varepsilon)M = \operatorname{Ker}(\varepsilon).$$

Consequently,

$$M = \varepsilon M \oplus (1-\varepsilon)M.$$

The obtained decomposition holds for every idempotent ε of the ring $\operatorname{End}(M)$. More generally, if $M = A_1 \oplus \cdots \oplus A_n$ is a direct decomposition of the module M, then we denote by ε_i the projection $M \to A_i$ with kernel $\bigoplus_{j\neq i} A_j$. Then $A_i = \varepsilon_i M$, $i = 1,\dots,n$, and $\{\varepsilon_i \mid i = 1,\dots,n\}$ is a complete orthogonal system of idempotents of the ring $\operatorname{End}(M)$. Let $\varkappa_i \colon A_i \to M$ be embeddings, $i = 1,\dots,n$. The embeddings $\varkappa_i$ and the projections ε_i are also called *coordinate embeddings* and *coordinate projections*.

Proposition 2.3. *Let M be a module and let $\{\varepsilon_i \mid i = 1,\dots,n\}$ be a complete orthogonal system of idempotents of the ring* $\operatorname{End}(M)$. *Then the correspondence*

$$M = \varepsilon_1 M \oplus \cdots \oplus \varepsilon_n M \to \operatorname{End}(M) = \operatorname{End}(M)\varepsilon_1 \oplus \cdots \oplus \operatorname{End}(M)\varepsilon_n$$

between finite direct decompositions of the module M and decompositions of the ring $\operatorname{End}(M)$ *into finite direct sums of left ideals is one-to-one.*

Proof. It has been proved that for a direct decomposition $M = A_1 \oplus \cdots \oplus A_n$, there exists a complete system $\{\varepsilon_i \mid i = 1,\dots,n\}$ of orthogonal idempotents of the ring $\operatorname{End}(M)$ such that $A_i = \varepsilon_i M$ for all i. This system induces the decomposition

$$\operatorname{End}(M) = \operatorname{End}(M)\varepsilon_1 \oplus \cdots \oplus \operatorname{End}(M)\varepsilon_n$$

of the ring $\operatorname{End}(M)$ into a direct sum of left ideals. Conversely, assume that $\operatorname{End}(M) = L_1 \oplus \cdots \oplus L_n$, where L_i are left ideals of the ring $\operatorname{End}(M)$. Then

$$1 = \varepsilon_1 + \cdots + \varepsilon_n \quad \text{with} \quad \varepsilon_i \in L_i$$

and we have a complete orthogonal system $\{\varepsilon_i \mid i = 1,\dots,n\}$ of idempotents of the ring $\operatorname{End}(M)$. It is directly verified that

$$M = \varepsilon_1 M \oplus \cdots \oplus \varepsilon_n M.$$

The considered correspondence is one-to-one. □

For direct sums with infinite number of summands and direct products of modules, coordinate embeddings and projections are defined similarly.

We consider several standard interrelations between a module and the endomorphism ring of the module related to idempotent endomorphisms. The following properties directly follow from Proposition 2.3.

(a) If ε is an idempotent of the ring $\operatorname{End}(M)$, then εM is an indecomposable direct summand of the module M if and only if ε is a primitive idempotent. The last condition is equivalent to the property that the left ideal $\operatorname{End}(M)\varepsilon$ is an indecomposable $\operatorname{End}(M)$-module.

(b) Let ε and τ be two idempotents of the ring $\operatorname{End}(M)$. There exist the canonical group isomorphism $\operatorname{Hom}(\varepsilon M, \tau M) \cong \varepsilon \operatorname{End}(M)\tau$ and the ring isomorphism $\operatorname{End}(\varepsilon M) \cong \varepsilon \operatorname{End}(M)\varepsilon$. Here we have

$$\varepsilon \operatorname{End}(M)\tau = \{\varepsilon\alpha\tau \mid \alpha \in \operatorname{End}(M)\}.$$

Let $\varphi\colon \varepsilon M \to \tau M$ be a homomorphism. Assuming that $\bar{\varphi}$ annihilates the complement $(1-\varepsilon)M$ to εM, we extend φ to an endomorphism $\bar{\varphi}$ of the module M. Using the correspondence $f\colon \varphi \to \varepsilon\bar{\varphi}\tau$, we obtain the required isomorphism

$$f\colon \operatorname{Hom}(\varepsilon M, \tau M) \to \varepsilon \operatorname{End}(M)\tau.$$

Indeed, if $\varepsilon\psi\tau \in \varepsilon \operatorname{End}(M)\tau$ for some $\psi \in \operatorname{End}(M)$, then $\varepsilon\psi\tau_{|\varepsilon M}$ is a homomorphism $\varepsilon M \to \tau M$ and $f\colon \varepsilon\psi\tau_{|\varepsilon M} \to \varepsilon\psi\tau$. If $\varepsilon = \tau$, then we have an isomorphism $\operatorname{End}(\varepsilon M) \cong \varepsilon \operatorname{End}(M)\varepsilon$ which is a ring isomorphism.

Let $M = A \oplus B$ and let $\varepsilon\colon M \to A$ be the projection with kernel B. We can assume that $\operatorname{End}(A)$ is a subring of the ring $\operatorname{End}(M)$ if we identify $\operatorname{End}(A)$ and $\varepsilon \operatorname{End}(M)\varepsilon$ using the isomorphism constructed in (b).

We consider two basic results about interrelations between isomorphisms of modules and isomorphisms of their endomorphism rings.

(c) If two modules M and N are isomorphic to each other, then their endomorphism rings $\operatorname{End}(M)$ and $\operatorname{End}(N)$ also are isomorphic to each other. More precisely, every module isomorphism $\varphi\colon M \to N$ induces the ring isomorphism $\psi\colon \operatorname{End}(M) \to \operatorname{End}(N)$ defined as

$$\psi\colon \alpha \to \varphi^{-1}\alpha\varphi, \qquad \alpha \in \operatorname{End}(M).$$

If $\alpha, \beta \in \operatorname{End}(M)$, then we have

$$\psi(\alpha\beta) = \varphi^{-1}\alpha\beta\varphi = (\varphi^{-1}\alpha\varphi)(\varphi^{-1}\beta\varphi) = \psi(\alpha)\psi(\beta)$$

and

$$\psi(\alpha + \beta) = \psi(\alpha) + \psi(\beta), \qquad \psi(1_M) = 1_N.$$

Consequently, ψ is a ring homomorphism. If $0 \neq \alpha \in \mathrm{End}(M)$, then it is clear that $\varphi^{-1}\alpha\varphi \neq 0$. Now let $\gamma \in \mathrm{End}(N)$. Then

$$\varphi\gamma\varphi^{-1} \in \mathrm{End}(M) \quad \text{and} \quad \psi(\varphi\gamma\varphi^{-1}) = \gamma,$$

i.e., ψ is a ring isomorphism.

(d) We assume that we have modules $M = A_1 \oplus A_2$ and N. If $\psi\colon \mathrm{End}(M) \to \mathrm{End}(N)$ is a ring isomorphism, then the module N has a decomposition $N = C_1 \oplus C_2$ and ψ induces isomorphisms $\mathrm{End}(A_i) \to \mathrm{End}(C_i)$ for $i = 1, 2$.

We denote by ε the projection $M \to A_1$ with kernel A_2. Then $\psi(\varepsilon)$ is an idempotent of the ring $\mathrm{End}(N)$. We denote it by τ. We have $N = C_1 \oplus C_2$, where $C_1 = \mathrm{Im}(\tau)$, $C_2 = \mathrm{Ker}(\tau)$. The isomorphism ψ induces the ring isomorphism

$$\varepsilon \,\mathrm{End}(M)\varepsilon \to \tau \,\mathrm{End}(N)\tau.$$

Consequently, ψ induces the ring isomorphism

$$\mathrm{End}(A_1) \to \mathrm{End}(C_1)$$

(see property (b) and the remark after it). The second isomorphism follows from a similar argument.

In linear algebra, the matrix representation of linear operators is well known. Using direct decompositions, we can obtain a similar representation of module endomorphisms by some matrices.

Let $M = \bigoplus_{i=1}^n A_i$ be a direct sum of modules. We consider a square matrix of order n $(\alpha_{ij})_{i,j=1,\dots,n}$ with elements $\alpha_{ij} \in \mathrm{Hom}(A_i, A_j)$. For such matrices, it is possible to define usual matrix operations of addition and multiplication. It is directly verified that the addition and the multiplication are always possible and the result are matrices of the same form. As a result, we obtain the ring of matrices of mentioned form.

Proposition 2.4. *For a module $M = \bigoplus_{i=1}^n A_i$, the endomorphism ring of M is isomorphic to the ring of all matrices (α_{ij}) of order n, where $\alpha_{ij} \in \mathrm{Hom}(A_i, A_j)$.*

Proof. Let $\{\varepsilon_i \mid i = 1, \dots, n\}$ be coordinate projections corresponding to a given decomposition of the module M. Every element $a \in M$ is equal to the sum $\sum_{i=1}^n \varepsilon_i a$. For $\alpha \in \mathrm{End}(M)$, we have

$$\alpha(a) = \sum_{i=1}^n \alpha(\varepsilon_i a) = \sum_{i,j=1}^n (\varepsilon_i \alpha \varepsilon_j) a.$$

With the endomorphism α, we associate the matrix $(\alpha_{ij})_{i,j=1,\dots,n}$ with $f\colon \alpha \to (\alpha_{ij})$, where $\alpha_{ij} = \varepsilon_i \alpha \varepsilon_j$. By (b), we identify $\varepsilon_i \,\mathrm{End}(M)\varepsilon_j$ with

$$\mathrm{Hom}(\varepsilon_i M, \varepsilon_j M) = \mathrm{Hom}(A_i, A_j).$$

If $\beta \in \mathrm{End}(M)$ and (β_{ij}) is the corresponding matrix with $\beta_{ij} = \varepsilon_i \beta \varepsilon_j$, then the difference $(\alpha_{ij} - \beta_{ij})$ and the product $(\sum_{k=1}^{n} \alpha_{ik}\beta_{kj})$ of the matrices (α_{ij}) and (β_{ij}) are the matrices corresponding to $\alpha - \beta$ and $\alpha\beta$. Consequently, f is a ring homomorphism. It is clear that only the zero endomorphism of the module M corresponds to the zero matrix. Conversely, let $(\alpha_{ij})_{i,j=1,\dots,n}$ be a matrix with entries $\alpha_{ij} \in \varepsilon_i \,\mathrm{End}(M)\varepsilon_j$. We define $\alpha \in \mathrm{End}(M)$ by the relation $\alpha(a) = \sum_{i,j=1}^{n} \alpha_{ij}(a)$ for every $a \in M$. Then $f\colon \alpha \to (\alpha_{ij})$. Consequently, f is a ring isomorphism. □

In Proposition 2.4, if $\mathrm{Hom}(A_i, A_j) = 0$ for all distinct subscripts i and j, then endomorphisms of the module M are represented by diagonal matrices. In other words, we have

$$\mathrm{End}(M) \cong \mathrm{End}(A_1) \times \cdots \times \mathrm{End}(A_n).$$

Corollary 2.5. (1) *Let M be an R-module and let $S = \mathrm{End}_R(M)$. Then*

$$\mathrm{End}_R(M^n) \cong S_n \quad \textit{for every } n \geq 1.$$

(2) *For every $n \geq 1$, we have $\mathrm{End}_R(R^n) \cong R_n$.*

Now we consider induced exact sequences for the group Hom. We assume that M, N, and L are left R-modules and $f\colon L \to N$, $g\colon L \to M$ are two homomorphisms. We have *induced homomorphisms* of Abelian groups

$$f_*\colon \mathrm{Hom}(M, L) \to \mathrm{Hom}(M, N) \quad \text{and} \quad g^*\colon \mathrm{Hom}(M, N) \to \mathrm{Hom}(L, N),$$

where $f_*(\varphi) = \varphi f$ for all $\varphi \in \mathrm{Hom}(M, L)$ and $g^*(\psi) = g\psi$ for all $\psi \in \mathrm{Hom}(M, N)$.

A sequence of left R-modules and homomorphisms

$$0 \to A \xrightarrow{f} B \xrightarrow{g} C \to 0$$

is called a *short exact sequence* if f is a monomorphism, g is an epimorphism, and $\mathrm{Im}(f) = \mathrm{Ker}(g)$. We have the following exact sequences of Abelian groups which are called *induced sequences*:

$$0 \to \mathrm{Hom}(M, A) \xrightarrow{f_*} \mathrm{Hom}(M, B) \xrightarrow{g_*} \mathrm{Hom}(M, C),$$

$$0 \to \mathrm{Hom}(C, M) \xrightarrow{g^*} \mathrm{Hom}(B, M) \xrightarrow{f^*} \mathrm{Hom}(A, M),$$

where f_*, g_*, f^*, and g^* are induced mappings. If M, A, B, and C are bimodules and all groups Hom are considered as modules by the above method, then we have two exact sequences of modules. Using the extension group Ext, more long exact sequences are constructed in Section 21. A sequence of modules and homomorphisms

$$\cdots \xrightarrow{\alpha_{-3}} A_{-2} \xrightarrow{\alpha_{-2}} A_{-1} \xrightarrow{\alpha_{-1}} A_0 \xrightarrow{\alpha_0} A_1 \xrightarrow{\alpha_1} A_2 \longrightarrow \cdots$$

is said to be *exact* if it is exact at every place. For example, the exactness at the place A_1 means that the relation $\mathrm{Im}(\alpha_0) = \mathrm{Ker}(\alpha_1)$ holds.

We touch on some properties of the tensor product. Let N be a right R-module and let M be a left R-module. In such a case we can form the tensor product $N \otimes_R M$ of these modules. The *tensor product* $N \otimes_R M$ is an Abelian group with generator system elements $n \otimes m$ for all $n \in N$, $m \in M$ and defining relations

$$(n_1 + n_2) \otimes m = (n_1 \otimes m) + (n_2 \otimes m),$$
$$n \otimes (m_1 + m_2) = (n \otimes m_1) + (n \otimes m_2),$$
$$nr \otimes m = n \otimes rm$$

for all $n_1, n_2, n \in N$, $m_1, m_2, m \in M$, and $r \in R$. An arbitrary element $x \in N \otimes_R M$ is a finite sum $x = \sum n_i \otimes m_i$ for some $n_i \in N$ and $m_i \in M$. Similar to the case of homomorphism groups, the Abelian group $N \otimes_R M$ can be canonically turned into some bimodule. Let ${}_TN_R$ and ${}_RM_S$ be two bimodules. Then the group $N \otimes_R M$ is a T-S-bimodule with module multiplications defined by the relations

$$t\Big(\sum n_i \otimes m_i\Big) = \sum tn_i \otimes m_i,$$
$$\Big(\sum n_i \otimes m_i\Big)s = \sum n_i \otimes m_i s$$

for all $n_i \in N$, $m_i \in M$, $t \in T$, and $s \in S$. In particular, $R \otimes_R M$ is a left R-module which is isomorphic to M under the correspondence $r \otimes m \to rm$, where $r \in R$ and $m \in M$. If we have a T-R-bimodule N and a family of R-S-bimodules M_i ($i \in I$), then there exists the canonical isomorphism of T-S-bimodules

$$N \otimes_R \Big(\bigoplus_{i \in I} M_i\Big) \cong \bigoplus_{i \in I} (N \otimes_R M_i).$$

An exact sequence of right R-modules $0 \to A \xrightarrow{f} B \xrightarrow{g} C \to 0$ induces the exact sequence of Abelian groups

$$A \otimes_R M \xrightarrow{f \otimes 1} B \otimes_R M \xrightarrow{g \otimes 1} C \otimes_R M \to 0,$$

where $f \otimes 1$ and $g \otimes 1$ are the induced homomorphisms. For example, $(f \otimes 1)(a \otimes m) = f(a) \otimes m$ for all $a \in A$ and $m \in M$. For a fixed module N and an exact sequence of left R-modules, we can present a similar exact sequence of tensor products.

It should be noted that the question about correctness of various definitions often arises in the work with tensor products. Above all, difficulties are related to the non-uniqueness of the representation of elements in the form of sums of generators. It is known that these difficulties are usually removed by the use of balanced mappings which are defined in each of cited in Section 1 books in the theory of rings and modules.

A module ${}_RM$ is said to be *flat* if for every monomorphism $f\colon A \to B$ of right R-modules, the mapping

$$f \otimes 1\colon A \otimes_R M \to B \otimes_R M$$

is a monomorphism. We note that if R is a discrete valuation domain, then M is a flat module if and only if M is a torsion-free module (Lambek [198, Section 5.4]). In addition, if $\operatorname{Im}(f)$ is a pure submodule in B (the purity is defined in Section 7), then $f \otimes 1$ is a monomorphism (see Fuchs [92, Theorem 60.4]).

Finally, we note that all left R-modules form the category $R\text{-}\mathrm{mod}$ whose morphisms are R-homomorphisms. We also obtain a category if we consider only modules in a certain class of modules. This is a full subcategory of the category $R\text{-}\mathrm{mod}$. In such module categories, functors defined by the group Hom and the tensor product play an especial role (see Section 13). Sometimes we assume that if a category is such that the class of objects consists of modules, then the set of morphisms of the category consists of some sets obtained from groups of module homomorphisms. As a rule, the structures of these sets are more simple than the structure of the initial group Hom. Such categories appear in Sections 24 and 29.

Exercise 1. Let M be a left R-module and let $S = \mathrm{End}_R(M)$. For any two endomorphisms α and β from S, we define the product $\alpha \circ \beta$ by the relation

$$(\alpha \circ \beta)a = \alpha(\beta a)$$

for all $a \in M$. With respect to this new multiplication, all endomorphisms of the module M form one more endomorphism ring S°. Prove that the rings S and S° are *anti-isomorphic* to each other, i.e., there exists an additive isomorphism $f\colon S \to S^\circ$ such that

$$f(\alpha\beta) = f(\beta) \circ f(\alpha)$$

for all $\alpha, \beta \in S$.

Exercise 2. Prove that the new endomorphism ring of the module ${}_RR$ defined in Exercise 1 is anti-isomorphic to the ring R.

Exercise 3. Let R be a ring and let e be an idempotent of R. Then $\mathrm{End}_R(Re) \cong eRe$, where $Re = \{re \mid r \in R\}$ is a left R-module.

Exercise 4. Let A and B be two left R-module. If $M = A \oplus B$ and T is the biendomorphism ring of the R-module M, then A and B are T-submodules of the module ${}_TM$.

Exercise 5. Let R be a ring. For every $n \geq 1$, R^n is a right R_n-module (see Corollary 2.5 (2)). Prove that $\mathrm{End}_{R_n}(R^n) \cong R$.

3 Discrete valuation domains

We define discrete valuation domains by two equivalent methods: the first method is formal and uses some conditions for an abstract ring, and the second method uses valuations of division rings. We present various examples of discrete valuation domains.

In this section and several other sections, elements of topological algebra are used. We need the definition and one property of the Jacobson radical of the ring. A more detailed information about the radical is contained in Section 14. The *Jacobson radical* of the ring R is denoted by $J(R)$. The radical $J(R)$ is equal to the intersection of all left (or right) maximal ideals of the ring R. An element $x \in R$ is contained in $J(R)$ if and only if $1 - yxz$ is an invertible element of R for all $y, z \in R$.

For a better understanding of the place of discrete valuation domains among other rings, we recall the definition of a local ring and present several simple results on local rings. A ring R is said to be *local* if the factor ring $R/J(R)$ is a division ring.

Proposition 3.1. *For a ring R, the following conditions are equivalent:*

(1) *R is a local ring;*

(2) *R has a unique maximal left ideal;*

(3) *R has a unique maximal right ideal;*

(4) *all non-invertible elements of the ring R form an ideal;*

(5) *for every element $r \in R$, at least one of the elements r, $1 - r$ is an invertible element.*

Proof. (1) $\Longrightarrow$ (2). Let M be a maximal left ideal of the ring R. Then $J(R) \subseteq M$ and $M/J(R)$ is a maximal left ideal of the division ring $R/J(R)$. Consequently, $M/J(R) = 0$ and $M = J(R)$.

(2) $\Longrightarrow$ (1). We obtain that the unique maximal left ideal coincides with $J(R)$. Consequently, the ring $R/J(R)$ does not have proper left ideals. Consequently, it is a division ring and R is a local ring.

The equivalence of (1) and (3) can be proved similarly.

(1) $\Longrightarrow$ (4). An element r of the ring R is invertible if and only if the residue class $r + J(R)$ is invertible in $R/J(R)$. Indeed, let

$$(r + J(R))(s + J(R)) = 1 + J(R).$$

There exist two elements $x, y \in J(R)$ such that $rs = 1 + x$ and $sr = 1 + y$. Therefore rs and sr are invertible elements. Consequently, the element r is also invertible. Now it is clear that the set of all non-invertible elements of the ring R coincides with $J(R)$.

(4) $\Longrightarrow$ (5). All non-invertible elements of the ring R form a unique maximal left ideal of the ring R, which coincides with $J(R)$. Consequently, if the element r is not invertible, then $r \in J(R)$. Then $1 - r$ is an invertible element.

(5) $\Longrightarrow$ (1). Let $r \notin J(R)$. There exists a maximal left ideal M of the ring R such that $r \notin M$. Then $M + Rr = R$ and $m + ar = 1$, where $m \in M$ and $a \in R$. Since m is a non-invertible element, we have that ar is an invertible element. Similarly, rb is an invertible element for some b. Consequently, r is an invertible element. Thus, $R/J(R)$ is a division ring and R is a local ring. □

A ring without zero-divisors is called a *domain*.

Definition 3.2. A ring R is called a *discrete valuation domain* (or a *discrete valuation ring*) if it satisfies the following conditions:

(1) R is a local ring;

(2) $J(R) = pR = Rp$, where p is some non-nilpotent element of the ring R;

(3) $\bigcap_{n \geq 1} (J(R))^n = 0$.

We will verify below that the ring R from Definition 3.2 really is a domain. Sometimes, condition (2) is formulated as follows: $J(R) = pR = Rp$, where either $p = 0$ or p is some non-nilpotent element of the ring R. For $p = 0$, we obtain that R is a division ring. This allows us to formally include division rings in the class of discrete valuation domains.

The element p from condition (2) is called a *prime element* of the discrete valuation domain R. A prime element is not uniquely defined. The elements of the form pv or vp, where v is an arbitrary invertible element of the ring R, also are prime elements.

Throughout the book, the symbol p denotes some fixed prime element of a discrete valuation domain. However, sometimes p is some prime integer; this case is especially specified.

We consider some properties of discrete valuation domains (see Baer [26] and Liebert [206]).

(1) $J(R)^n = p^nR = Rp^n$ for every positive integer n.

Since $pR = Rp$, we can directly verify that

$$J(R)^n = pR \cdots pR = p^nR = Rp^n.$$

(2) Every proper one-sided ideal of the ring R is an ideal and it is equal to p^nR for some $n \geq 1$.

Let A be a proper nonzero left ideal of the ring R. Then $A \subseteq J(R)$, since otherwise A contains an invertible element and A coincides with R. We choose the minimal positive integer n with $A \subseteq p^nR$. We take an element $a \in A$ such that $a \notin p^{n+1}R$. We have $a = vp^n$, where $v \notin Rp$. By Proposition 3.1, we have $p^nR = Rp^n \subseteq A$. Consequently, $A = p^nR$. For right ideals, the proof is similar to the proof in the left-side case.

(3) Every nonzero element r of the ring R has a unique representation in the form $r = p^nv = up^n$, where n is a non-negative integer and v, u are invertible elements of the ring R. (As usual, we assume that $p^0 = 1$.)

Let n be the minimal positive integer with $r \in p^nR$. Then $r = p^nv$, where $v \notin pR$. Therefore v is an invertible element. Since $p^nR = Rp^n$, we have $r = up^n$, where u is an invertible element. Now it is clear that p is a non-zero-divisor in R. Therefore, if $p^nv = p^mw$ ($up^n = wp^m$, respectively), then $n = m$ and $v = w$ ($u = w$, respectively).

(4) The ring R is a domain. We assume that $xy = 0$ for some nonzero elements x and y of the ring R. We have $x = p^nv$, $y = p^mu$, and $vp^m = p^mw$, where v, u, and w are invertible elements. We have

$$xy = p^nvp^mu = p^np^mwu = 0.$$

Therefore $p^{n+m} = 0$. This is impossible, since p is a non-nilpotent element.

(5) Let x and y be two nonzero elements of the ring R. There exist unique elements $x_1, y_1 \in R$ such that $xy = yx_1 = y_1x$. We take the elements v, u, and w considered in (4). Then we have

$$xy = (p^mu)x_1 = yx_1, \quad \text{where} \quad x_1 = u^{-1}p^nwu.$$

Using the same method, we obtain the element y_1.

(6) The elements x_1 and y_1 from (5) can be obtained with the use of some automorphism of the ring R. We fix a nonzero element y of the ring R. For

every $x \in R$, it follows from (5) that there exists the unique element x_1 such that $xy = yx_1$. We set $\sigma_y(x) = x_1, x \in R$. Here σ_y is a bijection from the ring R onto itself. The relations

$$\sigma_y(x + z) = \sigma_y(x) + \sigma_y(z) \quad \text{and} \quad \sigma_y(xz) = \sigma_y(x)\sigma_y(z)$$

also hold for any two elements $x, z \in R$. Consequently, σ_y is an automorphism of the ring R and $xy = y\sigma_y(x)$ for every $x \in R$. It is clear that $\sigma_y(y) = y$. For $y = 0$ as σ_0, we take the identity automorphism of the ring R. Furthermore, it follows from $yx = x\sigma_x(y)$ that $\sigma_x^{-1}(y)x = xy$. Therefore, we have $xy = y\sigma_y(x) = \sigma_x^{-1}(y)x$ for any two elements x and y of the ring R.

A ring is called a *principal ideal ring* if every one-sided ideal of the ring is principal. A principal ideal ring without zero-divisors is called a *principal ideal domain*. A commutative ring R is a discrete valuation domain if and only if R is a principal ideal domain with unique nonzero proper prime ideal. In commutative principal ideal domains, prime elements coincide with elements generating nonzero proper prime ideals.

It follows from Properties (3) and (5) that elements of non-commutative discrete valuation domains and elements of commutative discrete valuation domains can be similarly handled. It is only necessary to do sometimes inessential specializations.

(7) It follows from (5) that the domain R satisfies the right and left Ore conditions. Consequently, there exists the division ring of fractions K of the domain R. We can assume that K consists of all expressions (which are called *fractions*) of the form r/p^n, where $r \in R$ and n is a non-negative integer. In addition, we assume that

$$r/p^n = s/p^m \iff rp^{m-n} = s \text{ for } n \leq m \text{ and } r = sp^{n-m} \text{ for } m \leq n.$$

We define addition and multiplication of fractions by the relations

$$\frac{r}{p^n} + \frac{s}{p^m} = \frac{rp^m + sp^n}{p^{n+m}},$$

$$\frac{r}{p^n}\frac{s}{p^m} = \frac{r\sigma_{p^n}(s)}{p^{n+m}},$$

where σ_{p^n} is the automorphism of the ring R defined in (6). It can be verified that K is a division ring which is the division ring of fractions of the domain R. In particular, if $r = p^m v \neq 0$ and the element v is invertible, then for the fraction r/p^n, the fraction $(p^n v^{-1})/p^m$ is the inverse fraction. The ring R is identified with the subring of the division ring K consisting of fractions $r/1$ for all possible $r \in R$.

(8) Every automorphism τ of the ring R can be extended to an automorphism τ of the division ring K by the relation $\tau(r/p^n) = (\tau(r)v_n^{-1})/p^n$, where $\tau(p^n) = p^n v_n$ and v_n is an invertible element in R. Then it follows from (6) that for every $y \in R$, there exists an automorphism σ_y of the division ring K which satisfies the relation $ky = y\sigma_y(k)$ for all $k \in K$. If R is a commutative domain, then σ_y is the identity automorphism.

We present examples of discrete valuation domains. For an element x of a commutative ring, the principal ideal generated by x is denoted by (x).

Let R be an arbitrary commutative principal ideal domain with field of fractions F. We take some proper prime ideal P of the ring R. We set $R_P = \{a/b \in F \mid b \notin P\}$. Then R_P is a discrete valuation domain with maximal ideal PR_P. The field of fractions of the ring R is the field F. The ring R_P is called the *localization* of the ring R with respect to the prime ideal P. If $P = (p)$, then the notation R_p is also used.

As R, we take R the ring of integers $\mathbb{Z}$. Let p be some prime integer. Then the localization $\mathbb{Z}_p$ is the ring of all rational numbers whose denominators are coprime to p. The field of rational numbers is the field of fractions of $\mathbb{Z}_p$.

Now we assume that F is a field and $F[x]$ is the polynomial ring in variable x over the field F. The ring $F[x]$ is a commutative principal ideal domain. The field of fractions of the ring $F[x]$ is called the *field of rational functions*. Prime elements of the ring $F[x]$ coincide with irreducible polynomials. Let $p(x)$ be some irreducible polynomial. Then $(p(x))$ is a prime ideal and the localization of the ring $F[x]$ with respect to $(p(x))$ is a discrete valuation domain.

We consider examples of another type.

Let p be a fixed prime integer. We take sequences of positive integers $(k_1, k_2, \dots, k_n, \dots)$ such that $0 \leq k_n < p^n$ and $k_{n+1} \equiv k_n \pmod{p^n}$ for $n \geq 1$. We define the sum and the product of two such sequences using the component-wise addition and multiplication and the subsequent factorization modulo p^n. As a result, all considered sequences form a commutative domain. This domain is called the *ring of p-adic integers*; it is denoted by $\widehat{\mathbb{Z}}_p$. In this ring, the sequence $(0, 0, \dots, 0, \dots)$ is the zero and the sequence $(1, 1, \dots, 1, \dots)$ is the identity element. There is a ring embedding $\mathbb{Z} \to \widehat{\mathbb{Z}}_p$ such that an integer s corresponds to the sequence $s \cdot 1$, where 1 is the identity element of the ring $\widehat{\mathbb{Z}}_p$. If $(s, p) = 1$, then $s \cdot 1$ is an invertible element in $\widehat{\mathbb{Z}}_p$. Every fraction s/t in $\mathbb{Z}_p$ corresponds to the element $(s \cdot 1)(t \cdot 1)^{-1}$ of the ring $\widehat{\mathbb{Z}}_p$. In such a manner, we obtain an embedding of the ring $\mathbb{Z}_p$ in $\widehat{\mathbb{Z}}_p$. By identifying, we assume that $\mathbb{Z}_p$ is a subring in $\widehat{\mathbb{Z}}_p$. Proper ideals of the ring $\widehat{\mathbb{Z}}_p$ have the form (p^n) with $n \geq 1$. Consequently, $\widehat{\mathbb{Z}}_p$ is a commutative discrete valuation domain with prime element p. A sequence $(k_1, k_2, \dots, k_n, \dots)$ is invertible in $\widehat{\mathbb{Z}}_p$ if and only if $k_1 \neq 0$. The field of fractions of the ring $\widehat{\mathbb{Z}}_p$ is called the *field of p-adic numbers*; it consists of all fractions π/p^n, where π is a

p-adic integer and n is a non-negative integer.

It is known that p-adic numbers can be represented in the form of a formal infinite series. In the real use of the ring of p-adic integers (for example, in the study of modules over the ring), it is not necessary to know the concrete form of elements and operations in the ring.

Again, let F be a field. Over the field F, we can define not only polynomials, but and (formal) power series

$$a_0 + a_1x + a_2x^2 + \cdots + a_kx^k + \cdots = \sum_{k=0}^{\infty} a_kx^k, \quad a_k \in F,$$

in variable x. We can directly extend definitions of operations from polynomials to power series:

$$\sum_{k=0}^{\infty} a_kx^k + \sum_{k=0}^{\infty} b_kx^k = \sum_{k=0}^{\infty}(a_k + b_k)x^k,$$

$$\sum_{k=0}^{\infty} a_kx^k \cdot \sum_{l=0}^{\infty} b_lx^l = \sum_{m=0}^{\infty} c_mx^m,$$

where $c_m = \sum_{k+l=m} a_kb_l$. Similar to the case of polynomials, it can be verified that all power series with the mentioned addition and multiplication form a commutative domain. The domain is called the *power series ring* in variable x over the field F; it is denoted by $F[[x]]$.

A series f with nonzero constant term has the inverse series. Indeed, we represent f in the form $f = a(1 - g)$, where $a \in F$, $a \neq 0$, and g is some series. It is easy to verify that the series $1 + g + g^2 + \cdots$ exists. Then $f^{-1} = a^{-1}(1 + g + g^2 + \cdots)$. More explicitly, let

$$f = \sum_{k=0}^{\infty} a_kx^k, \quad \text{where} \quad a_0 \neq 0.$$

Then $f^{-1} = \sum_{k=0}^{\infty} b_kx^k$, where the coefficients b_k can be obtained from the relations

$$1 = a_0b_0, \qquad 0 = a_0b_1 + a_1b_0, \qquad 0 = a_0b_2 + a_1b_1 + a_2b_0, \qquad \ldots .$$

Therefore, all nonzero ideals in $F[[x]]$ form the chain

$$(1) \supset (x) \supset (x^2) \supset \cdots .$$

Therefore $F[[x]]$ is a discrete valuation domain.

We describe the field of fractions of the domain $F[[x]]$. We define a (formal) Laurent series in variable x over the field F as the expression

$$a_n x^n + a_{n+1} x^{n+1} + \cdots + a_k x^k + \cdots = \sum_{k=n}^{\infty} a_k x^k, \quad a_k \in F,$$

where n is any (possibly, negative) integer. Therefore, a Laurent series in a general case contains a finite number of terms with negative degrees of the variable x and infinitely many terms with positive degrees of x. The addition and multiplication of Laurent series are similar to the addition and multiplication of power series. As a result, we obtain a commutative domain consisting of all Laurent series. This domain is called the *Laurent series ring* in variable x over the field F; it is denoted by $F\langle x\rangle$. It is clear that $F[[x]]$ is a subring in $F\langle x\rangle$. Let f be a nonzero Laurent series. We choose some power x^n (n is some integer) such that

$$f \cdot x^n = b_0 + b_1 x + b_2 x^2 + \cdots,$$

where $b_0 \neq 0$. As it was shown above, the right series is invertible. The series x^n is also invertible. Consequently, f is an invertible series. Therefore $F\langle x\rangle$ is a field. For a series $f \notin F[[x]]$, there exist a positive integer n and a power series g such that $f = (x^n)^{-1} \cdot g$. This means that $F\langle x\rangle$ is the field of fractions for the domain $F[[x]]$.

In the considered examples, discrete valuation domains are commutative. We present examples of non-commutative discrete valuation domains. Let D be some division ring. We can similarly define the polynomial ring $D[x]$, the power series ring $D[[x]]$, and the Laurent series ring $D\langle x\rangle$ in variable x over the division ring D. We assume that the variable x can be permuted with elements of the division ring D, i.e., $ax = xa$ for all $a \in D$. We can similarly obtain that $D[[x]]$ is a discrete valuation domain and $D\langle x\rangle$ is the division ring of fractions for $D[[x]]$. If D is a non-commutative division ring, then $D[[x]]$ is a non-commutative domain.

We can obtain new (non-commutative) discrete valuation domains if we define other operations of multiplication of power series and Laurent series. Let α be some automorphism of the division ring D. We preserve the previous addition on the set of all power series in variable x over the division ring D. We define multiplication of power series by the relation $xa = \alpha(a)x$ ($a \in D$) and corollaries of this relation. Precisely,

$$\sum_{k=0}^{\infty} a_k x^k \cdot \sum_{l=0}^{\infty} b_l x_l = \sum_{m=0}^{\infty} c_m x^m$$

where $c_m = \sum_{k+l=m} a_k \alpha^k(b_l)$. The set of all power series with given operations of the addition and multiplication forms the ring $D[[x, \alpha]]$. The ring is called the *skew power series ring*. As above, it is verified that $D[[x,\alpha]]$ is a discrete valuation domain. Even if D is a field, but α is not the identity automorphism, then $D[[x,\alpha]]$ is a non-commutative domain. For the domain $D[[x,\alpha]]$, the division ring of skew Laurent series $D\langle x,\alpha\rangle$ is the division ring of fractions. Elements of the division ring are Laurent series in variable x over the division ring D. The division ring $D\langle x,\alpha\rangle$ and the division ring $D\langle x\rangle$ have the same additive structure. The multiplication is induced by the relation $x^k a = \alpha^k(a)x^k$, where k is an integer and $a \in R$. We note that

$$x^{-1}\alpha(a) = \alpha^{-1}(\alpha a)x^{-1} = ax^{-1}.$$

Therefore $\alpha(a) = xax^{-1}$ for every $a \in D$. Consequently, the automorphism α can be considered as the restriction to the subring D of the conjugation by the element x of the division ring $D\langle x,\alpha\rangle$.

We return to an arbitrary discrete valuation domain R. One quite useful topology can be naturally defined in R. We take the ideals p^nR with $n = 0,1,2,\dots$ as a basis of neighborhoods of zero. We recall that p denotes some fixed prime element of the domain R. This turns R into a topological space. The mapping $(r,s) \to r+s$ from the topological space $R \times R$ onto the topological space R is continuous. The mapping $r \to -r$ from the space R onto R is continuous. Consequently, R is a topological Abelian group. The mapping $(r,s) \to rs$ from $R \times R$ into R is also continuous. Therefore R is a topological ring. The defined topology is called the *p-adic topology* of the ring R. Since $\bigcap_{n\geq 1} p^nR = 0$, this topology is Hausdorff. We assume that discrete valuation domains have the p-adic topology and all topological notions are related to this topology.

To define complete discrete valuation domains, we consider Cauchy sequences and convergent sequences. For a discrete valuation domain R, a sequence $\{r_i\}_{i\geq 1}$ of elements of R is called a *Cauchy sequence* if for every positive integer n, there exists a positive integer k (depending on n) such that $r_i - r_j \in p^nR$ for all $i,j \geq k$. We say that the sequence $\{r_i\}_{i\geq 1}$ *converges* to the limit $r \in R$ if for every positive integer n, there exists a positive integer k (depending on n) such that $r - r_i \in p^nR$ for all $i \geq k$. Convergent sequences are Cauchy sequences. If a sequence converges, then every infinite subsequence of the sequence also converges and has the same limit. Therefore, it is clear that we can mainly consider so-called pure Cauchy sequences and purely convergent sequences. A sequence $\{r_i\}$ is called a *pure Cauchy sequence* if for every positive integer n, we have $r_i - r_n \in p^nR$ for all $i \geq n$. We say that a sequence $\{r_i\}$ *purely converges* to the element $r \in R$ if $r - r_i \in p^nR$ for all $i \geq n$. If a pure Cauchy sequence converges to the limit r, then this sequence purely converges to r.

A discrete valuation domain R is called a *complete discrete valuation domain* if R is a complete topological ring with respect to the p-adic topology. This means that every Cauchy sequence of elements in the ring R has the limit in R. For completeness of the ring R, it is sufficient the existence of the limit in R for every pure Cauchy sequence of elements in the ring R.

We consider the question about an embedding of a discrete valuation domain in a complete discrete valuation domain.

Proposition 3.3. *If R is a discrete valuation domain, then there exists a complete discrete valuation domain $\widehat{R}$ containing R as a dense subring.*

Proof. We set $R_{p^n} = R/p^nR\,(n \geq 1)$ and form the ring product $\bar{R} = \prod_{n\geq 1} R_{p^n}$. We have the canonical ring homomorphism $\pi_n \colon R_{p^{n+1}} \to R_{p^n}$ for every n, where

$$\pi_n(r + p^{n+1}R) = r + p^nR \quad \text{for all} \quad r \in R.$$

Let $\widehat{R}$ be the set of all elements $(\bar{r}_n)_{n\geq 1}$ in $\bar{R}$, where $\bar{r}_n = r + p^nR$, such that $\pi_n(\bar{r}_{n+1}) = \bar{r}_n$ for all n. Then $\widehat{R}$ is a subring in $\bar{R}$ and $\widehat{R}$ is a domain. Furthermore, the mapping $R \to \widehat{R}$ with $r \to (r + p^nR)_{n\geq 1}$ is a ring embedding. We identify the ring R with the image of this embedding; in particular, let $p = (p + p^nR)_{n\geq 1}$. If $(\bar{r}_n) = (r_1 + p^nR) \in \widehat{R}$ and $\bar{r}_1 \neq 0$, then $(\bar{r}_n)$ is an invertible in $\widehat{R}$ element. We obtain that all non-invertible elements of the ring $\widehat{R}$ form an ideal and $\widehat{R}$ is a local ring. An element $(\bar{r}_n)$ of the ring $\widehat{R}$ is divisible by p^k if and only if $\bar{r}_1 = \cdots = \bar{r}_k = 0$. Therefore

$$J(\widehat{R}) = p\widehat{R} = \widehat{R}p \quad \text{and} \quad \bigcap_{n\geq 1} J(R)^n = 0.$$

Consequently, $\widehat{R}$ is a discrete valuation domain.

Let

$$(\bar{r}_n^{(1)}), (\bar{r}_n^{(2)}), \ldots, (\bar{r}_n^{(i)}), \ldots$$

be a pure Cauchy sequence of elements in $\widehat{R}$. Then for every $n \geq 1$, we have $\bar{r}_n^{(n)} = \bar{r}_n^{(n+1)} = \cdots$. We denote this element by $\bar{r}_n$. Then $(\bar{r}_n) \in \widehat{R}$ and the given pure sequence converges to this element.

We prove that the subring R is dense in $\widehat{R}$. Since $p^n\widehat{R} \cap R = p^nR$ for every n, we have that the p-adic topology of the ring R coincides with the induced topology. Let $(\bar{r}_n) \in \widehat{R}$, where for every n, we have $\bar{r}_n = r_n + p^nR$ and $r_n \in R$. We take the sequence of elements $(\bar{s}_n^{(1)}), (\bar{s}_n^{(2)}), \ldots, (\bar{s}_n^{(i)}), \ldots$ of the ring R, where $(\bar{s}_n^{(i)}) = (\bar{r}_i, \bar{r}_i, \ldots, \bar{r}_i, \ldots)$ for every i. Then the element $(\bar{r}_n)$ is the limit of this sequence. Consequently, R is a dense subring in $\widehat{R}$. □

In the proof of Proposition 3.3, we have used the construction of the inverse limit. The constructed complete discrete valuation domain $\widehat{R}$ is called the *p-adic completion* or merely the *completion* of the domain R. It is unique in some sense (see Exercise 2, and also Corollary 11.14).

We denote by $\widehat{K}$ the division ring of fractions of the domain $\widehat{R}$. We have inclusions $R \subseteq \widehat{R} \subset K$, $R \subset K \subseteq \widehat{K}$, and ring equalities $\widehat{R} \cap K = R$, $\widehat{K} = \widehat{R}K$, where $\widehat{R}K$ consists of products rk for all $r \in R$ and $k \in K$.

There are two typical examples of complete discrete valuation domains. The ring $\widehat{\mathbb{Z}}_p$ of p-adic integers is a complete discrete valuation domain and it is the completion of the ring $\mathbb{Z}_p$. The ring $\mathbb{Z}_p/p^n\mathbb{Z}_p$ is the residue ring modulo p^n. In fact, we constructed $\widehat{\mathbb{Z}}_p$ as the completion of the ring $\mathbb{Z}_p$ with the use the method from Proposition 3.3.

The skew power series ring $D[[x, \alpha]]$ is a complete discrete valuation domain. For every pure sequence $f_1(x), f_2(x), \dots$ of series in $D[[x, \alpha]]$, it is easy to construct the series which is the limit of this sequence. For example, if F is a field and S is the localization of the ring $F[x]$ with respect to the prime ideal (x), then $F[[x]]$ is the completion for S.

Let R be a discrete valuation domain. For every positive integer n, we set $R_{p^n} = R/p^nR$. Since R is dense in the completion $\widehat{R}$, we have $R + p^n\widehat{R} = \widehat{R}$. Therefore, we obtain

$$\widehat{R}/p^n\widehat{R} = (R + p^n\widehat{R})/p^n\widehat{R} \cong R/(R \cap p^n\widehat{R}) = R/p^nR \quad \text{and} \quad \widehat{R}_{p^n} \cong R_{p^n}.$$

For $n = 1$, we have the division ring R_p. It is called the *residue division ring* of the domain R. We also have the division ring of fractions K of the domain R. Thus, there are two division rings connected to the discrete valuation domain. If R is a commutative discrete valuation domain, then we deal with the *residue field* R_p and the *field of fractions* K. The characteristics of these division rings satisfy the following condition. Either the characteristic of the division ring of fractions is equal to zero and the characteristic of the residue division ring is finite or these characteristics are equal to each other. In the case of commutative complete discrete valuation domains, the rings $\widehat{\mathbb{Z}}_p$ and $F[[x]]$ can be considered as typical examples. Properties of commutative complete discrete valuation domains with distinct characteristics of the field of fractions and of the residue field resemble the properties of the ring $\widehat{\mathbb{Z}}_p$. On the other hand, the following theorem of Teichmuller is known: if R is a commutative complete discrete valuation domain such that the field of fractions K and the residue field R_p have the same characteristics, then R is isomorphic to the power series ring over the field R_p.

We briefly consider another approach to the definition of discrete valuation domains, which explains the origin of the word "valuation" in the definition.

Let $(\Gamma, <)$ be a linearly ordered group. This means that the linear order on the set Γ satisfies the following condition: for all $x, y, z \in \Gamma$, if $x \leq y$, then $zx \leq zy$ and $xz \leq yz$.

We assume that D is a division ring and D^* is the multiplicative group of D. A function $\nu \colon D^* \to \Gamma$ is called a *valuation* of the division ring D with values in the group Γ if the following conditions hold:

(1) $\nu(ab) = \nu(a)\nu(b)$ for any two elements $a, b \in D^*$;

(2) $\nu(a + b) \geq \min(\nu(a), \nu(b))$ for any two elements $a, b \in D^*$ such that $a + b \neq 0$.

For the valuation $\nu \colon D^* \to \Gamma$, we set $R = \{a \in D^* \mid \nu(a) \geq e\} \cup \{0\}$, where e is the identity element of the group Γ. We verify that R is a subring in D. Since ν is a group homomorphism, we have $\nu(1) = e$. Therefore

$$e = \nu(1) = \nu(-1)\nu(-1) = \nu(-1)^2.$$

Consequently, $\nu(-1) = e$, since Γ is a torsion-free group as a linearly ordered group. Now if $b \in R$, then

$$\nu(-b) = \nu(b)\nu(-1) \geq \nu(-1) = e \quad \text{and} \quad -b \in R.$$

In addition, we have

$$\nu(a + b) \geq e \quad \text{and} \quad \nu(ab) = \nu(a)\nu(b) \geq \nu(b) \geq e$$

for all $a, b \in R$. We obtain that R is a subring in D.

We prove some properties of the ring R.

(1) R is a local ring. Let $A = \{a \in R \mid \nu(a) > e\} \cup \{0\}$. For $a, b \in A$, we have

$$\nu(a - b) \geq \min(\nu(a), \nu(-b)) > e \quad \text{and} \quad a - b \in A.$$

If $r \in R$ and $r \neq 0$, then

$$\nu(ra) = \nu(r)\nu(a) \geq \nu(a) > e \quad \text{and} \quad ra \in A.$$

Similarly, we have $ar \in A$. Consequently, A is an ideal of the ring R. If $a \in R$ and $a \notin A$, then $\nu(a) = e$. Therefore $\nu(a^{-1}) = e$, $a^{-1} \in R$, and a is an invertible element. Let a be an invertible element of the ring R. Then

$$\nu(a), \nu(a^{-1}) \geq e \quad \text{and} \quad \nu(a)\nu(a^{-1}) = \nu(1) = e.$$

Consequently, $\nu(a) = e$ and $a \notin A$. We obtain that the ideal A consists of all non-invertible elements of the ring R. Consequently, R is a local ring.

(2) For every element g of the group Γ with $g \geq e$, the sets

$$X_g = \{a \in R \mid \nu(a) > g\} \quad \text{and} \quad Y_g = \{a \in R \mid \nu(a) \geq g\}$$

are ideals of the ring R and every nonzero one-sided ideal of the ring R contains one from the ideals Y_g.

It is directly verified that X_g and Y_g are ideals. For any two nonzero elements $a, b \in R$ with $\nu(b) \geq \nu(a)$, we have $b = ca$, where $c = ba^{-1}$. Consequently,

$$\nu(c) = \nu(b)\nu(a^{-1}) \geq e \quad \text{and} \quad c \in R.$$

Therefore $Ra \supseteq Y_g$ for every g with $g \geq \nu(a)$. Since $a \in Y_g$, we have $aR \subseteq Y_g$. Therefore, we obtain property (3).

(3) Every left ideal of the ring R is an ideal. Similarly, we have that every right ideal is an ideal.

(4) For every element $a \in D^*$, we have that either $a \in R$ or $a^{-1} \in R$, where both inclusions are possible. If $\nu(a) \geq e$, then $a \in R$. Otherwise, $\nu(a) \leq e$. Then

$$e = \nu(aa^{-1}) = \nu(a)\nu(a^{-1}) \leq \nu(a^{-1}) \text{ and } a^{-1} \in R.$$

(5) For every element $a \in D^*$, we have $aRa^{-1} = R$. Let $ara^{-1} \in aRa^{-1}$, where $r \neq 0$. It follows from $\nu(r) \geq e$ that

$$\nu(ara^{-1}) = \nu(a)\nu(r)\nu(a^{-1}) \geq \nu(a)\nu(a^{-1}) = e, \quad ara^{-1} \in R,$$

and $aRa^{-1} \subseteq R$. Similarly, we have $a^{-1}Ra \subseteq R$. Consequently, $R \subseteq aRa^{-1}$ and $aRa^{-1} = R$ or $aR = Ra$.

(6) Any two ideals of the ring R are comparable with respect to inclusion. Let B and C be two ideals of the ring R and let $C \not\subseteq B$. We choose an element $a \in C \setminus B$. Let $b \in B$ and $b \neq 0$. If $ab^{-1} \in R$, then $a = (ab^{-1})b \in RB = B$. This is impossible. Therefore $ab^{-1} \notin R$. By (4),

$$ba^{-1} \in R \quad \text{and} \quad b = (ba^{-1})a \in RC = C.$$

Now we assume that R is a subring of the division ring D such that conditions (4) and (5) hold for R. We prove that there exists a linearly ordered group $(\Gamma, <)$ and a valuation $\nu\colon D^* \to \Gamma$ such that $R = \{a \in D^* \mid \nu(a) \geq e\} \cup \{0\}$. For this purpose, we denote by U the group of all invertible elements of the ring R. Then U is a normal subgroup in D^*, i.e., $aUa^{-1} = U$ for every $a \in D^*$. We have $aRa^{-1} = R$ and $a^{-1}Ra = R$. Let $ava^{-1} \in aUa^{-1}$. Then $ava^{-1} \in R$ and the element ava^{-1} is invertible in R, since it has the converse $av^{-1}a^{-1} \in R$. Consequently, $ava^{-1} \in U$. Thus, $aUa^{-1} \subseteq U$. Similarly, we have $a^{-1}Ua \subseteq U$. Therefore $U \subseteq aUa^{-1}$ and $aUa^{-1} = U$.

We take the factor group D^*/U as the group Γ. We define a relation $\leq$ on the group Γ as follows. Let $g, h \in \Gamma$, $g = aU$, and let $h = bU$, where $a, b \in D^*$. We assume that

$$h \leq g \Longleftrightarrow aR \subseteq bR.$$

Our definition does not depend on the choice of representatives a and b. It is easy to verify that $\leq$ is a reflexive transitive relation. Let $h \leq g$ and $g \leq h$. Therefore

$aR \subseteq bR$, $bR \subseteq aR$, and $aR = bR$. Therefore $b^{-1}a, a^{-1}b \in R$. Now we have $a = b(b^{-1}a)$, where $b^{-1}a \in U$, since $(b^{-1}a)^{-1} = a^{-1}b$. Consequently, $a \in bU$ and $b \in aU$. Therefore $aU = bU$ and $g = h$. We have the order $\leq$ on the group Γ. Similar to (6), we can prove that either $aR \subseteq bR$ or $bR \subseteq aR$. Therefore, either $h \leq g$ or $g \leq h$, i.e., the order $\leq$ is linear. Now we assume that $h \leq g$ and $f = cU$. It follows from $aR \subseteq bR$ that $caR \subseteq cbR$. However,

$$fh = cU \cdot bU = cbU \quad \text{and} \quad fg = caU.$$

Therefore $fh \leq fg$. Similarly, we have $hf \leq gf$. Thus, $\langle \Gamma, \leq \rangle$ is a linearly ordered group.

Let $\nu \colon D^* \to \Gamma$ be the canonical homomorphism, i.e., $\nu \colon a \to aU$, $a \in D^*$. We prove that ν is a valuation. We need to verify only property (2) from the definition of a valuation. We take arbitrary elements $a, b \in D^*$ with $a + b \neq 0$ and prove the relation $\nu(a + b) \geq \min(\nu(a), \nu(b))$. For definiteness, assume that $\nu(a) \leq \nu(b)$ (therefore, we have $bR \subseteq aR$). Then $\nu(a^{-1}b) \geq \nu(1)$. Consequently,

$$a^{-1}bR \subseteq 1 \cdot R \quad \text{and} \quad a^{-1}b \in R.$$

Therefore

$$a^{-1}b + 1 \in R \quad \text{and} \quad (a + b)R \subseteq aR.$$

Consequently,

$$\nu(a + b) \geq \nu(a) = \min(\nu(a), \nu(b))$$

and ν is a valuation. Finally, we prove that the initial ring R coincides with $\{a \in D^* \mid \nu(a) \geq e\} \cup \{0\}$. Let $a \in R$, $a \neq 0$. Then $aR \subseteq 1 \cdot R$ and $\nu(a) \geq e$. If $a \in D^*$, $a \neq 0$, and $\nu(a) \geq e = \nu(1)$, then $aR \subseteq 1 \cdot R = R$ and $a \in R$.

The ring R constructed by the first or second method (these methods are equivalent) is called the *invariant ring of the valuation* ν of the division ring D. For the field D, property (5) holds automatically. The invariant ring of a valuation of the field D is called a *valuation ring* of D (see Exercise 6). An additional information about valuations of fields is contained in Section 26.

Let Γ be an infinite cyclic group with generator g. We assume that the group Γ is naturally linearly ordered:

$$g^n \leq g^m \iff n \leq m$$

(n and m are arbitrary integers). In this case, every valuation $\nu \colon D^* \to \Gamma$ is said to be *discrete*.

Let ν be a discrete valuation of the division ring D and let R be an invariant ring of the valuation ν, i.e.,

$$R = \{a \in D^* \mid \nu(a) \geq e\} \cup \{0\}.$$

We prove that R is a discrete valuation domain in the sense of Definition 3.2. By property (1), the ring R is local and

$$J(R) = A = \{a \in R \mid \nu(a) > e\} \cup \{0\}.$$

Let p one of the elements of the group D^* with $\nu(p) = g$. It is clear that $p \in R$ and p is a non-nilpotent element. We verify that $J(R) = pR = Rp$. The second relation follows from property (5). Since $\nu(p) = g > e$, we have that $p \in A$ and $pR \subseteq A$. On the other hand, for every nonzero $a \in A$, there exists a positive integer n such that $\nu(a) = \nu(p^n)$. Therefore $a = p^n v$, where v is some invertible element in R. Consequently, $a \in pR$ and $A = pR$. Let $a \in \bigcap_{n\geq 1}(pR)^n$ and $a \neq 0$. Then $a = p^n x_n$ for every n, where $x_n \in R$. We obtain $\nu(a) = \nu(p^n)\nu(x_n) = g^n \nu(x_n)$. This is impossible. Consequently,

$$\bigcap_{n\geq 1}(pR)^n = 0.$$

Thus, R is a discrete valuation domain.

Now we assume that R is a discrete valuation domain and K is the division ring of fractions of R. Let Γ be an infinite cyclic group with generator p, where p is a prime element of the domain R. Every element of K^* can be uniquely written in the form vp^n, where v is an invertible element of the ring R and n is some integer. We define the mapping $\nu\colon K^* \to \Gamma$ by the relation $\nu(vp^n) = p^n$. Then ν is a valuation of the division ring K (it is said to be *canonical*) and R coincides with invariant ring of the valuation ν, i.e.,

$$R = \{a \in K^* \mid \nu(a) \geq e\} \cup \{0\}.$$

Let U be the group of all invertible elements of the ring R. It is clear that the so-defined valuation ν coincides with the valuation if we define it as above, starting from the factor group K^*/U.

We can conclude that discrete valuation domains are certain invariant valuation rings of division rings.

We present examples of discrete valuations. The corresponding valuation rings are those discrete valuation domains which were directly defined before. An infinite cyclic group with generator g is denoted by $\langle g \rangle$.

(1) Let p be a fixed prime integer. Every nonzero element of the field of rational numbers $\mathbb{Q}$ has the form $(m/n)p^k$, where m and n are nonzero integers which are not divisible by p, k is a integer. The mapping

$$\nu_p\colon \mathbb{Q}^* \to \langle p \rangle \quad \text{with} \quad \nu_p((m/n)p^k) = p^k$$

is a (discrete) valuation of the field $\mathbb{Q}$. The valuation ring of the valuation is $\mathbb{Z}_p$. The valuation ν_p is called the *p-adic valuation.*

(2) Every nonzero Laurent series $f(x)$ over the field F is equal to $g(x)x^k$, where $g(x)$ is a power series with nonzero constant term and k is a integer. The mapping

$$\nu_x \colon F\langle x\rangle^* \to \langle x\rangle$$

is a valuation of the Laurent series field $F\langle x\rangle$ with valuation ring $F[[x]]$. Instead of x, we take some irreducible polynomial $p(x)$ over the field F. By the same method, we can obtain the valuation $\nu_{p(x)}$. The ring of the valuation $\nu_{p(x)}$ coincides with the completion of the localization of the ring $F[x]$ with respect to the prime ideal $(p(x))$.

Two above examples have the following generalization.

(3) Let R be a commutative principal ideal domain, F be the field of fractions of R, and let p be some prime element of the domain R. Every element a of F^* can be represented in the form $(r/s)p^k$, where r and s are elements of R which are not divisible by p and k is a integer. We can also note that k is the greatest degree of p included in the decomposition of the element a into the product of positive and negative powers of prime elements (or into the decomposition of the fractional ideal aR in the product of powers of prime ideals). We define the mapping $\nu_p \colon F^* \to \langle p\rangle$ by the relation $\nu_p(a) = p^k$. We obtain the valuation of the field F with valuation ring which coincides with the localization of the ring R with respect to the prime ideal (p).

We can define a similar valuation ν_P provided R is a Dedekind domain (see Exercise 3) and P is a proper nonzero prime ideal. We have $\nu_P(a) = P^k$ for $a \in F^*$, where k is the greatest degree of the ideal P included in the decomposition of the fractional ideal aR into the product of powers of prime ideals.

Exercise 1. We assume that R is a discrete valuation domain, D is the division ring of fractions of R, $U(R)$ is the group of invertible elements of the domain R, and D^* is the group of invertible elements of the division ring D. Prove that the group D^* is a semidirect product of the subgroups $U(R)$ and $\langle p\rangle$. This means that $U(R)$ is a normal subgroup in D^*, $D^* = U(R)\langle p\rangle$, and $U(R) \cap \langle p\rangle = \langle 1\rangle$.

Exercise 2. The p-adic completion $\widehat{R}$ of a discrete valuation domain R is unique in the following sense. If S is one more completion of the domain R, then the identity mapping of the ring R can be extended to a ring isomorphism between $\widehat{R}$ and S.

Exercise 3. A commutative domain R with field of fractions F is called a *Dedekind domain* if for every nonzero R-submodule A in F, there exists an R-submodule B in F with $AB = R$. Prove that a commutative domain R is a

Dedekind domain if and only if the ring R is Noetherian and R_P is a discrete valuation domain for all maximal ideals P of the ring R. Here, R_P is the localization of the ring R with respect to P.

Exercise 4. Let F be a field of algebraic numbers (i.e., F is a finite extension of the field of rational numbers). The integral closure of the ring of integers $\mathbb{Z}$ in F is called the *ring of algebraic integers* of the field F. Prove that the ring of algebraic integers is a Dedekind domain.

Exercise 5. Let z_0 be a point in the complex plane and let R be the ring of all functions which are holomorphic in some disk centered of z_0. Then R is a discrete valuation domain and the maximal ideal of R consists of all functions which are equal to zero in z_0. The function $z - z_0$ is a prime element of the ring.

Exercise 6. Let R be a commutative domain and let F be the field of fractions of R. The ring R is called the *valuation ring* (of the field F) if either $a \in R$ or $a^{-1} \in R$ for every nonzero element a in F. Prove that R is a valuation ring if and only if any two ideals of R are comparable with respect to inclusion.

Exercise 7. Let R be a commutative principal ideal domain, F be the field of fractions of R, and let S be a valuation ring of the field F such that S contains R and $S \neq F$. Prove that S is the localization of the ring R with respect to some prime ideal of the ring R. (This can be also applied to the ring of integers and the polynomial ring over any field.)

Exercise 8. Every valuation ring of the field of rational numbers $\mathbb{Q}$ is either equal to $\mathbb{Q}$ or has the form $\mathbb{Z}_p$ for some p.

4 Primary notions of module theory

The symbol R denotes a discrete valuation domain. We present some beginning properties of modules over discrete valuation domains.

We consider the structure of cyclic modules. Let $M = Ra$ be a cyclic module with generator a. We have the epimorphism $\varphi\colon R \to M$, where $\varphi(r) = ra$ for every element $r \in R$. If $\mathrm{Ker}(\varphi) = 0$, then $M \cong R$. Otherwise, $\mathrm{Ker}(\varphi) = p^n R$ for some positive integer n. Then $M \cong R/p^n R$. The cyclic module $R/p^n R$ is denoted by $R(p^n)$. Every module which is isomorphic to $R(p^n)$ is also denoted by $R(p^n)$. The left module R satisfies the same property. Since $p^n R$ is an ideal of the ring R, we have that $R(p^n)$ is a right R-module, i.e., it is an R-R-bimodule. We have that $R(p^n)$ also is an R_{p^n}-R_{p^n}-bimodule, where R_{p^n} denotes the factor ring R/p^nR (see Section 3). It follows from this section that for every element $y \in R$, there exists a group automorphism σ_y of the module $R(p^n)$ such that

$$\sigma_y(rc) = \sigma_y(r)\sigma_y(c) \quad \text{and} \quad cy = y\sigma_y(c)$$

for any two elements $r \in R$ and $c \in R(p^n)$.

Thus, each cyclic R-modules either is isomorphic to the module R or is isomorphic to the module $R(p^n)$ with $n \geq 1$. Therefore, we obtain that every simple module is isomorphic to $R(p)$. (A nonzero module M is said to be *simple* if M does not have proper nonzero submodules.) Proper submodules of the module R have the form p^nR, $n \geq 1$. If $R(p^n) = Ra$, then the submodules $R(pa), R(p^2a), \dots, R(p^{n-1}a)$ are proper submodules of the module Ra. Consequently, submodules of cyclic modules are cyclic modules. Again, let $Ra = R(p^n)$. We take some submodule $R(p^ka)$, where $1 \leq k \leq n-1$. Then

$$Ra/R(p^ka) \cong R(p^k).$$

The next three important modules are $\widehat{R}$, K, and K/R, where $\widehat{R}$ is the completion of the ring R in the p-adic topology and K is the division ring of fractions of the domain R (see Section 3). Modules which are isomorphic to $\widehat{R}$, K, and K/R, respectively, have the same notations. These three modules are R-R-bimodules. We will show below that every proper submodule of the module K is isomorphic to R. The structure of the factor module $\widehat{R}/R$ is presented in Exercise 2. The factor module K/R is of especial interest. It has elements $1/p + R, 1/p^2 + R, \dots, 1/p^n + R, \dots$ such that

$$p\left(\frac{1}{p} + R\right) = 0 \quad \text{and} \quad p\left(\frac{1}{p^{n+1}} + R\right) = \frac{1}{p^n} + R$$

for every $n \geq 1$. The set of these elements is a generator system of the module K/R. We assume that some module C has a countable generator system $c_1, c_2, \dots, c_n, \dots$ such that $pc_1 = 0$ and $pc_{n+1} = c_n$ for every $n \geq 1$. A module C is the union of an ascending chain of the submodules

$$Rc_1 \subset Rc_2 \subset \cdots \subset Rc_n \subset \cdots,$$

where

$$Rc_n \cong R(p^n), \quad n \geq 1.$$

Every element of the module C is equal to vc_n for some integer n and an invertible element v of the ring R. The mapping $vc_n \to v(1/p^n + R)$ is an isomorphism of modules C and K/R. Sometimes, we denote the module K/R and every module which is isomorphic to K/R by $R(p^\infty)$; it is called the *quasicyclic* module. Every proper submodule of the quasicyclic module C is equal to Rc_n for some $n \geq 1$ and $C/Rc_n \cong R(p^\infty)$. It follows from Section 3 that for every element $y \in R$, there exists a group automorphism σ_y of the module $R(p^\infty)$ such that

$$\sigma_y(rc) = \sigma_y(r)\sigma_y(c) \quad \text{and} \quad cy = y\sigma_y(c)$$

for any two elements $r \in R$ and $c \in R(p^\infty)$.

Direct sums of cyclic modules, copies of the quasicyclic module, and copies of the module K are very important. Such sums are studied in the next chapter.

Let a be a nonzero element of a module M. The element a is called a *torsion* element if $ra = 0$ for some nonzero element r of the ring R. The element a is a torsion element if and only if $p^n a = 0$ for some positive integer n. Indeed, let $ra = 0$, where $r \in R$ and $r \neq 0$. We have $r = vp^n$, where v is an invertible element of the ring R and n is a positive integer. Then $vrp^n a = 0$. Therefore $p^n a = 0$, and it is sufficient to consider only the powers of the element p in questions related to the annihilation of elements of the module by elements of the ring. As a rule, this also holds for other questions related to elements of modules or submodules, since multiplication by invertible elements of the ring does not change the whole given module.

Let a be a torsion element of the module M and let n be the least positive integer with $p^n a = 0$. The element p^n is called the *order* of the element a and the order of the module Ra; it is denoted by $o(a)$. We also say that a is an *element of finite order*. If a is an element of the module M such that $p^n a \neq 0$ for every $n \geq 1$, then a is called an *element of infinite order* (we denote $o(a) = \infty$ in this case). We assume that the order of the zero element is equal to zero.

In relation to orders of elements, we formulate the following remark on cyclic modules. We assume that $a \in R(p^n)$, $o(a) = p^k$, and $ra = 0$, where $r \in R$. Then r is divisible by p^k.

A module M is called a *primary* module or a *torsion* module if every element of the module M is of finite order. If all nonzero elements of the module M have an infinite order, then M is called a *torsion-free* module. If the module contains nonzero elements of finite order and elements of infinite order, then it is said to be *mixed*. Thus, every nonzero module is contained in the one of the following three classes of modules: the class of primary modules, the class of torsion-free modules, and the class of mixed modules.

Theorem 4.1. *Let M be a mixed module. The set T of all elements of finite order of the module M is a submodule of the module M. In addition, T is a primary module and the factor module M/T is a torsion-free module.*

Proof. Let $a, b \in T$ and let $p^m a = p^n b = 0$ for some positive integers m and n. For definiteness, let $m \leq n$. Then $p^n(a - b) = 0$ and $a - b \in T$. Let $r \in R$

and $r = p^k u$, where k is a non-negative integer and u is an invertible element of the ring R. We have $p^m u = vp^m$ for some invertible element v of the ring R. Now we obtain $p^m(ra) = p^k v p^m a = 0$ and $ra \in T$. Consequently, T is a submodule in M. We assume that a residue class $a + T$ has the finite order p^n. Then $p^n(a + T) = p^n a + T = 0$. Therefore $p^n a \in T$. Therefore, there exists a number k such that $p^k p^n a = 0$. Therefore $a \in T$ and $a + T = 0$. Consequently, M/T is a torsion-free module. □

Let M be a mixed module. The submodule T is called the *torsion* (or *primary*) submodule of the module M; it is denoted by $t(M)$. If M is a mixed module such that $M = t(M) \oplus F$ for some submodule F (it is clear that F is a torsion-free module), then M is called a *split* mixed module. An example of a nonsplit mixed module is contained in Exercise 1. If M is a primary module, then we set $t(M) = M$. For the torsion-free module M, we set $t(M) = 0$. The modules $R(p^n)$, $n \geq 1$, and $R(p^\infty)$ are primary modules, and R, K, $\widehat{R}$, and $\widehat{R}/R$ are torsion-free modules.

Let R be an incomplete discrete valuation domain. The rings R and $\widehat{R}$ have the same prime element p ($\widehat{R}$ is the p-adic completion of the domain R defined in Section 3). Since R is a subring in $\widehat{R}$, every $\widehat{R}$-module is an R-module. We assume that M is a primary R-module, $a \in M$, and $r \in \widehat{R}$. We have that for some positive integer k, the relation $p^k a = 0$ holds and $r = \lim r_i$ ($r_i \in R$ and $i \geq 1$) is the limit in the p-adic topology. There exists a positive integer n such that $r_i - r_j \in p^k R = Rp^k$ for all $i, j \geq n$. Therefore, the elements r_i have the same action on the element a for $i \geq n$. We assume that ra is equal to this element $r_i a$ for $i \geq n$. By this method, M is turned into a primary $\widehat{R}$-module. We conclude that primary R-modules and primary $\widehat{R}$-modules can be identified. In particular, primary cyclic R-modules and quasicyclic R-modules coincide with primary cyclic and quasicyclic $\widehat{R}$-modules, respectively. It can be directly verified that there exist isomorphisms

$$R(p^n) \to \widehat{R}(p^n) \quad \text{and} \quad R(p^\infty) \to \widehat{R}(p^\infty).$$

Indeed, we have

$$R/p^n R = R/(R \cap p^n \widehat{R}) \cong (R + p^n \widehat{R})/p^n \widehat{R} = \widehat{R}/p^n \widehat{R}.$$

In addition, the mapping of generators $1/p^n + R \to 1/p^n + \widehat{R}$ induces the isomorphism $K/R \cong \widehat{K}/\widehat{R}$.

Let M and N be two primary $\widehat{R}$-modules. Then every R-homomorphism $\varphi \colon M \to N$ is an $\widehat{R}$-homomorphism. We have to prove that $\varphi(ra) = r\varphi(a)$ for every element $a \in M$ and $r \in \widehat{R}$. If $r = \lim r_i$, $r_i \in R$, then $ra = r_i a$

and $r\varphi(a) = r_i\varphi(a)$ for quite large subscripts i (consider that if $p^k a = 0$, then $p^k\varphi(a) = 0$). We have

$$\varphi(ra) = \varphi(r_i a) = r_i\varphi(a) = r\varphi(a).$$

These results for primary modules are also related to Exercises 5 and 7.

We explain as commutative discrete valuation domains appear in the theory of modules over commutative principal ideal domains. We assume that R is a commutative principal ideal domain, p is a prime element of R, $P = (p)$, and R_P is the localization of the ring R with respect to the prime ideal P. We know that R_P is a discrete valuation domain (see Section 3). Let M be some R-module which is primary with respect to p. This means that for every $x \in M$, the relation $p^k x = 0$ holds for some positive integer k. There exists a natural method of the construction a module over R_P from M. This leads to the question about the definition of the product $b^{-1}x$, where $x \in M$ and $b \in R \setminus P$. We assume that $p^k x = 0$. There exists an element c in R such that $bc = 1 + p^k r$ for some $r \in R$. We assume that $b^{-1}x = cx$. In addition, M is a primary R_P-module. Therefore, primary R_P-modules and R-modules primary with respect to p can be assumed the same objects.

We specialize these properties for the ring of integers $\mathbb{Z}$. Let $\mathbb{Z}_p$ be the ring of all rational numbers whose denominators are coprime to a given prime integer p, i.e., the localization of the ring $\mathbb{Z}$ with respect to (p). Then $\mathbb{Z}_p$ is a discrete valuation domain and the p-adic completion of $\mathbb{Z}_p$ is the ring of p-adic integers $\widehat{\mathbb{Z}}_p$ (this topic was considered in Section 3). $\mathbb{Z}_p$-modules coincide with Abelian groups G such that $qG = G$ for all prime numbers $q \neq p$. Such Abelian groups are called *p-local* groups. Primary $\mathbb{Z}_p$-modules, primary $\widehat{\mathbb{Z}}_p$-modules, Abelian p-groups, and primary groups are the same objects. Submodules and homomorphisms of primary $\widehat{\mathbb{Z}}_p$-modules coincide with subgroups and group homomorphisms. $\widehat{\mathbb{Z}}_p$-modules are also called p-adic modules. Such modules play an important role in several sections of the theory of Abelian groups. For an Abelian p-group G, we present the exact method of defining the structure of a p-adic module on G (see Exercise 7 on the uniqueness of such a structure). If $\alpha = (k_1, k_2, \dots, k_n, \dots) \in \widehat{\mathbb{Z}}_p$ and the element $a \in G$ has the order p^n, then we set $\alpha a = k_n a$.

There are quite many works on Abelian p-local groups. The content of these works can be extended to modules over discrete valuation domains with the use of small unsophisticated corrections.

Since R is a subring of the division ring of fractions K of R, every left vector K-space is a torsion-free R-module. Let V and W be two left K-spaces. Every R-homomorphism $\varphi\colon V \to W$ is a linear operator of these spaces. We verify that $\varphi((r/p^n)x) = (r/p^n)\varphi(x)$ for any two elements $x \in V$ and $r/p^n \in K$. We have the relations

$$\varphi(rx) = r\varphi(x) \quad \text{and} \quad p^n\varphi\left(\left(\frac{r}{p^n}\right)x\right) = p^n\left(\frac{r}{p^n}\right)\varphi(x).$$

Therefore the required relation holds.

Let M be a module. For an element $r \in R$, we set $rM = \{rx \mid x \in M\}$. We have $r = p^k v$, where k is a non-negative integer and v is an invertible element of the ring R. Then $rM = p^k M$. We have that $p^n M$ is a submodule in M for every non-negative integer n. Indeed, we represent the element $r \in R$ in the form $r = p^k v$, as above. In addition, let $vp^n = p^n w$, where w is an invertible element in R. Then we obtain

$$r(p^n M) = p^k v(p^n M) = p^n(p^k wM) \subseteq p^n M.$$

We note that $p^0 M = M$.

A module M is said to be *bounded* if $p^n M = 0$ for some positive integer n. Any bounded module is primary. Let M be a bounded module and $p^n M = 0$ for some n. Since $(p^n R)M = 0$, the module M is turned into a left module over the ring R_{p^n} (where $R_{p^n} = R/p^n R$) if we set $\bar{r}m = rm$ for all $m \in M$ and $\bar{r} \in R_{p^n}$, where $\bar{r} = r + p^n R$. Conversely, if M is a left R_{p^n}-module, then using the same relations $rm = \bar{r}m$, we obtain the R-module M. Since $p^n \bar{1} = 0$ in R_{p^n}, we have

$$p^n M = (p^n 1)M = (p^n \bar{1})M = 0,$$

i.e., M is a bounded module. Consequently, R_{p^n}-modules and bounded R-modules M with $p^n M = 0$ can be identified. We have the especial case if $pM = 0$. Then M is a left vector space over the residue division ring R_p. The structure of R_{p^n}-modules is presented in Theorem 4.8, and bounded R-modules are considered in Theorem 7.1. Let M and N be bounded modules with $p^n M = 0 = p^n N$. Then R-homomorphisms and R_{p^n}-homomorphisms from M into N coincide.

For every $n \geq 1$, the module $M/p^n M$ is bounded, since $p^n(M/p^n M) = 0$. We set

$$M[p^n] = \{a \in M \mid p^n a = 0\}.$$

Let $r = vp^k \in R$, where v is an invertible element, $k \geq 0$. We have $p^n v = up^n$, where u is an invertible element of the ring R. Then

$$p^n ra = p^n vp^k a = up^n p^k a = 0 \quad \text{and} \quad ra \in M[p^n].$$

Consequently, $M[p^n]$ is a submodule in M. Since $p^n M[p^n] = 0$, we have that $M[p^n]$ is a bounded module. The modules M/pM and $M[p]$ are R_p-spaces. We often use this property. The submodule $M[p]$ is called the *socle* of the module M.

The dimension of the R_p-space M/pM is called the *p-rank* of the module M. We denote it by $r_p(M)$.

We can compare the orders of elements of the module. For this purpose, we define an order on the set of powers of the element p. We assume that

$$p^m \leq p^n \qquad \Longleftrightarrow \qquad m \leq n$$

(where m and n are non-negative integers). We also assume that $0 < p^n$ and $p^n < \infty$ for every $n \geq 1$.

A submodule G of the module M is said to be *essential* if G has the nonzero intersection with every nonzero submodule of the module M. We present one method of constructing essential submodules. Let A and C be two submodules of the module M with $A \cap C = 0$. The set of all submodules B of the module M such that $C \subseteq B$ and $A \cap B = 0$ is inductive. Consequently, it contains a maximal submodule B_0. Then $A + B_0$ is an essential submodule in M. Indeed, let H be a submodule in M such that $(A + B_0) \cap H = 0$. Then $A \cap (B_0 + H) = 0$. By the choice of the submodule B_0, we have $B_0 + H = B_0$ and $H \subseteq B_0$. Therefore $H = 0$. Setting $C = 0$, we obtain the following property. For every submodule A, there exists a submodule B such that $A \cap B = 0$ and $A + B$ is an essential submodule in M.

Proposition 4.2. *For a module M, a submodule G of M is essential if and only if $M[p] \subseteq G$ and M/G is a primary module.*

Proof. Let G be an essential submodule of the module M. If a is a nonzero element of the socle $M[p]$, then $Ra \cong R(p)$. Consequently, Ra is a simple module, $Ra \subseteq G$ and $M[p] \subseteq G$. If a is an arbitrary nonzero element of the module M, then $Ra \cap G \neq 0$. Consequently, there exists a $r \in R$ such that $ra \in G$ and $ra \neq 0$. This means that M/G is a primary module.

Now we assume that $M[p] \subseteq G$ and M/G is a primary module. It is sufficient to prove that G has the nonzero intersection with every nonzero cyclic submodule Ra of the module M. If $o(a) = p$, then $Ra \subseteq G$. If $o(a) = p^k$ with $k > 1$, then $Rp^{k-1} \subseteq G$. Let $o(a) = \infty$. Since M/G is a primary module, $p^n a \in G$ for some positive integer n. □

Let M be a module. We take the tensor product $K \otimes_R M$ (main properties of the tensor product are presented in Section 2). Every element x is equal to a finite sum $\sum_{i=1}^{s} k_i \otimes a_i$, where $k_i \in K$ and $a_i \in M$, $i = 1, \dots, s$. For an arbitrary element $k \in K$, we set $kx = \sum_{i=1}^{s} kk_i \otimes a_i$. With respect to this outer multiplication, $K \otimes_R M$ is a left vector space over the division ring of fractions K; in particular, $K \otimes_R M$ is an R-module. We have $k_i = r_i/p^{n_i}$, where $r_i \in R$, $n_i \geq 0$ and

$t_i = n - n_i$, where n is the greatest integer among the integers n_i, $i = 1, \dots, s$. Then

$$x = \sum_{i=1}^{s} 1/p^n \otimes p^{t_i} r_i a = 1/p^n \otimes \sum_{i=1}^{s} p^{t_i} r_i a_i.$$

Thus, every element of the K-space $K \otimes_R M$ is equal to $1/p^n \otimes a$, where $n \geq 0$, $a \in M$.

We denote by KM the set of all formal expressions (fractions) $(1/p^n)a$ (where $n \geq 0$ and $a \in M$) which satisfy the following condition: $(1/p^n)a = (1/p^k)b$ if and only if there exist nonzero $r, s \in R$ such that $ra = sb$ and $rp^n = sp^k$. This equality relation is an equivalence relation and the set KM is decomposed to classes of equal fractions. Now we assume that KM is the set classes of equal fractions. Naturally, we deal with representatives of classes. We can add fractions by a usual rule of addition of fractions. We obtain the left vector K-space KM. Let $(1/p^n)a = (1/p^k)b$. We take the above elements r and s. We set $t = rp^n = sp^k$. Then

$$1/p^n = t^{-1}r \quad \text{and} \quad 1/p^k = t^{-1}s.$$

Now we have

$$1/p^n \otimes a = t^{-1}r \otimes a = t^{-1} \otimes ra \quad \text{and} \quad 1/p^k \otimes b = t^{-1} \otimes sb.$$

Since $ra = sb$, we have $1/p^n \otimes a = 1/p^k \otimes b$. Therefore, the mapping

$$\gamma\colon KM \to K \otimes_R M, \qquad (1/p^n)a \to 1/p^n \otimes a, \qquad n \geq 0, \qquad a \in M$$

is correctly defined. It is clear that γ is a K-isomorphism. Therefore $KM \cong K \otimes_R M$.

We have the canonical R-module homomorphism $\varphi\colon M \to K \otimes_R M$, where $\varphi(a) = 1 \otimes a$ for every $a \in M$.

Lemma 4.3. *The kernel of the homomorphism φ coincides with the torsion submodule $t(M)$ of the module M. In particular, M is a torsion-free module if and only if φ is a monomorphism.*

Proof. Let $a \in M$. Then $1 \otimes a = 0$ in $K \otimes_R M$ (i.e., $a = 0$ in KM) if and only if there exists a nonzero element $r \in R$ such that $ra = 0$, i.e., $a \in t(M)$. Consequently, $\mathrm{Ker}(\varphi) = t(M)$. □

A torsion-free module M can be identified with the image of the monomorphism φ (for every $m \in M$, we identify m with $1 \otimes m$). Thus, we can assume that M is a submodule in $K \otimes_R M$.

Every module contains sets of elements with properties which are similar to the properties of bases of vector spaces.

A system (or a set) $\{a_i\}_{i\in I}$ of nonzero elements of the module M is said to be *linearly independent* if for all elements $a_1, \dots, a_k$, the relation $r_1a_1+\cdots+r_ka_k = 0$ with $r_i \in R$, implies the relation $r_1a_1 = \cdots = r_ka_k = 0$.

Lemma 4.4. *A system $\{a_i\}_{i\in I}$ is linearly independent if and only if the submodule A generated by this system is equal to the direct sum $\bigoplus_{i\in I} Ra_i$.*

Proof. We assume that the mentioned system is linearly independent. The submodule A is equal to the sum $\sum_{i\in I} Ra_i$. It is necessary only to prove that $Ra_i \cap \sum_{j\neq i} Ra_j = 0$ for every $i \in I$. Let

$$ra_i = r_1a_1 + \cdots + r_ka_k, \quad \text{where} \quad r, r_1, \dots, r_k \in R \quad \text{and} \quad i \neq 1, \dots, k.$$

Since the system is linearly independent, we have $ra_i = 0$. Conversely, if $A = \bigoplus_{i\in I} Ra_i$, then the linear independence of the system $\{a_i\}_{i\in I}$ directly follows from properties of a direct sum. □

A nonzero module M always has linearly independent systems. For example, if $a \in M$ and $a \neq 0$, then $\{a\}$ is a linearly independent system. An ordered set of linearly independent systems is inductive. Therefore, it contains a *maximal linearly independent system.* Let $\{a_i\}_{i\in I}$ be a maximal linearly independent system of elements of the module M. Then for every element $a \in M$ there exist elements $a_1, \dots, a_k$ and $r, r_1, \dots, r_k \in R$ such that

$$ra = r_1a_1 + \cdots + r_ka_k \quad \text{and} \quad ra \neq 0.$$

Consequently, we obtain $ra \in \bigoplus_{i\in I} Ra_i$. Therefore, $\bigoplus_{i\in I} Ra_i$ is an essential submodule in M. The converse assertion also holds. If $\{a_i\}_{i\in I}$ is a linearly independent system and $\bigoplus_{i\in I} Ra_i$ is an essential submodule in M, then $\{a_i\}_{i\in I}$ is a maximal linearly independent system.

Linearly independent systems of elements are especially useful for torsion-free modules. Let M be a torsion-free module. A system of nonzero elements $a_1, \dots, a_k$ from M is linearly independent if and only if the relation

$$r_1a_1 + \cdots + r_ka_k = 0 \quad \text{with} \quad r_i \in R$$

implies that $r_1 = \cdots = r_k = 0$. In contrast to vector spaces, in a general case, elements of the module are not necessarily linear combinations of elements from a maximal linearly independent system. However, we have the result presented below. We identify the module M with the submodule of the left K-space $K \otimes_R M$ consisting of elements $1 \otimes a$ for all $a \in M$.

Lemma 4.5. *A system $\{a_i\}_{i\in I}$ of elements of the torsion-free module M is a maximal linearly independent if and only if, when it is a basis of the K-space $K \otimes_R M$.*

Proof. Let $\{a_i\}_{i\in I}$ be a maximal linearly independent system of elements of the module M. We assume that

$$(r_1/p^{n_1})a_1 + \cdots + (r_k/p^{n_k})a_k = 0, \quad \text{where} \quad r_i/p^{n_i} \in K,\ i = 1, \ldots, k.$$

We multiply both parts of this relation by p^m, where m is the greatest integer among the integers $n_1, \ldots, n_k$. We obtain

$$p^{t_1} r_1 a_1 + \cdots + p^{t_k} r_k a_k = 0, \quad \text{where} \quad t_i = m - n_i,\ i = 1, \ldots, k.$$

Consequently,

$$p^{t_1} r_1 = \cdots = p^{t_k} r_k = 0 \quad \text{and} \quad r_1 = \cdots = r_k = 0.$$

Now we assume that

$$1/p^n \otimes a \in K \otimes_R M, \quad \text{where} \quad n \geq 0 \quad \text{and} \quad a \in M.$$

There exist elements $r, r_1, \ldots, r_k \in R$ with $r \neq 0$ such that $ra = r_1 a_1 + \cdots + r_k a_k$. Therefore

$$(1/p^n) r \otimes a = 1/p^n r_1 (1 \otimes a_1) + \cdots + 1/p^n r_k (1 \otimes a_k).$$

We can use the automorphism σ_r of the division ring K (see Subsection 8) in Section 3) to represent the element $(1/p^n)r$ in the form $q(1/p^n)$ for some $q \in K$. Therefore it is clear that the element $1/p^n \otimes a$ is linearly expressed over K in terms of $1 \otimes a_1, \ldots, 1 \otimes a_k$.

Let $\{a_i\}_{i\in I}$ be a basis of the space $K \otimes_R M$. This is obvious that this system is linearly independent. If $a \in M$ and $a \neq 0$, then

$$a = (r_1/p^{n_1})a_1 + \cdots + (r_k/p^{n_k})a_k, \quad \text{where} \quad r_i/p^{n_i} \in K.$$

There exists a positive integer m such that

$$p^m a = s_1 a_1 + \cdots + s_k a_k \quad \text{and} \quad p^m a \neq 0$$

for some $s_1, \ldots, s_k \in R$. Consequently, $\{a_i\}_{i\in I}$ is a maximal linearly independent system of elements of the module M. □

We prove that any two maximal linearly independent systems of elements of the module M have equal cardinalities. Let $\{b_j\}_{j\in J} \cup \{c_l\}_{l\in L}$ be some maximal linearly independent system of elements of the module M, where every element b_j is of finite order, and every element c_l is of infinite order. Then $\{b_j\}_{j\in J}$ is a maximal linearly independent system of elements of the module $t(M)$. Let $o(b_j) = p^{n_j}$,

$n_j \geq 1$. The system of elements $\{p^{n_j-1}b_j\}_{j\in J}$ of the socle $M[p]$ is also linearly independent. The socle $M[p]$ is an R_p-space. This is obvious that this system is a basis of the R_p-space. Consequently, the cardinality $|J|$ is the dimension of the space $M[p]$ and it is uniquely defined.

We take a system $\{c_l\}_{l\in L}$. We prove that $\{\bar{c}_l\}_{l\in L}$ is a maximal linearly independent system of elements of the module $M/t(M)$, where $\bar{c}_l = c_l + t(M), l \in L$. Let

$$r_1\bar{c}_1 + \cdots + r_k\bar{c}_k = 0, \qquad r_i \in R.$$

Then $r_1c_1 + \cdots + r_kc_k \in t(M)$. There exists a nonzero element $r \in R$ with $rr_1 = \cdots = rr_k = 0$. Therefore $r_1 = \cdots = r_k = 0$. Now we assume that $\bar{x} = x + t(M) \in M/t(M)$. There exist elements $r, r_1, \ldots, r_k \in R$ with $r \neq 0$ such that

$$rx = r_1c_1 + \cdots + r_kc_k \quad \text{and} \quad r\bar{x} = r_1\bar{c}_1 + \cdots + r_k\bar{c}_k.$$

Consequently, $\{\bar{c}_l\}_{l\in L}$ is a maximal linearly independent system of elements of the module $M/t(M)$. We note that the converse assertion also holds. By Lemma 4.5, the cardinality $|L|$ is equal to the dimension of the K-space $K \otimes_R M/t(M)$, and it is uniquely defined. Now it is clear that any two maximal linearly independent systems of elements of the module M have the same cardinalities. Therefore, the following definition is meaningful.

For a nonzero module M, the cardinality of every maximal linearly independent system of elements of M is called the *rank* $r(M)$ of M. If $M = 0$, then we set $r(M) = 0$. It follows from the above argument that $r(M) = r(t(M)) + r(M/t(M))$. The rank $r(M/t(M))$ is also denoted by $r_0(M)$; it is called the *torsion-free rank* of the module M. The ranks $r(M)$, $r(t(M))$, and $r_0(M)$ are *invariants* of the module M. Invariants of the module usually are positive integers, cardinal numbers, or other easy described objects (for example, matrices). The invariants should be uniquely determined by the module.

It follows from Lemma 4.5 that the module M is a torsion-free module of finite rank if and only if M is isomorphic to the submodule of some finite-dimensional left vector space over K. It is easy to list all modules of rank 1.

Proposition 4.6. *A module M has rank 1 if and only if M is isomorphic to one of the following modules*: $R(p^n)$, $n \geq 1$, $R(p^\infty)$, R, *and* K.

Proof. The above modules have rank 1. We assume that $r(M) = 1$, i.e., M has a maximal linearly independent system of elements consisting of one element. In this case, either M is a primary module or M is a torsion-free module.

Let M be a primary module. Then $M[p]$ is an one-dimensional R_p-space. We assume that M has an element a such that the order p^n of a is the greatest order among the orders of all elements from M. We prove that $M = Ra$. It follows from properties of spaces that $M[p] \subseteq Ra$. Let $b \in M$ and $o(b) > p$. By the induction hypothesis, $pb = ra$, where $r \in R$. We have $r = p^k v$, where v is an invertible element of the ring R and $k \geq 1$ by the choice of the element a. Consequently,

$$pb = p^k va, \quad p(b - p^{k-1}va) = 0, \quad b - p^{k-1}va \in M[p] \subseteq Ra \quad \text{and} \quad b \in Ra.$$

Therefore $M = Ra = R(p^n)$. We assume that M does not have an element of greatest order. Then we have a properly ascending chain of the submodules

$$M[p] \subset M[p^2] \subset \cdots \subset M[p^n] \subset \cdots.$$

By the proved property, $M[p^n] = R(p^n)$ for every n. Therefore M is a quasicyclic module $R(p^\infty)$.

Let M be a torsion-free module with $pM = M$. We fix a nonzero element $a \in M$. For every $n \geq 0$, there exists the unique element $b_n \in M$ such that $a = p^n b_n$. We define the mapping $\varphi\colon K \to M$ as $\varphi(r/p^n) = rb_n$ for every $r/p^n \in K$. Then φ is an isomorphism of R-modules and $M \cong K$. If $pM \neq M$, then we choose an element $a \in M$ with $a \notin pM$. Let $b \in M$. Then $rb = sa$, where $r, s \in R$ and $r \neq 0$. We have $r = p^m u$ and $s = p^n v$, where $m, n \geq 0$, u and v are invertible elements of the ring R, and $m \leq n$. Consequently,

$$p^m(ub - p^{n-m}va) = 0, \quad ub - p^{n-m}va = 0, \quad \text{and} \quad b = u^{-1}p^{n-m}va \in Ra.$$

We obtain $M = Ra$, where $Ra \cong R$. □

We return to the R-module K and prove that every proper nonzero submodule A of K is cyclic, i.e., $A \cong R$. Since A is a torsion-free module of rank 1, it follows from Proposition 4.6 that either $A \cong K$ or $A \cong R$. If $A \cong K$, then A is a K-space and the embedding $A \to K$ is an embedding of the K-spaces. Consequently, A is a proper subspace in K. This is impossible. Therefore $A \cong R$.

Let M be a module. A submodule H is called a *fully invariant* submodule of the module M if $\alpha H \subseteq H$ for every endomorphism α of the module M. For all $n \geq 1$, the modules $p^n M$, $M[p^n]$, 0, and M are fully invariant submodules of M.

Lemma 4.7. *Let $M = A \oplus B$ and let H be a fully invariant submodule of M. Then $H = (H \cap A) \oplus (H \cap B)$.*

Proof. If $c \in H$, then $c = \pi_1(c) + \pi_2(c)$, where π_1 and π_2 are the coordinate projections related to decomposition $M = A \oplus B$. Since H is a fully invariant submodule, we have that $\pi_1(c) \in H \cap A$ and $\pi_2(c) \in H \cap B$. □

In the end of the section, we consider modules over the ring R_{p^n}, where n is some positive integer. Let $\bar{p} = p + p^n R$ (we recall that $R_{p^n} = R/p^n R$). Then

$$\bar{p}^k R_{p^n} = R_{p^n} \bar{p}^k \quad \text{for} \quad k = 1, \dots, n-1.$$

Every proper left or right ideal of the ring R_{p^n} is an ideal and it is equal to $\bar{p}^k R_{p^n}$ for some k. If $r \in R_{p^n}$, then $r = \bar{p}^k v = u \bar{p}^k$, where $k \in \{0, 1, \dots, n\}$ and v, u are invertible elements of the ring R_{p^n}.

Theorem 4.8. *Every left R_{p^n}-module is a direct sum of cyclic modules.*

Proof. Let M be a left R_{p^n}-module and let $S = \{x \in M \mid \bar{p}x = 0\}$. Then S is a left vector space over the residue division ring R_p. We construct a basis $\{e_i\}_{i \in I}$ of the R_p-space S as follows. We take some basis of the space $S \cap \bar{p}^{n-1} M$. Then we extend the basis to a basis of the space $S \cap \bar{p}^{n-2} M$. We continue in such a manner. For every subscript $i \in I$, if $e_i \in S \cap \bar{p}^k M$ and $e_i \notin S \cap \bar{p}^{k+1} M$, then we choose an element $a \in M$ with $e_i = \bar{p}^k a_i$. We assert that $M = \bigoplus_{i \in I} R_{p^n} a_i$.

Let

$$r_1 a_1 + \dots + r_j a_j = 0, \quad \text{where} \quad r_1, \dots, r_j \in R_{p^n} \quad \text{and} \quad r_1 a_1, \dots, r_j a_j \neq 0.$$

We multiply this relation by some power of the element $\bar{p}$. As a result, we obtain $s_1 e_1 + \dots + s_j e_j = 0$, where at least one element s_i is invertible in R_{p^n}. This means that $\bar{s}_1 e_1 + \dots + \bar{s}_j e_j = 0$, where $\bar{s}_i \in R_p$ and not all $\bar{s}_i$ are equal to zero. However, this is impossible. We obtain that the sum of all submodules $R_{p^n} a_i$ is a direct sum. We note the following property of elements e_i. If

$$v_1 e_1 + \dots + v_n e_n \in \bar{p}^m M \quad \text{with} \quad m \geq 1,$$

then $e_1, \dots, e_n \in \bar{p}^m M$.

Let $N = \bigoplus_{i \in I} R_{p^n} a_i$. If $x \in M$ and $\bar{p}x = 0$, then it is clear that $x \in N$. We assume that for elements $x \in M$ such that $\bar{p}^s x = 0$ and $s > 1$, we have $x \in N$. We also assume $y \in M$, $\bar{p}^{s+1} y = 0$, and $\bar{p}^s y \neq 0$. Then $\bar{p}^s y = v_1 e_1 + \dots + v_n e_n$, where $v_1, \dots, v_n$ are invertible elements of the ring R_{p^n}. By the above remark, we obtain $e_i = \bar{p}^{t_i} a_i$, where $t_i \geq s$, $i = 1, \dots, n$. Let $v_i \bar{p}^{t_i} = \bar{p}^{t_i} w_i$, where w_i is an invertible element. Then

$$\bar{p}^s (y - \bar{p}^{m_1} w_1 a_1 - \dots - \bar{p}^{m_n} w_n a_n) = 0,$$

where $m_i \geq 0$, $i = 1, \dots, n$. By the induction hypothesis, the element $y - \bar{p}^{m_1} w_1 a_1 - \dots - \bar{p}^{m_n} w_n a_n$ is contained in N. Therefore $y \in N$ and the relation $M = \bigoplus_{i \in I} R_{p^n} a_i$ has been proved. □

Exercise 1. The product $\prod_{n\geq 1} R(p^n)$ is a nonsplit mixed module.

Exercise 2. The factor module $\widehat{R}/R$ is isomorphic to the direct sum of some number of copies of the module K.

Exercise 3. For a module M and every positive integer n, we have the isomorphism $R(p^n) \otimes_R M \cong M/p^n M$.

Exercise 4. Let M be a left $\widehat{R}$-module and let N be a right $\widehat{R}$-module. Then there exists an epimorphism of Abelian groups $N \otimes_R M \to N \otimes_{\widehat{R}} M$ such that $a \otimes_R b \to a \otimes_{\widehat{R}} b$ for all $a \in N$ and $b \in M$. The epimorphism $\widehat{R} \otimes_R M \to \widehat{R} \otimes_{\widehat{R}} M$ is split, i.e., the kernel of the epimorphism is a direct summand in $\widehat{R} \otimes_R M$.

Exercise 5. Let M be an $\widehat{R}$-module. The following conditions are equivalent.

(1) The canonical mapping $N \otimes_R M \to N \otimes_{\widehat{R}} M$ is an isomorphism for every right $\widehat{R}$-module N.

(2) The canonical mapping $M \to \widehat{R} \otimes_R M$, where $a \to 1 \otimes a$ for all $a \in M$, is an isomorphism.

(3) $\widehat{R}/R \otimes_R M = 0$.

(4) M is a primary module.

Exercise 6. There is the canonical isomorphism of left $\widehat{R}$-modules

$$\widehat{R} \otimes_R \widehat{R} \cong \widehat{R} \oplus (\widehat{R} \otimes_R \widehat{R}/R)$$

(see Exercise 4).

Exercise 7. Let M be a primary $\widehat{R}$-module. We assume that there exists another module multiplication $\circ$ on M such that M is a left $\widehat{R}$-module and $rm = r \circ m$ for all elements $r \in R$ and $m \in M$. Then $rm = r \circ m$ for all $r \in \widehat{R}$ and $m \in M$.

Exercise 8. Let V be a left K-space and let W be a right K-space. Then the canonical mapping $W \otimes_R V \to W \otimes_K V$ is an isomorphism (cf., Exercise 4).

Exercise 9 (Levy [204]). Let R be an arbitrary ring and let M be a left R-module. An element $a \in M$ is called a *torsion* element if $ra = 0$ for some non-zero-divisor r of the ring R. Prove that the set of all torsion elements of every left R-module is a submodule if and only if R has the left classical ring of fractions.

Remarks. A general theory of valuations and valuation rings are due to Krull (1932). The classical theory of valuations of fields is presented, for example, in the books of Bourbaki [35] and Zariski–Samuel [327]. In the papers of Vidal [293, 294, 295] and Roux [266, 267, 268, 269, 270] some delicate questions of the theory of non-commutative discrete valuation domains are studied. It is difficult to describe the history of notions and results of Section 4. These notions and results are not automatically generalized to modules over an arbitrary ring (cf., Theorem 4.1 and Exercise 9). Kulikov began to systematically study Abelian p-local groups and p-adic modules in [189, 190].

Chapter 2

Basic facts

In Chapter 2, we consider the following topics:

- free modules (Section 5);
- divisible modules (Section 6);
- pure submodules (Section 7);
- direct sums of cyclic modules (Section 8);
- basic submodules (Section 9);
- pure-projective and pure-injective modules (Section 10);
- complete modules (Section 11).

We define and study several classes of modules which play a base role in the theory. We also consider such important submodules as pure submodules and basic submodules. In the main part of the text, we assume that a ring R is a (not necessarily commutative) discrete valuation domain. The case of a complete domain is specified. In exercises, other rings are considered; this is always specified.

5 Free modules

A module F is said to be *free* if F is isomorphic to the direct sum of some number of copies of the module R. Therefore, if F is a free module, then $F \cong \bigoplus_s R$ for some cardinal number s. It is clear that s coincides with the rank $r(F)$ of the module F defined in Section 4. Since the rank is an invariant of the module, s is uniquely defined and s is also an invariant of the module F. Thus, the isomorphism $\bigoplus_s R \cong \bigoplus_t R$, where t is some cardinal number, implies that $s = t$.

Proposition 5.1. *A module F is free if and only if F has a generator system $\{a_i\}_{i \in I}$ such that every element $x \in F$ can be uniquely represented in the form $x = \sum_{i \in I} r_i a_i$ for some $r_i \in R$, where almost all r_i are equal to zero.*

Proof. Let the module F be free. Then $F = \bigoplus_{i \in I} R_i$, where $R_i \cong R$ for all i. If an element a_i of the module R_i corresponds under this isomorphism to the identity element of the ring R, then $\{a_i\}_{i \in I}$ is the required generator system of the module F.

Now we assume that F has a generator system $\{a_i\}_{i \in I}$ with the property mentioned in the proposition. For every $i \in I$, we take the homomorphism $\varphi_i \colon R \to Ra_i$ such that $\varphi_i(r) = ra_i$ for all $r \in R$. If $ra_i = 0$ for some r, then it follows from the relation $ra_i = 0 = 0 \cdot a_i$ that $r = 0$. Consequently, φ_i is an isomorphism and $Ra_i \cong R$ for every i. It follows from the uniqueness of the representation of elements of the module F in the form $\sum r_i a_i$ that the sum $\sum_{i \in I} Ra_i$ is a direct sum. Consequently, $F = \bigoplus_{i \in I} Ra_i$, where $Ra_i \cong R$ for every i. □

Let F be a free module. The generator system $\{a_i\}_{i \in I}$ from Proposition 5.1 is called a *free basis* of the module F. In this case, $F = \bigoplus_{i \in I} Ra_i$.

The following theorem provides a complete information about the structure of submodules of free modules.

Theorem 5.2. *Every submodule of a free module is a free module.*

Proof. Let F be a free module. Then $F = \bigoplus_{i \in I} Ra_i$, where $\{a_i\}_{i \in I}$ is a free basis of the module F. We can assume that I is the set of all ordinal numbers which are less than some ordinal number τ. Let G be some submodule of the module F. For every ordinal number $\sigma \leq \tau$, we set

$$F_\sigma = \bigoplus_{\rho < \sigma} Ra_\rho \quad \text{and} \quad G_\sigma = G \cap F_\sigma.$$

Then we have

$$G_{\sigma+1}/G_\sigma = G_{\sigma+1}/(G \cap F_\sigma) \cong (G_{\sigma+1} + F_\sigma)/F_\sigma \subseteq F_{\sigma+1}/F_\sigma.$$

Since $F_{\sigma+1}/F_\sigma \cong Ra_\sigma \cong R$, we have $G_{\sigma+1}/G_\sigma \cong R$ for $G_{\sigma+1} \neq G_\sigma$. We have $G_{\sigma+1}/G_\sigma = R(b_\sigma + G_\sigma)$, where $b_\sigma \in G_{\sigma+1}$ (if $G_{\sigma+1} = G_\sigma$, then $b_\sigma = 0$). Then $G_{\sigma+1} = G_\sigma \oplus Rb_\sigma$. The sum $\sum_{\sigma \leq \tau} Rb_\sigma$ is a direct sum. In addition, $G = \bigoplus_{\sigma \leq \tau} Rb_\sigma$, since the module G is the union of the submodules G_σ. □

The following result shows the universality property of free modules.

Proposition 5.3. *Every module is isomorphic to a factor module of some free module.*

Proof. Let M be an arbitrary module and let $\{m_i\}_{i \in I}$ be some generator system of M. We take a free module F of rank $|I|$. Then $F = \bigoplus_{i \in I} Ra_i$, where $\{a_i\}_{i \in I}$ is a free basis of the module F. We define the homomorphism $\varphi \colon F \to M$ as follows. For any elements $r_1, \ldots, r_k \in R$, we set

$$\varphi(r_1 a_1 + \cdots + r_k a_k) = r_1 m_1 + \cdots + r_k m_k.$$

Since $\{m_i\}_{i \in I}$ is a generator system of the module M, we have that φ is an epimorphism. Consequently, $M \cong F/\operatorname{Ker}(\varphi)$. □

In other words, every module is a homomorphic image of some free module.

A module P is said to be *projective* if for every epimorphism $\pi\colon A \to B$ between arbitrary modules A, B and each homomorphism $\varphi\colon P \to B$, there exists a homomorphism $\psi\colon P \to A$ such that $\varphi = \psi\pi$.

Theorem 5.4. *For a module P, the following conditions are equivalent.*

(1) *P is projective.*

(2) *P is free.*

(3) *For every module M and each epimorphism $\pi\colon M \to P$, there exists a decomposition $M = \operatorname{Ker}(\pi) \oplus P'$, where P' is some module which is isomorphic to P.*

Proof. (1) $\Longrightarrow$ (3). Let M be a module and let $\pi\colon M \to P$ be an epimorphism. By (1), there exists a homomorphism $\psi\colon P \to M$ such that $\psi\pi = 1_P$. Then $\pi\psi$ is an idempotent endomorphism of the module M, i.e., $(\pi\psi)^2 = \pi\psi$. Consequently,

$$M = \operatorname{Ker}(\pi\psi) \oplus \operatorname{Im}(\pi\psi) = \operatorname{Ker}(\pi) \oplus P'.$$

Then $P' \cong M/\operatorname{Ker}(\pi) \cong P$.

(3) $\Longrightarrow$ (2). There exist a free module F and an epimorphism $\pi\colon F \to P$ (Proposition 5.3). By (3), $F = \operatorname{Ker}(\pi) \oplus P'$, where $P' \cong P$. By Theorem 5.2, P is a free module.

(2) $\Longrightarrow$ (1). Let $\pi\colon A \to B$ be an epimorphism and let $\varphi\colon P \to B$ be a homomorphism. We have $P = \bigoplus_{i\in I} Rc_i$, where $\{c_i\}_{i\in I}$ is a free basis of the free module P. Let $a_i \in A$ and let $\pi(a_i) = \varphi(c_i)$ for all $i \in I$. For elements $r_1, \dots, r_k \in R$, we set

$$\psi(r_1c_1 + \dots + r_kc_k) = r_1a_1 + \dots + r_ka_k.$$

Then ψ is a homomorphism from P in A and $\varphi = \psi\pi$. Consequently, P is a projective module. □

We can also define a projective module P as a module such that for every exact sequence of modules $0 \to C \to A \to B \to 0$, the induced sequence of Abelian groups

$$0 \to \operatorname{Hom}(P, C) \to \operatorname{Hom}(P, A) \to \operatorname{Hom}(P, B) \to 0$$

is exact (about induced sequences see Section 2).

Free and projective modules over an arbitrary ring are defined similarly.

Exercise 1. A module M is said to be *skew-projective* if every endomorphism of every factor module of the module M can be lifted to an endomorphism of the module M. The quasicyclic module K/R is skew-projective. The module K is skew-projective if and only if R is a complete domain.

Exercise 2. The ring of 2×2 matrices over a division ring has a projective module which is not free.

Exercise 3. Let $P = \bigoplus_{i \in I} P_i$. We assume that all submodules of the modules P_i are projective for every $i \in I$. Then every submodule M of the module P is isomorphic to the direct sum $\bigoplus_{i \in I} M_i$, where $M_i \subseteq P_i$ for all i.

The ring R is said to be *left hereditary* if every left ideal of R is a projective left module over R.

Exercise 4. Over a left hereditary ring R, every submodule of a free left R-module is isomorphic to the direct sum of some left ideals of the ring R.

Exercise 5 (Dual Basis Lemma). Let P be a left module over the ring R. A module P is projective if and only if there exist two sets $\{x_i\}_{i \in I}$ from P and $\{f_i\}_{i \in I}$ from $\mathrm{Hom}_R(P, R)$ such that every element $a \in P$ is the sum $a = \sum f_i(a)x_i$, where $f_i(a) = 0$ for almost all $i \in I$.

Exercise 6 (Kaplansky [167]). Over a local ring, every projective module is a free module.

6 Divisible modules

A module D is said to be *divisible* if for any element $a \in D$ and each nonzero element $r \in R$, there exists an element $b \in D$ with $a = rb$. Since $r = p^k v$, where k is a non-negative integer, v is an invertible element of the ring R, we have that the module D is divisible if and only if for every $a \in D$ there exists an element $b \in D$ with $a = pb$. This is equivalent to the relation that $pD = D$.

We present several simple properties of divisible modules.

(1) If A is a submodule of a divisible module D, then D/A is a divisible module. Indeed, $p(D/A) = (pD + A)/A = D/A$. Consequently, any homomorphic image of a divisible module is a divisible module.

(2) A direct sum or a direct product is a divisible module if and only if every summand is a divisible module. Let $D = A \oplus B$. Then $pD = pA \oplus pB$. The same property holds for infinite direct sums and direct products.

If D is a submodule of a module M and D is a divisible module, then we say that D is a divisible submodule.

(3) If D_i with $i \in I$ are divisible submodules of the module M, then $\sum D_i$ is a divisible submodule.

We have $p(\sum D_i) = \sum pD_i = \sum D_i$.

Theorem 6.1 (Baer [25]). *Let M be a module and let D be a divisible submodule of M. Then D is a direct summand of M.*

Proof. We consider the set of all submodules of the module M which have the zero intersection with D. The set is nonempty, since it contains the zero submodule and it is inductive. By the Zorn lemma, this set has a maximal element A. We assume that $M \neq D \oplus A$. Let $b \in M$ and $b \notin D \oplus A$. If $Rb \cap (D + A) = 0$, then $D \cap (A + Rb) = 0$; this contradicts to the choice of A. Consequently, $Rb \cap (D + A) \neq 0$. Let $rb \in D + A$, where $r \in R$, $b \in B$, and $rb \neq 0$. Then $p^k b \in D + A$ for some non-negative integer k. We assume that k is the least number with this property. We have $p^k b = d + a$, where $d \in D$ and $a \in A$. Since D is a divisible module, $d = p^k c$, where $c \in D$. We state that $D \cap (A + R(b - c)) = 0$. We note that $b - c \notin A$. Let $d_1 \in D$ and $d_1 = a_1 + r(b - c)$, where $a_1 \in A$ and $r \in R$. Then $rb = d_1 - rc - a_1 \in D + A$. We have $r = vp^m$, where v is an invertible element, m is a non-negative integer. It is clear that $p^m b \in D + A$. Therefore $k \leq m$ and $rb = r_1 p^k$, where $r_1 \in R$. We have

$$d_1 = a_1 + r_1 p^k b - r_1 p^k c = a_1 + r_1 d + r_1 a - r_1 d = a_1 - r_1 a.$$

Therefore $d_1 \in D \cap A = 0$ and $d_1 = 0$. By the choice of the submodule A, we have $b - c \in A$ and $b \in D + A$; this is impossible. Therefore $M = D \oplus A$. □

Remark. We assume that we consider Theorem 6.1 and B is a submodule of the module M with $D \cap B = 0$. In the decomposition $M = D \oplus A$, we can assume that the submodule A satisfies the relation $B \subseteq A$. For this, in the beginning of the proof of the theorem, we can take the set of all submodules of the module M which have the zero intersection with D and contain B. After this, we repeat the remaining part of the proof.

A module is said to be *reduced* if it does not have nonzero divisible submodules.

Corollary 6.2. *Every module M is equal to the direct sum $D \oplus A$, where D is a divisible module and A is a reduced module. Here the submodule D is unique and the submodule A is unique up to an isomorphism.*

Proof. Let D be the direct sum of all divisible submodules of the module M. Then D is a divisible module (property (3)). By Theorem 6.1, we have $M = D \oplus A$, where the module A does not have (nonzero) divisible submodules, i.e., A is a reduced module. The module D is unique by construction. We assume that $M = D \oplus A = D \oplus B$ for some submodule B. Then $A \cong M/D \cong B$. □

We consider two examples of divisible modules.

The module K is divisible. This is clear, since $K = \{r/p^n \mid r \in R, n \text{ is a non-negative integer}\}$.

The module $R(p^\infty)$ is divisible. Since $R(p^\infty) = K/R$, the result follows from property (1).

It is sufficient to have the modules $R(p^\infty)$ and K for the construction of every divisible module.

Theorem 6.3. *A divisible module D is a direct sum of some number of modules isomorphic to $R(p^\infty)$ and some number of modules isomorphic to K.*

Proof. It is clear that the torsion submodule T of the module D is a divisible module. By Theorem 6.1, $D = T \oplus C$ for some submodule C. Since $C \cong D/T$, we have that C is a divisible torsion-free module.

We consider the modules T and C separately. Let $T[p] = \{x \in T \mid px = 0\}$. Then $(pR)T[p] = 0$. Therefore $T[p]$ is a vector space over the residue division ring R_p, where $R_p = R/pR$. We choose some basis $\{a_i\}_{i \in I}$ of this space. Since the module T is divisible, for every i there exist in T elements $a_{i_1}, a_{i_2}, \dots, a_{i_n}, \dots$ such that

$$a_{i_1} = a_i, \quad pa_{i_2} = a_{i_1}, \quad \dots, \quad pa_{i_{n+1}} = a_{i_n}, \quad \dots .$$

Let T_i be the submodule in T generated by the elements $a_{i_1}, a_{i_2}, \dots$. Then $T_i \cong R(p^\infty)$. The sum of all submodules T_i is a direct sum (use the property that the sum $\sum_{i \in I} R_p a_i$ is a direct sum) and $\bigoplus_{i \in I} T_i$ is a divisible module. However, $\bigoplus_{i \in I} T_i$ contains the socle $T[p]$. Therefore $T = \bigoplus_{i \in I} T_i$, where $T_i \cong R(p^\infty)$ for every i.

We take the module C. It follows from Lemma 4.3 that the mapping $\varphi\colon C \to KC$, $x \to 1 \cdot x$, $x \in C$ is a monomorphism. Let $(1/p^n)x \in KC$. Since C is divisible, there exists an element $y \in C$ with $x = p^n y$. Now we obtain $\varphi(y) = 1 \cdot y = (1/p^n)x$. Thus, φ is an isomorphism and C is a K-space. Consequently, C is the direct sum of some number of copies of the module K. □

It follows from the proof of the theorem that divisible torsion-free R-modules and left vector spaces over the division ring K are the same objects. In addition, homomorphisms of R-modules coincide with linear operators of K-spaces; this is shown in Section 4.

Let D be a divisible module, n_p be the dimension of the R_p-space $D[p]$ (it coincides with $r(t(D))$), and let n_0 be the dimension of the K-space $D/t(D)$ (it coincides with $r(D/t(D))$). In addition, if D is a torsion-free module, then we assume that $n_p = 0$, and if D is a primary module, then we assume that $n_0 = 0$. There is a decomposition

$$D \cong \bigoplus_{n_p} R(p^\infty) \oplus \bigoplus_{n_0} K.$$

We can put the pair of cardinal numbers (n_p, n_0) into correspondence with the module D; the pair is a system of invariants of the module D. If two divisible modules have the same systems of invariants, then the modules are isomorphic to each other. We say that (n_p, n_0) is a complete system of invariants. This is obvious that for every pair of cardinal numbers, there exists a divisible module such that the system of invariants of the module coincides with the prescribed system. In such a case, we say that the module has an independent system of invariants. Therefore, a divisible module has a *complete and independent system of invariants*.

Corollary 6.2 and Theorem 6.3 reduce the problem of the description of the structure of R-modules to the case of reduced modules.

Divisible modules are universal in some sense with respect to other modules.

Theorem 6.4. *Every module is a submodule of some divisible module.*

Proof. Let M be a module. By Proposition 5.3, $M \cong F/A$ for some free module F and a submodule A of F. The module F is equal to the direct sum of some number of copies of the module R. Consequently, F can be embedded in a direct sum of some number of copies of the module K, i.e., F can be embedded in a divisible module D. Therefore M can be embedded in the divisible module D/A. □

A module M is said to be *injective* if for every monomorphism $\varkappa\colon B \to A$ of arbitrary modules A and B and any homomorphism $\varphi\colon B \to M$, there exists a homomorphism $\psi\colon A \to M$ such that $\varphi = \varkappa\psi$. This is equivalent to the property that for every module A and any submodule B of A, each homomorphism $B \to M$ can be extended to a homomorphism $A \to M$.

Theorem 6.5. *A module D is injective if and only if it is divisible.*

Proof. We assume that D is injective. We can assume that D is a submodule of a divisible module E (Theorem 6.4). The identity mapping of the module D can be extended to a homomorphism $\pi\colon E \to D$. If we consider π as an endomorphism of the module E, then we obtain $\pi^2 = \pi$. Therefore $E = \operatorname{Im}(\pi) \oplus \operatorname{Ker}(\pi)$. Here we have $\operatorname{Im}(\pi) = D$ and the module D is divisible, since it is a direct summand of the divisible module E.

Let D be a divisible module, $\varkappa\colon B \to A$ be a monomorphism, and let $\varphi\colon B \to D$ be a homomorphism. We take the direct sum $D \oplus A$ and the submodule N of the sum, where $N = \{(\varphi(b), -\varkappa(b)) \mid b \in B\}$. Let $G = (D \oplus A)/N$. We define homomorphisms $\alpha\colon D \to G$ and $\beta\colon A \to G$ such that $\alpha(d) = (d, 0) + N$ for all $d \in D$ and $\beta(a) = (0, a) + N$ for all $a \in A$. It is clear that $\varkappa\beta = \varphi\alpha$. Since $\varkappa$ is a monomorphism, we have that α is a monomorphism. Consequently, $\operatorname{Im}(\alpha)$ is a divisible module, and it is a direct summand of the module G by Theorem 6.1. Let

π be the projection G onto $\operatorname{Im}(\alpha)$ and $\psi = \beta\pi\alpha^{-1}\colon A \to D$. Then the relation

$$\varkappa\psi = \varkappa\beta\pi\alpha^{-1} = \varphi\alpha\pi\alpha^{-1} = \varphi$$

holds. □

Let M be a module. A divisible module E containing M is called the *divisible* (or *injective*) *hull* of the module M if E does not have proper divisible submodules containing M.

Proposition 6.6. *A divisible module E is the divisible hull of the module M if and only if E is an essential extension of the module M (i.e., M is an essential submodule in E).*

Proof. Let E be the divisible hull of the module M. If M is not an essential submodule of the module E, then E has a cyclic submodule C with $M \cap C = 0$. Here we have $C \cong R(p^k)$ for some positive integer k or $C \cong R$. Then C can be embedded in a divisible submodule B of the module E such that $B \cong R(p^\infty)$ or $B \cong K$ and $M \cap B = 0$. By Theorem 6.1 and the remark after the theorem, $E = D \oplus B$ for some divisible submodule D containing M. This is impossible, since E is the divisible hull of the module M. Consequently, E is an essential extension M.

Using Theorem 6.1, we can similarly prove the converse assertion. □

Sometimes it is more convenient to define the divisible hull of the module M as a divisible module E which has a monomorphism $M \to E$ such that the image of the monomorphism is an essential submodule in E.

Theorem 6.7. *Every module M has the divisible hull which is unique in the following sense. If E_1 and E_2 are two divisible hulls of the module M, then the identity mapping of the module M can be extended to an isomorphism between E_1 and E_2.*

Proof. Let D be a divisible module containing the module M. Such a module exists by Theorem 6.4. We consider the set of all divisible submodules of the module D which have the zero intersection with M. The set is inductive and contains a maximal element G. If such divisible submodules do not exist, then $G = 0$. By Theorem 6.1, $D = E \oplus G$ for some submodule E containing M (see the remark after Theorem 6.1). The module E is divisible. Let H be a divisible module such that $M \subseteq H \subseteq E$. Then $E = H \oplus X$ for some module X. Since $(X + G) \cap M = 0$ and G is maximal, we have $X = 0$; therefore $H = E$. Therefore, E is the divisible hull of the module M.

If E_1 and E_2 are two divisible hulls of the module M, then the identity mapping of the module M can be extended to a homomorphism $\gamma\colon E_1 \to E_2$ (we note that the module E_2 is injective by Theorem 6.5). Since $\mathrm{Ker}(\gamma) \cap M = 0$, it follows from Proposition 6.6 that $\mathrm{Ker}(\gamma) = 0$ and γ is a monomorphism. The image $\mathrm{Im}(\gamma)$ is a divisible submodule containing M; therefore, it coincides with E_2. Therefore γ is an isomorphism which does not move every element of the module M. □

We describe the structure of the divisible hull of the module M. Let n_p be the dimension of the R_p-space $M[p]$. There exists an embedding $M[p] \to \bigoplus_{n_p} R(p^\infty)$. Let $\{a_i\}_{i\in I}$ be some maximal linearly independent system of elements of infinite order of the module M and $n_0 = |I|$. There exists an embedding $\bigoplus_{i\in I} Ra_i \to \bigoplus_{n_0} K$. As a result, we have a monomorphism from $M[p] \oplus \bigoplus_{i\in I} Ra_i$ into the divisible module

$$\bigoplus_{n_p} R(p^\infty) \oplus \bigoplus_{n_0} K.$$

It can be extended to a monomorphism from the module M, since $M[p] \oplus \bigoplus_{i\in I} Ra_i$ is an essential submodule in M. The module $\bigoplus_{n_p} R(p^\infty) \oplus \bigoplus_{n_0} K$ is the divisible hull of the module M. For torsion-free modules, the following result holds.

Proposition 6.8. *Let M be a torsion-free module. Then $K \otimes_R M$ is the divisible hull of the module M.*

Proof. The R-module $K \otimes_R M$ is a divisible module since it is a K-space. We can assume that the module M is a submodule of the module $K \otimes_R M$ if we identify elements $m \in M$ and $1 \otimes m$, as in Section 4. Let $\{a_i\}_{i\in I}$ be a maximal linearly independent system of elements of the module M. It is a system with the same property for the module $K \otimes_R M$. Consequently, $\bigoplus_{i\in I} Ra_i$ is an essential submodule in $K \otimes_R M$; therefore M is also essential in $K \otimes_R M$. By Theorem 6.1, $K \otimes_R M$ is a divisible hull of the module M. □

Exercise 1. Two divisible primary modules M and N are isomorphic to each other if and only if $M[p] \cong N[p]$.

Exercise 2. A module M is said to be *quasiinjective* if every homomorphism from any submodule of the module M in M can be extended to an endomorphism of the module M. Prove that every quasiinjective R-module is either an injective module or a direct sum of isomorphic primary cyclic modules.

Exercise 3. An $\widehat{R}$-module M is a reduced module if and only if for every $\widehat{R}$-module N, each R-homomorphism $N \to M$ is an $\widehat{R}$-homomorphism.

Exercise 4. Let M be a reduced $\widehat{R}$-module. We assume that there exists one more module multiplication $\circ$ on M such that M is a left $\widehat{R}$-module and $rm = r \circ m$ for all $r \in R$ and $m \in M$. Then $rm = r \circ m$ for all $r \in \widehat{R}$ and $m \in M$.

In the remaining exercises, R is an arbitrary ring. The definition of an injective R-module coincides with the corresponding definition in the case of discrete valuation domains. A left R-module M is said to be *divisible* if $rM = M$ for every non-zero-divisor r of the ring R.

Exercise 5 (Levy [204]). We assume that R has the two-sided classical ring of fractions S. Then every divisible left R-module is injective if and only if S is an Artinian semiprimitive ring and R is a left hereditary ring (such rings are defined before Exercise 4 in Section 5).

Exercise 6. For a commutative domain R, the following conditions are equivalent.

(1) Every divisible torsion-free R-module is injective.

(2) Every divisible R-module is injective.

(3) R is a Dedekind domain.

7 Pure submodules

We introduce the notion of the height of an element of a module. Let M be a module, $a \in M$, and let k be a non-negative integer. In M, we can consider equations of the form $p^k x = a$ and pose the question about the solvability of such equations. If $b \in M$ and $p^k b = a$, then element b is called a *solution* of the equation $p^k x = a$. We say also that a is *divisible* by p^k. The equation $p^k x = a$ is soluble if and only if $a \in p^k M$. The greatest non-negative integer n, such that the equation $p^n x = a$ has a solution in M, is called the *height* $h(a)$ of the element a in the module M. Consequently, $h(a) = n$ if $a \in p^n M$ and $a \notin p^{n+1} M$. If the equation $p^n x = a$ has a solution for every n, then a is called an *element of infinite height*, $h(a) = \infty$. Therefore, the height of an element can be either a non-negative integer or the symbol ∞. It is clear that $h(0) = \infty$. A module M is divisible if and only if every element of the module M has infinite height.

If the element a is contained in a submodule A of the module M, then we can define two heights for a. As necessary, we write $h_A(a)$ and $h_M(a)$ for the height of the element a in A and M, respectively. We always have $h_A(a) \leq h_M(a)$, where we assume that every non-negative integer is less than ∞.

A submodule A of the module M is said to be *pure* if every equation $p^k x = a$ with $a \in A$, which has a solution in M, also has a solution in A. The submodule

A is pure if and only if $h_A(a) = h_M(a)$ for every $a \in A$, and if and only if $p^k A = A \cap p^k M$ for every non-negative integer k.

We present main properties of pure submodules. The proofs are easy and will be omitted.

(1) The zero submodule and the module itself are pure submodules.

(2) Every direct summand is a pure submodule.

(3) If A is pure in B and B is pure in M, then A is pure in M.

(4) The union of an ascending chain of pure submodules is a pure submodule.

(5) The torsion submodule of the module is a pure submodule.

(6) If A is a submodule in M such that M/A is a torsion-free module, then A is a pure submodule. If A is a pure submodule of the torsion-free module M, then M/A is a torsion-free module.

(7) In torsion-free modules, intersections of pure submodules are also pure. Indeed, in a torsion-free module the equation $p^k x = a$ has at most one solution.

Let X be a subset of a torsion-free module M. Then the intersection A of all pure submodules containing X is the least pure submodule containing X. A submodule A is called the *pure submodule* generated by X. It consists of elements a such that $ra = r_1 x_1 + \cdots + r_n x_n$ for some elements $x_1, \ldots, x_n \in X$, $r, r_1, \ldots, r_n \in R$, and $r \neq 0$. If B is a submodule in M, then the pure submodule generated by B is equal to $\{a \in M \mid ra \in B$ for some nonzero element $r \in R\}$.

We recall that a module M is said to be *bounded* if $p^n M = 0$ for some positive integer n. In this case, M is an R_{p^n}-module (see Section 4). From Theorem 4.8, we obtain the following result.

Theorem 7.1. *A bounded module is a direct sum of cyclic modules.*

Any direct summand always is a pure submodule (property (2)). The following theorem distinguish a very important case of the positive solution of the converse question.

Theorem 7.2. *A bounded pure submodule is a direct summand.*

Proof. Let M be a module and let A be a bounded pure submodule of M. By Theorem 7.1, A is a direct sum of cyclic modules. First, assume that all these cyclic summands are isomorphic to the module $R(p^k)$ for a fixed integer k. Since A is a pure submodule, we have $A \cap p^k M = p^k A = 0$. Let B be a submodule in M such that $p^k M \subseteq B$, $A \cap B = 0$, and B is maximal among submodules with such properties. Then $A \oplus B$ is an essential submodule in M (see Section 4). If $M \neq A \oplus B$, then $M/(A \oplus B)$ is a primary module by Proposition 4.2. Let $m \in M, m \notin A \oplus B$, and let $pm = a + b$, where $a \in A$ and $b \in B$. Then $p^k m = p^{k-1} a + p^{k-1} b$, where $p^k m \in B$. Consequently, $p^{k-1} a = 0$. By considering the structure of the module A, we obtain $a = pa_1$, where $a_1 \in A$ (see Exercise 2). In addition,

$$b = pm - a = p(m - a_1) \in B \quad \text{and} \quad m - a_1 \notin B.$$

Let $X = Rx + B$, where $x = m - a_1$. Since B is maximal, the submodule $A \cap (Rx + B)$ has a nonzero element y. Then $y = rx + b_1$, where $r \in R$, $b_1 \in B$. If $r = r_1 p$ for some $r_1 \in R$, then $y = r_1 px + b_1 \in A \cap B = 0$, since $y \in A$ and $px \in B$; this is impossible. Therefore r is an invertible element. However, then we have

$$x = r^{-1}(y - b_1) \in A \oplus B \quad \text{and} \quad m \in A \oplus B.$$

This is a contradiction. Therefore $M = A \oplus B$.

Let A be an arbitrary bounded pure submodule. Then $A = A_1 \oplus \cdots \oplus A_k$, where the summand A_i is a direct sum of pairwise isomorphic cyclic modules ($i = 1, \ldots, k$). It is clear that A_i is a pure in M submodule. Let $A = A_1 \oplus C$, where $C = A_2 \oplus \cdots \oplus A_k$. By the proved property,

$$M = A_1 \oplus M_1 \quad \text{and} \quad A = A_1 \oplus C_1,$$

where $C_1 = A \cap M_1$, $C_1 \cong A/A_1 \cong C$, and C_1 is a pure submodule in M_1. By repeating a similar argument for C_1 and M_1, we obtain

$$M = A_1 \oplus \cdots \oplus A_k \oplus M_k = A \oplus M_k,$$

after several steps. □

We consider two main corollaries of the proved theorem.

Corollary 7.3. *Let M be a reduced module which is not torsion-free. Then M has a direct summand which is isomorphic to $R(p^n)$ for some positive integer n.*

Proof. Let T be the torsion submodule of the module M. We have that T is a reduced primary module. We assume that all elements of order p of the module T are of infinite height. Using the induction on the order, we prove that all elements of the module T are of infinite height. Let $a \in T$ and $h(a) = k < \infty$. By the induction hypothesis, $h(pa) = \infty$. Then $pa = pb$, where $h(b) > k$. We take element $a - b$. We have $h(a - b) = k$ (see Exercise 1) and $p(a - b) = 0$; this is impossible. Therefore, all elements of the module T are of infinite height, i.e., T is a divisible module. This is a contradiction.

Thus, M contains an element a of order p of finite height $n - 1$, where n is a positive integer. Let $b \in M$ and let $p^{n-1}b = a$. Then the cyclic submodule Rb is isomorphic to $R(p^n)$. Indeed, the homomorphism $R \to Rb$ with $r \to rb$ ($r \in R$) has the kernel $p^n R$. Consequently, $Rb \cong R/p^n R = R(p^n)$. We prove that Rb is a pure submodule in M. Let $vp^k b \in Rb$, where v is an invertible element of the ring R. Then $h_B(vp^k b) = k$. We assume that $vp^k b = p^{k+1}x$, where $x \in M$. Then $p^k b = p^{k+1}y$ for some $y \in M$. We have

$$a = p^{n-k-1}(p^k b) = p^{n-1}b = p^{n-k-1}p^{k+1}y = p^n y,$$

i.e., $h(a) \geq n$; this is impossible. Consequently, $h_M(vp^k b) = k$ and Rb is a pure in M submodule. By Theorem 7.2, Rb is a direct summand of M. □

Corollary 7.4. *An indecomposable module M is either a primary module or a torsion-free module. If M is primary, then either M is isomorphic to $R(p^n)$ or M is isomorphic to $R(p^\infty)$.*

In Section 11 we will return to the question about the structure of indecomposable modules.

Using the proof of Corollary 7.3, we can prove the following result.

Corollary 7.5. *Every element of order p and of finite height can be embedded in a direct summand which is isomorphic to $R(p^n)$.*

To develop Corollary 7.5, we consider the following definition. A module M is called a module *without elements of infinite height* if M does not contain nonzero elements of infinite height.

Corollary 7.6. *Let M be a primary module without elements of infinite height. Then every finite set of elements of the module M can be embedded in a direct summand of the module M which is a direct sum of a finite number of cyclic modules.*

Proof. We assume that every element of the module M can be embedded in a direct summand with properties mentioned in the corollary. Now we assume that $a, b \in M$. Then $M = A \oplus C$, where $a \in A$ and A is a direct sum of a finite number of cyclic modules. Let $b = a_1 + c$, where $a_1 \in A$ and $c \in C$. The element c can be embedded in a direct summand B of the module C which is a direct sum of a finite number of cyclic modules. Then the set $\{a, b\}$ is contained in the direct summand $A \oplus B$ of the module M.

Therefore, it is sufficient to prove the assertion for a single element of the module M. We prove this property by induction on the order of the elements. For elements of order p, the required property follows from Corollary 7.5. Let $a \in M$ and $o(a) = p^k$, where $k > 1$. By Corollary 7.5, we have $M = C \oplus B$, where C is a cyclic module and $p^{k-1}a \in C$. Let $a = c + b$, where $c \in C$ and $b \in B$. Then $p^{k-1}b = 0$ and $o(b) < p^k$. By the induction hypothesis, the element b can be embedded in a direct summand A of the module B, where A is a direct sum of a finite number of cyclic modules. Then the element a is contained in the direct summand $C \oplus A$ of the module M. □

Finally, we present some useful criterion of purity of the submodule.

Proposition 7.7. *For a module M, a submodule A of M is pure in M if and only if every residue class $\overline{m}$ of M with respect to A contains an element m such that the order of m is equal to the order of $\overline{m}$.*

Proof. Let A be a pure submodule in M and $y + A \in A/M$. If the order of the residue class $y + A$ is equal to ∞, then every element in $y + A$ has the order ∞. We assume that the order of $y + A$ is equal to p^n. Then $p^n y \in A$. Since A is pure, we have $p^n y = p^n a$ for some $a \in A$. Then

$$p^n(y - a) = 0, \qquad o(y - a) = p^n, \quad \text{and} \quad y - a \in y + A.$$

Conversely, let $a = p^n b \in A$ and $b \in M$. In the residue class $b + A$, we choose an element x whose order is equal to the order of $b + A$. Then

$$p^n x = 0, \qquad b - x \in A, \quad \text{and} \quad p^n(b - x) = a.$$

Consequently, A is a pure submodule in M. □

Exercise 1. The inequality $h(a + b) \geq \min(h(a), h(b))$ always holds for heights. If $h(a) \neq h(b)$, then the relation $h(a + b) = \min(h(a), h(b))$ holds.

Exercise 2. Let $a \in R(p^n)$ and let $o(a) = p^k$. Then $k + h(a) = n$.

Exercise 3. A discrete valuation domain R is a pure submodule in the p-adic completion $\widehat{R}$ of R.

Exercise 4. If T is the torsion submodule of the module M and A is a pure submodule in M, then $T + A$ is a pure submodule in M.

Exercise 5. Let A be a pure submodule in a module M and $p^k A = 0$. Then $(A + p^k M)/p^k M$ is a pure submodule in $M/p^k M$.

Exercise 6. Let M be a primary module and A is a submodule in M. We assume that $h_A(a) = h_M(a)$ for every element a of the submodule A of order p. Then A is a pure submodule in M.

Exercise 7. Let M be a module, A and B be two submodules of M, and let $A \subseteq B$. Then the following assertions hold.

(1) If B is pure in M, then B/A is pure in M/A.

(2) If A is pure in M and B/A is pure in M/A, then B is pure in M.

Exercise 8. Let M be a primary module and let A be a pure submodule in M with $M[p] \subseteq A$. Then $A = M$.

Exercise 9. A submodule A of the module M is said to be *pure in the sense of Cohn* if for every right R-module N, the induced mapping $N \otimes_R A \to N \otimes_R M$ is a monomorphism. Prove that the submodule A is pure in the sense of Cohn if and only if A is a pure submodule in the sense of this section.

8 Direct sums of cyclic modules

Direct sums of cyclic modules form a very important class of modules. Such sums appear in the study of almost all other classes of modules. We recall that direct sums of copies of the cyclic module R are called free modules which were considered in Section 5. A bounded module is a direct sum of cyclic modules (this follows from Theorem 7.1). In the following theorems, we consider two additional classes of modules which are direct sums of cyclic modules.

Theorem 8.1. *Every finitely generated module is a direct sum of cyclic modules.*

Proof. First, we assume that M is a finitely generated torsion-free module. Using the induction on the number of generators, we prove that M is a free module. We take some finite generator system of the module M. If it consists of a single element a, then M is a cyclic module, $M = Ra$, and $M \cong R$. If the system contains more than one element, then we take some element a of the system. Let A be a pure submodule in M generated by the element a. Then M/A is a finitely generated torsion-free module by property (6) from Section 7. By the induction hypothesis, M/A is a free module. Therefore $M = A \oplus F$ by Theorem 5.4, where $F \cong M/A$ and F is a finitely generated the module. The module A has rank 1. Therefore, either $A \cong R$ or $A \cong K$. The module K is not finitely generated. Consequently, $A \cong R$ and M is a free module.

We assume that M is a finitely generated primary module and $a_1, \dots, a_k$ is some generator system of M. We choose a positive integer n such that $p^n a_i = 0$ for $i = 1, \dots, k$. Let $a \in M$ and let

$$a = r_1 a_1 + \cdots + r_k a_k, \qquad \text{where } r_i \in R.$$

For every i, there exists an element $s_i \in R$ with $p^n r_i = s_i p^n$. It is clear that $p^n a = 0$. Consequently, $p^n M = 0$ and M is a bounded module. By Theorem 7.1, M is a direct sum of cyclic modules.

Let M be a mixed finitely generated module. Then the finitely generated torsion-free module $M/t(M)$ is a free module. Consequently, $M = t(M) \oplus F$, where F is a free module and $t(M)$ is a finitely generated primary module. Therefore $t(M)$ and M are direct sums of cyclic modules. □

A module is said to be *countably generated* if it has a countable generator system.

Theorem 8.2. *A countably generated primary module M without elements of infinite height is a direct sum of cyclic modules.*

Proof. Let $\{a_1, a_2, \dots, a_n, \dots\}$ be some generator system of the module M. By Corollary 7.6, there exists a decomposition $M = A_1 \oplus M_1$, where A_1 is a direct sum of cyclic modules and $a_1 \in A_1$. We assume that for a positive integer $n > 1$, we have obtained the decomposition

$$M = A_1 \oplus \cdots \oplus A_k \oplus M_k,$$

where $k \leq n$, $A_1, \dots, A_k$ are direct sums of cyclic modules, and $a_1, \dots, a_n \in A_1 \oplus \cdots \oplus A_k$. Now we take the element a_{n+1}. Let $a_{n+1} = a + b$, where

$$a \in A_1 \oplus \cdots \oplus A_k, \qquad b \in M_k, \quad \text{and} \quad b \neq 0$$

(if $a_{n+1} \in A_1 \oplus \cdots \oplus A_k$, i.e., $b = 0$, then we pass to the element a_{n+2}, and so on). By Corollary 7.6, we have

$$M_k = A_{k+1} \oplus M_{k+1},$$

where A_{k+1} is a direct sum of cyclic modules and $b \in A_{k+1}$. It is easy to verify that $M = A_1 \oplus \cdots \oplus A_k \oplus \cdots$. □

For torsion-free modules of countable rank, there exists a convenient criterion of the property to be a free module.

Theorem 8.3. *A torsion-free module of countable rank is free if and only if every submodule of finite rank of the module is free.*

Proof. The necessity follows from Theorem 5.2. For the proof of the sufficiency, we assume that $a_1, a_2, \dots, a_n, \dots$ is some maximal linearly independent system of elements of the module M. Let A_n be the pure submodule generated by the elements $a_1, \dots, a_n$ for $n \geq 1$ (see Section 7). We have a properly ascending chain

$$A_1 \subset A_2 \subset \cdots \subset A_n \subset \cdots.$$

We set $C_1 = A_1$. Here we have $C_1 \cong R$. For every $n \geq 1$, we have that A_{n+1} is a free module of finite rank. Consequently, A_{n+1} and A_{n+1}/A_n are finitely generated torsion-free modules. By Theorem 8.1, A_{n+1}/A_n is a free cyclic module and $A_{n+1}/A_n \cong R$. By Theorem 5.4, we have $A_{n+1} = A_n \oplus C_{n+1}$, where $C_{n+1} \cong R$. It can be verified that $M = \bigoplus_{n\geq 1} C_n$. Consequently, M is a free module. □

Before the proof of the main result on direct sums of cyclic modules, we present one criterion for a primary module to decompose into a direct sum of cyclic modules.

Theorem 8.4. *A primary module M is a direct sum of cyclic modules if and only if M is the union of an ascending chain of submodules*

$$A_1 \subseteq A_2 \subseteq \cdots \subseteq A_n \subseteq \cdots,$$

where the heights of all nonzero elements in A_n are less than some integer (depending on n).

Proof. Let M be a direct sum of cyclic modules. Then $M = \bigoplus_{n\geq 1} B_n$, where B_n is a direct sum of cyclic modules $R(p^n)$. The submodules $A_n = B_1 \oplus \cdots \oplus B_n$ with $n \geq 1$ satisfy the conditions of the theorem, since the heights of nonzero elements of the submodule A_n are less than n.

For the proof of the sufficiency, assume that submodules A_n with $n \geq 1$ satisfy the conditions of the theorem. Let $S = M[p]$. Then S is a left vector R_p-space. Similar to the proof of Theorem 4.8, we choose a suitable basis of the space S. Let $H_n = A_n \cap S$ with $n \geq 1$. All H_n are subspaces in S,

$$H_1 \subseteq H_2 \subseteq \cdots \subseteq H_n \subseteq \cdots, \qquad \bigcup_{n\geq 1} H_n = S,$$

and the heights of nonzero elements included in H_n, are less than some integer. It follows from properties of spaces that for every integer $k \geq 1$, there exists an ascending chain of subspaces

$$X_{1k} \subseteq X_{2k} \subseteq \cdots \subseteq X_{nk}$$

such that there are direct decompositions of spaces

$$H_n \cap p^{k-1}M = X_{nk} \oplus (H_n \cap p^k M)$$

(for $k = 1$, we have $H_n \cap p^{k-1}M = H_n$). For every k, we set $Y_k = \bigcup_{n\geq 1} X_{nk}$. Then Y_k is a subspace in S and $Y_k \subseteq p^{k-1}M \setminus p^k M$. Consequently, we have the direct sum $\bigoplus_{k\geq 1} Y_k$. If $x \in S$, then $x \in H_n$ for some n. It follows from the condition for the height of elements in H_n that $H_n \cap p^{k+1}M = 0$ for some k. Therefore $H_n \cap p^k M = X$. We have decompositions

$$H_n = X_{n1} \oplus (H_n \cap pM), \ldots, H_n \cap p^{k-1}M = X_{nk} \oplus (H_n \cap p^k M),$$

and $H_n \cap p^k M = X_{n,k+1}$. Now it is clear that

$$x \in X_{n1} \oplus \cdots \oplus X_{n,k+1} \quad \text{and} \quad x \in Y_1 \oplus \cdots \oplus Y_{k+1}.$$

Therefore $S \subseteq \bigoplus_{k \geq 1} Y_k$.

We take some basis in each of the space Y_k, $k \geq 1$. Let $\{e_i\}_{i \in I}$ be the union of these bases. If $h(e_i) = m_i$, then we take an element $a_i \in M$ with $p^{m_i} a_i = e_i$. We assert that $M = \bigoplus_{i \in I} Ra_i$. The proof is similar to the proof of the analogous relation in the proof of Theorem 4.8. □

Using Theorem 8.4, we can present another proof of Theorem 7.1 and 8.2.

Theorem 8.5. *Let a module M be a direct sum of cyclic modules. Then every submodule N of M is also a direct sum of cyclic modules.*

Proof. The torsion submodule $t(M)$ is a direct sum of cyclic modules. By Theorem 8.4, $t(M)$ is equal to the union of an ascending chain of submodules

$$A_1 \subseteq A_2 \subseteq \cdots \subseteq A_n \subseteq \cdots,$$

where the heights of nonzero elements in the submodule A_n are less than, say, the number l_n. Then the submodule $t(N)$ is the union of an ascending chain of submodules

$$B_1 \subseteq B_2 \subseteq \cdots \subseteq B_n \subseteq \cdots, \qquad B_n = t(N) \cap A_n,$$

and the heights of nonzero elements of B_n in the module $t(N)$ are less than l_n. By the previous theorem, $t(N)$ is a direct sum of cyclic modules. Furthermore, we have

$$N/t(N) = N/(N \cap t(M)) \cong (N + t(M))/t(M) \subseteq M/t(M),$$

where $M/t(M)$ is a free module. By Theorem 5.2, the module $N/t(N)$ is free. By Theorem 5.4, $N = t(N) \oplus F$, where $F \cong N/t(N)$. Consequently, N is a direct sum of cyclic modules. □

We consider the question about the uniqueness of a decomposition of the module into a direct sum of cyclic modules. First, we present two familiar notions related to direct decompositions. Direct decompositions

$$M = \bigoplus_{i \in I} A_i \quad \text{and} \quad M = \bigoplus_{j \in J} C_j$$

are said to be *isomorphic* to each other if there exists a bijection $f\colon I \to J$ such that $A_i \cong C_{f(i)}$ for all $i \in I$. If $M = \bigoplus_i A_i$, where every module A_i is a direct sum, $A_i = \bigoplus_j A_{ij}$, then the decomposition $M = \bigoplus_i \bigoplus_j A_{ij}$ is called a *refinement* of the first decomposition.

Theorem 8.6. *Any two decompositions of a module into a direct sum of cyclic modules are isomorphic to each other. Any two direct decompositions of a direct sum of cyclic modules have isomorphic refinements.*

Proof. Let the module M be a direct sum of cyclic modules. We fix one of decompositions of the module M into a direct sum of cyclic modules. Then $M = \bigoplus_{n\geq 1} B_n \oplus F$, where B_n is a direct sum of cyclic modules $R(p^n)$, F is a direct sum of cyclic modules R.

We denote by t_n, $n \geq 1$, the cardinality of the set of direct summands $R(p^n)$ in B_n, i.e., $t_n = r(B_n)$, t_0 is the cardinality of the set of summands R in F, i.e., $t_0 = r(F)$. We set $S = M[p]$ and $S_n = \{x \in S \mid h(x) \geq n-1\}$ for $n \geq 1$. Then

$$S_n/S_{n+1} \cong B_n[p] \quad \text{and} \quad t_n = r(S_n/S_{n+1}).$$

The submodules S_n are uniquely defined; this means that the submodules S_n do not depend on a concrete direct decomposition of the module M. Therefore, the cardinal numbers t_n are uniquely defined by the module M. Thus, $t_0 = r(M/t(M))$ and t_0 is also uniquely defined by the module M. Now it is obvious that the first assertion holds.

We pass to the second assertion. We assume that we have two direct decompositions of the module M. By Theorem 8.5, every direct summand of the module M is a direct sum of cyclic modules. Replacing every direct summand in two given decompositions to a direct sum of cyclic modules, we obtain refinements of initial decompositions. They isomorphic to each other by the first assertion. □

It follows from the presented proof that the set of cardinal numbers $\{t_0, t_n\ (n \geq 1)\}$ is a complete independent system of invariants for a direct sum of cyclic modules.

Exercise 1. Let M be the torsion submodule of the product $\prod_{n\geq 1} R(p^n)$. Then M is a primary module without elements of infinite height and M is not a direct sum of cyclic modules.

Exercise 2. The torsion-free module $\prod_{n\geq 1} C_n$, where $C_n \cong R$ for all n, is not a free, but every submodule of finite rank of the module is free. Moreover, every finite set of elements of the module M can be embedded in a free direct summand of the module M (cf., Corollary 7.6).

9 Basic submodules

In every non-divisible module, there exist largest in some sense pure submodules which are direct sums of cyclic submodules. Such submodules are called basic submodules. They are invariants of the module and are very useful in the study of modules. We define basic submodules with the use of some systems of elements of the module.

Let M be a module. A system of nonzero elements $a_1, \dots, a_n$ of the module M is said to be *purely independent* if for every positive integer k, the inclusion $r_1a_1 + \cdots + r_na_n \in p^kM$ with $r_ia_i \neq 0$ and $r_i \in R$ implies that r_i are divisible by p^k, $i = 1, \dots, n$. An arbitrary system of nonzero elements of the module M is said to be purely independent if every finite subsystem of the system is purely independent.

Lemma 9.1. *For a module M, a system of nonzero elements of M is purely independent if and only if the system is linearly independent and it generates a pure submodule in M.*

Proof. We assume that a system $\{a_i\}_{i\in I}$ of nonzero elements is purely independent. Let $r_1a_1 + \cdots + r_na_n = 0$, where $r_ia_i \neq 0$, $i = 1, \dots, n$. Then all r_i are divisible by p^k for every positive integer k. Therefore $r_i = 0$, since the ring R is Hausdorff (see Section 3). Therefore $\{a_i\}_{i\in I}$ is a linearly independent system. Let A be the submodule generated by this system and let $x \in A \cap p^kM$. We have $x = r_1a_1 + \cdots + r_na_n$, where all $r_i \in R$. Since the system is purely independent, we have $r_i = p^ks_i$ for some $s_i \in R$, $i = 1, \dots, n$. Therefore

$$x = p^k(s_1a_1 + \cdots + s_na_n) \in p^kA.$$

Consequently, A is a pure submodule in M.

We assume that the system $\{a_i\}_{i\in I}$ is linearly independent and generates the pure submodule A in M. Let $r_1a_1 + \cdots + r_na_n \in p^kM$, where all $r_ia_i \neq 0$. Since A is pure, we have

$$r_1a_1 + \cdots + r_na_n = p^k(s_1a_1 + \cdots + s_na_n) \qquad \text{for some} \quad s_i \in R.$$

Since the system is independent, we have

$$r_ia_i = p^ks_ia_i \quad \text{and} \quad (r_i - p^ks_i)a_i = 0, \qquad i = 1, \dots, n.$$

If $o(a_i) = \infty$, then $r_i - p^ks_i = 0$ and $r_i = p^ks_i$. If $o(a_i) = p^m$, then $m > k$ (consider that $p^ks_ia_i \neq 0$). Therefore $r_i - p^ks_i$ is divisible by p^m and r_i is divisible by p^k. Thus, in any case all r_i are divisible by p^k and $\{a_i\}_{i\in I}$ is a purely independent system. □

A non-divisible module M necessarily contains purely independent systems of elements. We can assume that M is a reduced module. If M is not a torsion-free module, then it follows from Corollary 7.3 that M has a cyclic direct summand Ra. Then $\{a\}$ is a purely independent system by Lemma 9.1. Let M be a torsion-free module, $a \in M$, and let $h(a) = 0$. Then Ra is a pure submodule. By Lemma 9.1, $\{a\}$ is a purely independent system. If M is a divisible module, then it is convenient to assume that M has a unique purely independent system, namely $\{0\}$.

The set of all purely independent systems of the module M is inductive with respect to inclusion. By the Zorn lemma, it contains maximal systems.

Proposition 9.2. *Let M be a module, $\{a_i\}_{i\in I}$ be some system of nonzero elements of M, and let B be the submodule generated by this system. The system $\{a_i\}_{i\in I}$ is a maximal purely independent system if and only if the following conditions hold.*

(1) *B is a direct sum of cyclic modules; more precisely, $B = \bigoplus_{i\in I} Ra_i$.*

(2) *B is a pure submodule in M.*

(3) *The factor module M/B is a divisible module.*

Proof. We assume that the mentioned system is a maximal purely independent system. By Lemma 9.1, it is linearly independent and generates the pure submodule. Therefore, we have (1) (by Lemma 4.4) and (2). We pass to (3). We prove that $p(M/B) = M/B$ or $pM + B = M$. Using the induction on the order, we verify that every element of finite order in the module M is contained in $pM + B$. We assume that $x \in M$, $o(x) = p^l$, $l \geq 1$, and for all elements of order $< p^l$, the assertion holds. Since the system $\{a_i\}$ is maximal, there exist a positive integer k, elements $r_1, \dots, r_n, r \in R$, and $y \in M$ such that

$$r_1a_1 + \cdots + r_na_n + rx = p^k y, \quad r_ia_i \neq 0, \quad rx \neq 0,$$

and one of the elements $r_1, \dots, r_n, r$ is not divisible by p^k. It is clear that r is this element. Consequently, $r = p^m v$, where $0 \leq m < k$, v is an invertible element. Now we have

$$r_1a_1 + \cdots + r_na_n \in p^m M.$$

Since the system $\{a_i\}$ is purely independent, we have $r_i = p^m s_i$, where $s_i \in R$ and $i = 1, \dots, n$. Therefore

$$p^m(s_1a_1 + \cdots + s_na_n + vx - p^{k-m}y) = 0.$$

Since $p^m vx \neq 0$, we have $m < l$. By the induction hypothesis,

$$s_1a_1 + \cdots + s_na_n + vx - p^{k-m}y \in pM + B.$$

Therefore $x \in pM + B$, since $k - m \geq 1$. We obtain that the torsion submodule of the module M is contained in $pM + B$. If $x \in M$ and $o(x) = \infty$, then we can similarly obtain the relation

$$p^m(s_1a_1 + \cdots + s_na_n + vx - p^{k-m}y) = 0.$$

Since

$$s_1a_1 + \cdots + s_na_n + vx - p^{k-m}y \in pM + B,$$

we have $x \in pM + B$.

We assume that conditions (1)–(3) hold. We prove that $\{a_i\}_{i\in I}$ is a maximal purely independent system. The independence follows from (1) and Lemma 4.4. Consequently, the system is a purely independent by Lemma 9.1. We assume that the system $\{a_i\}$ is not maximal. There exists an element a such that $\{a_i\}_{i\in I} \cup \{a\}$ is a purely independent system and $a \neq a_i$ for all i. It follows from (3) that $pM + B = M$; consequently, $a = px + r_1a_1 + \cdots + r_na_n$, where $x \in M$, $r_i \in R$, or $a - r_1a_1 - \cdots - r_na_n = px$. The system $\{a_i\}_{i\in I} \cup \{a\}$ generates the pure submodule C by Lemma 9.1. Therefore

$$a - r_1a_1 - \cdots - r_na_n = pc, \quad \text{where} \quad c \in C.$$

By expressing c in terms of a and a_i, we obtain $a = pra$, where $r \in R$ (consider that the considered system is independent). However, the relation $a = pra$ contradicts to the property that the system $\{a\}$ is purely independent. Consequently, $\{a_i\}_{i\in I}$ is a maximal purely independent system. □

For a module M, every submodule of M satisfying conditions (1)–(3) of Proposition 9.2 is called a *basic submodule* of M. By considering this proposition and the remark before it, we obtain that every non-divisible module has basic submodules. For a divisible module, we assume that the zero submodule is a basic submodule. Any maximal purely independent system of elements of the module M is called a *p-basis* of the module M. If $\{a_i\}_{i\in I}$ is a p-basis of the module M, then $\bigoplus_{i\in I} Ra_i$ is a basic submodule of the module M and conversely.

Let B be a basic submodule of the nondivisible module M. Then

$$B = \bigoplus_{n\geq 0} B_n,$$

where B_0 is a direct sum of cyclic modules isomorphic to R and B_n is a direct sum of cyclic modules which are isomorphic to $R(p^n)$, $n \geq 1$. It is clear that some summands B_n can be missing. Since $B_1 \oplus \cdots \oplus B_n$ is a bounded pure submodule in M, we can use Theorem 7.2 to obtain the following result.

Corollary 9.3. *For every positive integer* n*, we have*

$$M = B_1 \oplus \cdots \oplus B_n \oplus M_n$$

for some submodule M_n.

We also present other important properties of a basic submodule B of the module M.

(1) $M = p^n M + B$ for every positive integer n. Indeed, M/B is a divisible module. Consequently,

$$(p^n M + B)/B = p^n(M/B) = M/B \quad \text{and} \quad p^n M + B = M.$$

(2) $M/p^n M \cong B/p^n B$ for every positive integer n. Considering (1) and the purity of M, we have

$$M/p^n M = (p^n M + B)/p^n M \cong B/(p^n M \cap B) = B/p^n B.$$

(3) Let N be a reduced module and let $\varphi\colon M \to N$ be some homomorphism with $\varphi B = 0$. Then $\varphi = 0$. We have

$$\operatorname{Im}(\varphi) \cong M/\operatorname{Ker}(\varphi), \quad \text{where} \quad B \subseteq \operatorname{Ker}(\varphi).$$

Therefore $M/\operatorname{Ker}(\varphi)$ and $\operatorname{Im}(\varphi)$ is a divisible modules. Therefore $\operatorname{Im}(\varphi) = 0$ and $\varphi = 0$.

(4) Let N be a pure submodule of the module M and let C be a basic submodule of the module N. Then there exists a basic submodule B of the module M such that $B = A \oplus C$ for some submodule A. In addition, $(A \oplus N)/N$ is a basic submodule of the module M/N.

Indeed, let $C = \bigoplus_{i \in I} Rc_i$, where $\{c_i\}_{i \in I}$ is a p-basis of the module C. The p-basis is a purely independent system of elements of the module M. We extend this system to a maximal purely independent system $\{c_i\}_{i \in I} \cup \{a_j\}_{j \in J}$ of elements of the module M. Let $A = \bigoplus_{j \in J} Ra_j$. By Proposition 9.2, $A \oplus C$ is a basic submodule of the module M.

Let $a \in A \cap N$. By (1), $N = p^k N + C$ for every $k \geq 1$. Consequently, $a = p^k x + c$, where $x \in N$, $c \in C$. Therefore

$$a - c = p^k x \quad \text{and} \quad a - c = p^k(a_1 - c_1),$$

since the submodule $A \oplus C$ is pure, where $a_1 \in A$ and $c_1 \in C$. Therefore $a = p^k a_1$ and $a \in p^k A$ for every $k \geq 1$. It is clear that $a = 0$. We obtain $A \cap N = 0$.

Now we show that $(A \oplus N)/N$ is a basic submodule for M/N. Since $(A \oplus N)/N \cong A$, we have that $(A \oplus N)/N$ is a direct sum of cyclic modules, i.e., condition (1) of Proposition 9.2 holds. Furthermore, it follows from

$$(M/N)/(A \oplus N)/N \cong M/(A \oplus N)$$

that condition (3) of this proposition holds. We verify that $(A \oplus N)/N$ is a pure submodule in M/N. Let $a + b = p^k x$, where $a \in A$, $b \in N$, and $x \in M$. Since C is a basic submodule of the module N, property (1) implies the relation $b = p^k y + c$, where $y \in N$ and $c \in C$. Thus,

$$a + b = a + p^k y + c = p^k x \quad \text{and} \quad a + c = p^k(x - y).$$

Since $A \oplus C$ is pure in M, we obtain

$$a + c = a + b - p^k y = p^k(a_1 + c_1).$$

Therefore $a + b = p^k a_1 + p^k(c_1 + y)$, where $c_1 + y \in N$ and $a_1 \in A$.

The converse in some sense assertion to property (4) is contained in Exercise 3. In the assumptions of property (4), we also have the following property. A basic submodule of the module M is isomorphic to the direct sum of basic submodules of modules N and M/N.

(5) If B is a basic submodule of the module M and $B = t(B) \oplus B_0$ for some submodule B_0, then $t(B)$ is a basic submodule for $t(M)$ and B_0 is isomorphic to a basic submodule of the module $M/t(M)$.

It is clear that $t(B)$ is a pure submodule in $t(M)$. We assume that it is not a basic submodule. There exists a submodule C in $t(M)$ such that $t(B) \oplus C$ is a basic submodule for $t(M)$ (property (4)). We prove that $t(B) \oplus C \oplus B_0$ is a pure submodule of the module M. Let $a + c + b = p^k x$, where $a \in t(B)$, $c \in C$, $b \in B_0$, $x \in M$, and k is some positive integer. There exists a positive integer n such that $p^n a = p^n c = 0$. Consequently, $p^n b = p^{n+k} x$. Since B_0 is a pure submodule in M, we have $p^n b = p^{n+k} b_1$, where $b_1 \in B_0$. In addition, $b = p^k b_1$, since B_0 is a torsion-free module. Therefore

$$a + c = p^k(x - b_1) = p^k(a_1 + c_1), \qquad a_1 \in t(B), \qquad c_1 \in C,$$

since $t(B) \oplus C$ is pure in M. As a result, we have $a + c + b = p^k(a_1 + c_1 + b_1)$ which implies the required purity. Now we have that $(t(B) \oplus C \oplus B_0)/(t(B) \oplus B_0)$ is a divisible module as a pure submodule of the divisible module $M/(t(B) \oplus B_0)$. The module is isomorphic to C. Therefore C is a divisible module; this is impossible. Consequently, $t(B)$ is a basic submodule of the module $t(M)$. Now the remaining assertions follow from (4).

We consider interrelations between p-bases of the module M and bases of the R_p-space M/pM.

(6) Let $\{a_i\}_{i \in I}$ be a p-basis of the module M. Then $\{a_i + pM\}_{i \in I}$ is a basis of the R_p-space M/pM. If M is a torsion-free module, then the converse assertion holds.

We assume that

$$\bar{r}_1\bar{a}_1 + \cdots + \bar{r}_n\bar{a}_n = 0, \qquad \bar{r}_i = r_i + pR, \qquad \bar{a}_i = a_i + pM, \qquad i = 1, \ldots, n.$$

Then $r_1a_1 + \cdots + r_na_n = px$ for some $x \in M$. Since the system $\{a_i\}$ is purely independent, all r_i are divisible by p, i.e., all $\bar{r}_i = 0$. Let $b \in M$. By Proposition 9.2 and property (1), we obtain $b = r_1a_1 + \cdots + r_na_n + py$, where $r_i \in R$, $i = 1, \ldots, n$, $y \in M$. Therefore $b + pM = \bar{r}_1\bar{a}_1 + \cdots + \bar{r}_n\bar{a}_n$. Therefore $\{a_i + pM\}_{i \in I}$ is a maximal linearly independent system of elements of the space M/pM, i.e., it is a basis.

Now we assume that M is a torsion-free module and $\{a_i + pM\}_{i \in I}$ is some basis of the space M/pM. We prove that $\{a_i\}_{i \in I}$ is a maximal purely independent system of elements of the module M. First, we prove that if for the system of elements $a_1, \ldots, a_n$, the inclusion $r_1a_1 + \cdots + r_na_n \in pM$, where $r_ia_i \neq 0$ and $r_i \in R$ $(i = 1, \ldots, n)$, implies that all r_i are divisible by p, then this system is purely independent. Indeed, let $r_1a_1 + \cdots + r_na_n \in p^kM$, $r_ia_i \neq 0$, and $r_i \in R$, $i = 1, \ldots, n$, where $k > 1$. Then $r_i = ps_i$ with $s_i \in R$, $i = 1, \ldots, n$, and $p(s_1a_1 + \cdots + s_na_n) = p^ky$. Since M is a torsion-free module,

$$s_1a_1 + \cdots + s_na_n = p^{k-1}y \in p^{k-1}M.$$

By repeating a similar argument, we obtain that all r_i are divisible by p^k. Therefore $a_1, \ldots, a_n$ is a purely independent system of elements.

Now we assume that $r_1a_1 + \cdots + r_na_n \in pM$. Then $\bar{r}_1\bar{a}_1 + \cdots + \bar{r}_n\bar{a}_n = 0$ in the space M/pM. Consequently, all elements $\bar{r}_i$ are equal to zero and all elements r_i are divisible by p. By considering the property proved above, we obtain that $\{a_i\}_{i \in I}$ is a purely independent system.

Let $x \in M$ and $x \notin pM$. Then

$$\bar{x} = x + pM = \bar{r}_1\bar{a}_1 + \cdots + \bar{r}_n\bar{a}_n,$$

where $\bar{r}_i = r_i + pR \in R_p$ and $\bar{r}_i \neq 0$, i.e., $r_i \notin pR$ for all $i = 1, \ldots, n$. In the module M, we have $r_1a_1 + \cdots + r_na_n - x \in pM$. Consequently, $\{a_i\}_{i \in I} \cup \{x\}$ is not a purely independent system. Therefore $\{a_i\}_{i \in I}$ is a maximal purely independent system of elements, i.e., it is a p-basis of the module M.

The p-rank $r_p(M)$ of the module M is defined in Section 4; it is equal to the dimension of the R_p-space M/pM. Let $B = t(B) \oplus B_0$ be a basic submodule of the module M. By Lemma 9.1, we have the inequality $r_p(M) \leq r(M)$. It follows from properties (6) and (5) that

$$r(B) = r_p(M), \qquad r(t(B)) = r_p(t(M)), \quad \text{and} \quad r(B_0) = r_p(M/t(M)).$$

A natural question about the uniqueness of basic submodules has a positive answer.

Theorem 9.4. *Every module contains basic submodules. Any two basic submodules of the given module are isomorphic to each other.*

Proof. It is noted after the proof of Proposition 9.2 that the first assertion holds. We prove the second assertion. If M is a divisible module, then the zero submodule is the unique basic submodule of M by definition.

Let M be a non-divisible module and let B be some basic submodule of M. We have $B = \bigoplus_{n\geq 0} B_n$, where $B_0 = \bigoplus R_i$ with $R_i \cong R$ and $B_n = \bigoplus R(p^n)$ with $n \geq 1$. We fix n and show that the number of summands $R(p^n)$ in the decomposition of the module B_n is an invariant of the module M, i.e., does not depend on the choice of the concrete basic submodule. For a positive integer $k > n$ we have

$$p^k B = p^k B_0 \oplus \bigoplus_{m>k} p^k B_m$$

and

$$B/p^k B \cong B_0/p^k B_0 \oplus B_1 \oplus \cdots \oplus B_k \oplus \bigoplus_{m>k} B_m/p^k B_m,$$

where $B_0/p^k B_0$ and all $B_m/p^k B_m$ are direct sums of copies of the module $R(p^k)$. Consequently, the cardinality of the set of summands $R(p^n)$ in B_n is equal to the cardinality of the set of summands $R(p^n)$ in $B/p^k B$. However, $B/p^k B \cong M/p^k M$ by property (2). We also note that this cardinality is an invariant of the module $M/p^k M$ by Theorem 8.6.

By the remark before the theorem, $r(B_0) = r_p(M/t(M))$. Consequently, the number of the summands R_i in B_0 is an invariant of the module M. □

Exercise 1. If B_i is a basic submodule of the module M_i, $i \in I$, then $\bigoplus B_i$ is a basic submodule of the module $\bigoplus M_i$.

Exercise 2. For a reduced torsion-free module M, a basic submodule of M is cyclic if and only if M is isomorphic to some pure submodule of the module $\widehat{R}$.

Exercise 3. Let M be a module M, N be a pure submodule of M, $\{a_i\}_{i\in I}$ be a p-basis of the module N, and let $\{\bar{b}_j\}_{j\in J}$ be a p-basis of the module M/N. We assume that the element b_j of the module M is contained in the residue class $\bar{b}_j$ and the order of b_j is equal to the order of $\bar{b}_j$ (the element b_j exists by Proposition 7.7). Then $\{a_i, b_j \mid i \in I, j \in J\}$ is a p-basis of the module M.

Exercise 4. Let M be a torsion-free module of finite rank, $a_1, \dots, a_n$ be some maximal linearly independent system of elements of the module M, and let $F = Ra_1 \oplus \dots \oplus Ra_n$. Then

$$M/F = C_1 \oplus \dots \oplus C_k \oplus D_1 \oplus \dots \oplus D_l,$$

where $k + l = r(M)$, $l = r(M) - r_p(M)$, C_i is a cyclic module of finite order or $C_i = 0$, $i = 1, \dots, k$, and $D_j \cong R(p^\infty)$, $j = 1, \dots, l$.

Exercise 5. A torsion-free module M of finite rank is free if and only if $r(M) = r_p(M)$.

10 Pure-projective and pure-injective modules

We define the modules mentioned in the title of the section and present a satisfactory description of the modules.

A module P is said to be *pure-projective* if for every epimorphism $\pi \colon A \to B$ between any two modules A and B such that $\operatorname{Ker}(\pi)$ is a pure submodule in A and each homomorphism $\varphi \colon P \to B$, there exists a homomorphism $\psi \colon P \to A$ with $\varphi = \psi\pi$.

A module Q is said to be *pure-injective* if for every monomorphism $\varkappa \colon B \to A$ such that $\operatorname{Im}(\varkappa)$ is a pure submodule in A and each homomorphism $\varphi \colon B \to Q$, there exists a homomorphism $\psi \colon A \to Q$ such that $\varphi = \varkappa\psi$. This is equivalent to the property that for every module A and any pure submodule B of A, every homomorphism from B into Q can be extended to a homomorphism from A in Q.

It is clear that every projective module is pure-projective and every injective module is pure-injective.

We formulate two results of general type.

Lemma 10.1. (1) *For every module M, there exist a direct sum G of cyclic modules and an epimorphism $\varphi \colon G \to M$ such that* $\operatorname{Ker}(\varphi)$ *is a pure submodule in G.*

(2) *For every module M, there exist a product H of quasicyclic modules and primary cyclic modules and a monomorphism $\gamma \colon M \to H$ such that* $\operatorname{Im}(\gamma)$ *is a pure submodule in H.*

Proof. (1) The proof is similar to the proof of Proposition 5.3. Let $\{m_i\}_{i \in I}$ be the set of all nonzero elements of the module M. For every i, let Ra_i be a cyclic module with generator a_i such that $Ra_i \cong Rm_i$. We set $G = \bigoplus_{i \in I} Ra_i$ and we define an epimorphism $\varphi \colon G \to M$ as follows. For any elements $r_1, \dots, r_k \in R$, let

$$\varphi(r_1a_1 + \dots + r_ka_k) = r_1m_1 + \dots + r_km_k.$$

We assume that $a \in \operatorname{Ker}(\varphi)$ and $p^n x = a$, where n is some positive integer, $x \in G$. If $\varphi(x) = m_i$, then

$$\varphi(x - m_i) = \varphi(x) - \varphi(a_i) = m_i - m_i = 0.$$

Therefore $x - a_i \in \operatorname{Ker}(\varphi)$. We have

$$p^n m_i = p^n \varphi(x) = \varphi(p^n x) = \varphi(a) = 0.$$

Since $Ra_i \cong Rm_i$, we have $p^n a_i = 0$. Consequently, $p^n(x - a_i) = a$, where $x - a_i \in \operatorname{Ker}(\varphi)$. Therefore, the submodule $\operatorname{Ker}(\varphi)$ is pure in M.

(2) Let A_i with $i \in I$ be the set of all submodules of the module M such that $M/A_i \cong R(p^\infty)$ or $M/A_i \cong R(p^n)$ for some positive integer n. We set

$$H_i = M/A_i, \qquad i \in I, \quad \text{and} \quad H = \prod_{i \in I} H_i.$$

There exists a homomorphism $\gamma\colon M \to H$ such that $\gamma(m) = (\ldots, m + A_i, \ldots)$ for every $m \in M$. For a fixed nonzero element m in M, there exists a submodule A of the module M such that $m \notin A$ and A is maximal among such submodules of the module M. We prove that M/A is isomorphic to $R(p^\infty)$ or $R(p^n)$ for some integer n. Let $N = M/A$ and $c = m + A$. Then N is a module such that every nonzero submodule of M contains the element c. This is obvious that N is an indecomposable module. If N is a torsion-free module, then it follows from $c \in Rpc$ that $c = rpc$, where $r \in R$. Therefore $1 - rp = 0$ in the ring R; this is impossible. Therefore, it follows from Corollary 7.4 that N is isomorphic to $R(p^\infty)$ or $R(p^n)$ for some integer n. We return to the element m and obtain that there exists a submodule A_i with $m \notin A_i$. Therefore $m + A_i \neq 0$ and $\gamma(m) \neq 0$, i.e., γ is a monomorphism.

We prove that $\operatorname{Im}(\gamma)$ is a pure submodule in H. Let $\gamma(m) = p^k x$, where $m \in M$, $x \in H$, and k is a positive integer. We assume that $m \notin p^k M$. There exists a submodule A in M such that $p^k M \subseteq A$, $m \notin A$, and A is a maximal among submodules of the module M with such properties. As it was proved above, $M/A \cong R(p^\infty)$ or $M/A \cong R(p^n)$ for some integer n. Since $p^k M \subseteq A$, we have that $M/A \cong R(p^n)$ and $n \leq k$. Let $A = A_i$, where $i \in I$. Then $m + A_i \notin p^k H_i$, where $H_i = M/A_i$ and $\gamma(m) \notin p^k H$. This contradicts to $\gamma(m) = p^k x$. Thus, $m \in p^k M$ and $m = p^k a$, where $a \in M$. Then $\gamma(m) = p^k \gamma(a)$ and the submodule $\operatorname{Im}(\gamma)$ is pure in H. □

The following result is similar to Theorem 5.4.

Theorem 10.2. *For a module P, the following conditions are equivalent.*

(1) *P is a pure-projective module.*

(2) *P is a direct sum of cyclic modules.*

(3) *For every module M and any epimorphism $\pi\colon M \to P$ such that $\operatorname{Ker}(\pi)$ is a pure submodule in M, the relation $M = \operatorname{Ker}(\pi) \oplus P'$ holds for some submodule P' which is isomorphic to P.*

Proof. The proof of implication (1) $\Longrightarrow$ (3) repeats the proof of the implication (1) $\Longrightarrow$ (3) of Theorem 5.4.

(3) $\Longrightarrow$ (2). Let G be a direct sum of cyclic modules and let $\pi\colon G \to P$ be an epimorphism such that $\operatorname{Ker}(\pi)$ is a pure submodule in G (see Lemma 10.1). By (3), $G = \operatorname{Ker}(\pi) \oplus P'$, where $P' \cong P$. By Theorem 8.5, P is a direct sum of cyclic modules.

(2) $\Longrightarrow$ (1). We assume that $\pi\colon A \to B$ is an epimorphism such that $\operatorname{Ker}(\pi)$ is a pure submodule in A and $\varphi\colon P \to B$ is some the homomorphism. We have $P = \bigoplus_{i \in I} Rc_i$. Since the submodule $\operatorname{Ker}(\pi)$ is pure in A, we have that for every $i \in I$, we can choose element a_i in A such that $\pi(a_i) = \varphi(c_i)$ and the order of the element a_i is equal to the order of the element $\varphi(c_i)$ (Proposition 7.7). It is necessary also to consider that $B \cong A/\operatorname{Ker}(\pi)$. For elements $r_1, \dots, r_k \in R$, we set

$$\psi(r_1c_1 + \cdots + r_kc_k) = r_1a_1 + \cdots + r_ka_k;$$

we obtain the homomorphism ψ from P into A with $\varphi = \psi\pi$. Consequently, P is a pure-projective module. □

We directly obtain the following result. If A is a pure submodule of some module M and the factor module M/A is a direct sum of cyclic modules, then A is a direct summand in M.

The following theorem is dual in some sense to Theorem 10.2. It has analogues in Section 6.

Theorem 10.3. *For a module Q, the following conditions are equivalent.*

(1) *Q is a pure-injective module.*

(2) *Q is a direct summand of a product of quasicyclic modules and primary cyclic modules.*

(3) *If Q is a pure submodule of a module M, then Q is a direct summand in M.*

Proof. (1) $\Longrightarrow$ (3). Let Q be a pure submodule of some module M. The identity mapping of the module Q can be extended to a homomorphism $\pi\colon M \to Q$. Then

$$\pi^2 = \pi \quad \text{and} \quad M = \operatorname{Im}(\pi) \oplus \operatorname{Ker}(\pi),$$

where $\operatorname{Im}(\pi) = Q$.

(3) $\Longrightarrow$ (2). By Lemma 10.1, we can assume that the module Q is a pure submodule of a product of modules $R(p^\infty)$ and $R(p^n)$ for different n. By (3) Q is a direct summand of this product.

(2) $\Longrightarrow$ (1). Let B be a pure submodule of some module A and let $\varphi\colon B \to Q$ be a homomorphism. We prove that φ can be extended to a homomorphism from A into Q. If $Q = R(p^\infty)$, then Q is a injective the module (see Section 6). Consequently, the required extension of the homomorphism φ exists. Let $Q = R(p^n)$ for some positive integer n and $N = \mathrm{Ker}(\varphi)$. It follows from the inclusion $p^n B \subseteq \mathrm{Ker}(\varphi)$ that B/N is a bounded module. It is easy to verify that B/N is a pure submodule in A/N (see Exercise 7 in Section 7). By Theorem 7.2, we have $A/N = B/N \oplus C/N$ for some submodule C of the module A with $N \subseteq C$. Therefore $A = B + C$. Now we define the homomorphism $\psi\colon A \to Q$. Let

$$a \in A, \qquad a = b + c, \qquad b \in B, \quad \text{and} \quad c \in C.$$

We set $\psi(a) = \varphi(b)$. This is obvious that $\psi(a)$ does not depend on the choice of the element b. We obtain the homomorphism ψ from A into Q which extends φ.

Let Q be an arbitrary direct summand of the product $\prod_{i\in I} Q_i$, where every module Q_i is isomorphic to $R(p^\infty)$ or $R(p^n)$ for some integer n. Let ρ be the projection from the product $\prod Q_i$ onto Q and let π_i $(i \in I)$ be the coordinate projections related to this product. It follows from the proved property that for every $i \in I$, there exists a homomorphism $\psi_i\colon A \to Q_i$ extending $\varphi\pi_i$. We define the homomorphism $\eta\colon A \to \prod Q_i$ by the relation $\eta(a) = \langle \psi_i(a)\rangle$ for every $a \in A$, where $\langle \psi_i(a)\rangle$ is the vector with coordinate $\psi_i(a)$ in ith position. Let $\psi = \eta\rho$. For every element $a \in A$, we have

$$\psi(a) = (\eta\rho)a = \rho(\langle \psi_i(a)\rangle) = \rho(\langle (\varphi\pi)a\rangle) = \rho(\varphi(a)) = \varphi(a).$$

Consequently, ψ extends φ. □

Corollary 10.4. *A product $\prod_{i\in I} Q_i$ is a pure-injective module if and only if every module Q_i is pure-injective.*

Corollary 10.5. *A reduced pure-injective module is a direct summand of the product of cyclic primary modules.*

Proof. By Theorem 10.3 (2), we have $Q \oplus M = C \oplus D$, where M is some module, C is the product of cyclic modules $R(p^n)$, D is the product of modules $R(p^\infty)$. Since D is a fully invariant summand, we have $D = (D \cap Q) \oplus (D \cap M)$ by Lemma 4.7. However, $D \cap Q = 0$, since the module Q is reduced. Consequently, $D \subseteq M$ and $Q \oplus M/D \cong (Q \oplus M)/D \cong C$. □

We obtain that a reduced pure-injective module does not have nonzero elements of infinite height.

Corollary 10.6. (1) *Every module is isomorphic to a pure submodule of some pure-injective module.*

(2) *A module M is isomorphic to a pure submodule of some reduced pure-injective module if and only if M does not have elements of infinite height.*

Proof. (1) The assertion follows from Lemma 10.1 and Theorem 10.3 (2).

(2) We assume that the module M is isomorphic to a pure submodule of a reduced pure-injective module Q. The module Q does not have elements of infinite height. Therefore, the module M does not have such elements.

Conversely, assume that M does not have elements of infinite height. By (1) and Theorem 10.3, M is a pure submodule of the module $C \oplus D$, where C and D are the submodules from the proof of Corollary 10.5. We assume that $m \in M \cap D$ and $m \neq 0$. Since D is divisible, we have that for every integer k, there exists an element $x \in D$ with $m = p^k x$. Since the submodule M is pure in $C \oplus D$, we have that $m = p^k a$ for some $a \in M$. Therefore $h(m) = \infty$; this is impossible. Consequently, $M \cap D = 0$. Therefore, if ρ is the restriction to the submodule M of the projection $C \oplus D \to C$, then ρ is a monomorphism and $M \cong \rho(M)$. It remains to prove that $\rho(M)$ is a pure submodule in C. Let $\rho(m) = p^k x$, where

$$m \in M, \qquad x \in C, \qquad m = c + d, \qquad c \in C, \quad \text{and} \quad d \in D.$$

Then $\rho(m) = c = p^k x$. Since D is a divisible module, we have $d = p^k b$, where $b \in D$. We have $m = p^k x + p^k b = p^k(x + b)$. Since M is pure, there exists an element $m_1 \in M$ with $m = p^k m_1$. Therefore $\rho(m) = p^k \rho(m_1)$ which means that $\rho(M)$ is pure in C. □

The study of pure-injective modules will be continued in the next section.

Exercise 1. Prove that

$$\prod_{\aleph_0} R(p^\infty) \cong \bigoplus_{c} (R(p^\infty) \oplus K),$$

where c is the continual cardinality.

Exercise 2. A discrete valuation domain R is a pure-injective R-module if and only if R is a complete domain.

Exercise 3. Let A be a pure submodule of a pure-injective module Q. Then Q/A is a pure-injective module.

Exercise 4. The exact sequence of modules

$$0 \to A \xrightarrow{\alpha} B \to C \to 0$$

is said to be *purely exact* if $\mathrm{Im}(\alpha)$ is a pure submodule in B. The above sequence is a purely exact sequence if and only if each of the following three induced sequences is exact for every n:

$$0 \to p^n A \to p^n B \to p^n C \to 0,$$
$$0 \to A[p^n] \to B[p^n] \to C[p^n] \to 0,$$
$$0 \to A/p^n A \to B/p^n B \to C/p^n C.$$

We can also formulate the main definitions of the section as follows. A module is pure-projective (resp., pure-injective) if the module is projective (resp., is injective) with respect to every purely exact sequence.

11 Complete modules

In a module, a topology can be defined by different methods. Similar to the case of discrete valuation domains, the p-adic topology is most important (see Section 3 for the domain case). Let M be a module. We obtain the p-adic topology if we take the submodules $p^n M$, $n = 0, 1, 2, \dots$, as a basis of neighborhoods of zero. Then M is a topological Abelian group. If we assume that the ring R is topological with respect to the p-adic topology, then the mapping $(r, m) \to rm$ of the module multiplication $R \times M \to M$ is continuous with respect to the p-adic topologies on R and M. Consequently, M is a topological module. We assume that all modules have the p-adic topology. All topological notions such as Cauchy sequences, limits, closures, density, completeness, and completions are considered in the p-adic topology.

The p-adic topology is Hausdorff if and only if $\bigcap_{n \geq 1} p^n M = 0$, i.e., M does not have elements of infinite height. This topology is discrete if and only if $p^n M = 0$ for some integer $n \geq 1$, i.e., M is a bounded module. Homomorphisms of modules are continuous with respect to the p-adic topologies. A submodule A of the module M is pure if and only if the p-adic topology on A coincides with the topology induced on A by the p-adic topology of the module M.

We present main notions related to sequences of elements. We partially repeat some material of Section 3. All properties of sequences of elements formulated in terms of the p-adic topology of the ring hold for sequences in modules. A sequence $\{a_i\}_{i \geq 1}$ of elements of the module M is called a *Cauchy sequence* if for every positive integer n, there exists a positive integer k, depending on n, such that

$a_i - a_j \in p^n M$ for all $i, j \geq k$. We say that a sequence $\{a_i\}_{i\geq 1}$ *converges to the limit* $a \in M$ if for every integer n, there exists a number k, depending on n, such that $a - a_i \in p^n M$, as only $i \geq k$. We denote it by $a = \lim a_i$.

A module M is said to be *complete* if M is Hausdorff and every Cauchy sequence of elements in the module M has the limit in M. If every pure Cauchy sequence of elements in M has the limit in M, then the module M is complete.

Every bounded module is discrete; consequently, it is complete.

Proposition 11.1. *A product of modules M_i with $i \in I$ is a complete module if and only if every module M_i is complete.*

Proof. Let M be a complete module, where $M = \prod_{i\in I} M_i$. Every Cauchy sequence $\{c_n\}_{n\geq 1}$ in the module M_i is a Cauchy sequence in the module M. Let a be the limit of the sequence in M. Then $\pi_i(a)$ is the limit of the sequence $\{c_n\}$ in M_i, where π_i is the coordinate projection $M \to M_i$, $i \in I$. Conversely, assume that all modules M_i are complete and $\{a_n\}_{n\geq 1}$ is a pure Cauchy sequence in M. Then $\{\pi_i(a_n)\}_{n\geq 1}$ is a pure Cauchy sequence in M_i, $i \in I$. We assume that an element $c_i \in M_i$ is the limit in M_i of this sequence. There exists an element $c \in M$ such that $\pi_i(c) = c_i$ for every $i \in I$. This c is the limit of the sequence $\{a_n\}$ in M. □

We denote by $\bar{A}$ the closure in the p-adic topology of the submodule A of the module M. We set $M^1 = \bigcap_{n\geq 1} p^n M$. The submodule M^1 is called the *first Ulm submodule* of the module M. It is equal to zero if and only if M does not have elements of infinite height. The relation $M^1 = M$ is equivalent to the property that M is a divisible module.

Lemma 11.2. *Let A and C be two submodules in M such that $A \subseteq C$ and $C/A = (M/A)^1$. Then $\bar{A} = C$.*

Proof. Assume that $x \in \bar{A}$, $x = \lim a_i$, and $a_i \in A$. Therefore, for every positive integer n, we have $x - a_i \in p^n M$ for some i. Consequently,

$$x + A \in p^n(M/A), \quad x + A \in (M/A)^1, \quad \text{and} \quad x \in C.$$

Conversely, let $x \in C$. Then $x + A \in (M/A)^1$ and $x + A \in p^n(M/A)$ for every n. Therefore $x - a_n \in p^n M$, where $a_n \in A$. Consequently, $x = \lim a_n$ and $x \in \bar{A}$. □

By considering remarks before the lemma and Proposition 9.2, we obtain the following results.

Corollary 11.3. (a) *Let A be a submodule in M. Then the following assertions hold.*

(1) *A is a closed submodule if and only if M/A does not have elements of infinite height.*

(2) *A is a dense submodule if and only if M/A is a divisible module.*

(b) *Every basic submodule is a dense submodule.*

Complete modules do not form a new class of modules.

Theorem 11.4. *A module M is a complete if and only if M is a pure-injective module without elements of infinite height.*

Proof. Let M be a pure-injective module without elements of infinite height. By Corollary 10.5, the module M is isomorphic to the direct summand of the product of cyclic primary modules. Since every cyclic primary module is bounded, it is complete. By Proposition 11.1, M is a complete module.

Let M be a complete module. Then M does not have elements of infinite height. By Corollary 10.6, we can assume that M is a pure submodule of a pure-injective module Q without elements of infinite height. Let E be a basic submodule of the module M. There exists a basic submodule B of the module Q such that $B = A \oplus E$ for some submodule A (see property (4) and the proof of it in Section 9). In addition, $A \cap M = 0$ and $(A + M)/M$ is a pure submodule in Q/M. Consequently, $A + M$ is a pure dense submodule in Q by Corollary 11.3. Therefore if $x \in Q$, then $x = \lim(a_i + b_i)$, where $a_i \in A$ and $b_i \in M$ for every $i \geq 1$. Since M is a complete module, there exists the limit $\lim b_i$ in M. Consequently, $x = a + b$ where $a \in \bar{A}$ and $b \in M$. We obtain $Q = \bar{A} + M$. Let $y \in \bar{A} \cap M$. Then $y = \lim a_i$, $a_i \in A$, and $y - a_i \in p^i Q$ for every $i \geq 1$. Since the submodule $A + M$ is pure, we have $y - a_i = p^i b_i + p^i c_i$, $b_i \in A$, $c_i \in M$, and $y = p^i c_i$ for every $i \geq 1$. Therefore $y = 0$. Consequently, $Q = \bar{A} \bigoplus M$ and the direct summand M of a pure-injective module is a pure-injective module. □

Corollary 11.5. *If C is a pure submodule of some module M and C is a complete module, then C is a direct summand of the module M.*

We continue the study of indecomposable modules from Section 7 (see Corollary 7.4). We note that if R is a complete domain, then R is a complete R-module.

Corollary 11.6. *Let R be a complete discrete valuation domain and let M be a reduced R-module. Then M has a cyclic direct summand.*

Proof. By Corollary 7.4, we can assume that M is a torsion-free module. Let $a \in M \backslash pM$. Then $Ra \cong R$ and Ra is a pure submodule in M. By Corollary 11.5, Ra is a direct summand of the module M. □

It is useful to compare the following result with Corollary 7.6.

Corollary 11.7. *Let M be a reduced torsion-free module over a complete discrete valuation domain. Then every finite set of elements of the module M can be embedded in a direct summand of the module M which is a free module. If M has finite or countable rank, then M is a free module.*

Proof. Let b be a nonzero element of M. If $h(b) = k$, then $b = p^k a$, where $a \notin pM$. Then $b \in Ra$. Similar to Corollary 7.6, Ra is a direct summand of the module M. We obtain the first assertion.

Let $r(M) = n$, where n is a positive integer. We take some maximal linearly independent system of elements $a_1, \dots, a_n$ of the module M. By the proved property, it is contained in some free direct summand F of the module M. It is clear that $M = F$. We assume that the rank of M is countable. Since every submodule of finite rank of the module M is free, M is a free module (Theorem 8.3). □

Corollary 11.8. *Let M be an indecomposable module over a complete domain R. Then M is isomorphic to one of the following four modules: $R(p^n)$ for some n, $R(p^\infty)$, R, and K.*

Example 11.9. *Let R be an incomplete domain and let M be a pure submodule in $\widehat{R}$ containing R. Then M is an indecomposable R-module.*

Proof. Since $\widehat{R}/p\widehat{R} \cong R/pR$, we have $r_p(\widehat{R}) = 1$. If $M = A \oplus B$, then $r_p(M) = r_p(A) + r_p(B)$, where $r_p(X)$ is the p-rank of the module X. Consequently, $r_p(A) = 0$ or $r_p(B) = 0$, i.e., $A = 0$ or $B = 0$. Therefore, we have indecomposable R-modules of rank $1, 2, \dots, r(\widehat{R})$. □

Let M be a complete R-module. We take arbitrary elements $a \in M$ and $r \in \widehat{R}$. We have $r = \lim r_i$, where $r_i \in R$ for all i. It is clear that $\{r_i a\}_{i \geq 1}$ is a Cauchy sequence in M. We assume that $ra = \lim r_i a$. This definition of the module multiplication ra does not depend on the choice of the Cauchy sequence $\{r_i\}$. We obtain the left $\widehat{R}$-module M. Thus, every complete R-module is an $\widehat{R}$-module (see also Exercise 4 from Section 6). In addition, R-homomorphisms of complete modules are $\widehat{R}$-homomorphisms. Indeed, let $\varphi\colon E \to C$ be an R-homomorphism of complete modules E and C. We take elements $x \in E$ and $r \in \widehat{R}$, where $r = \lim r_i$, $r_i \in R$. We have

$$\varphi(rx) = \varphi(\lim r_i x) = \lim \varphi(r_i x) = \lim(r_i \varphi(x)) = (\lim r_i)\varphi(x) = r\varphi(x).$$

Consequently, φ is an $\widehat{R}$-module homomorphism.

Theorem 11.10. *Let C be a complete module. Then the closure of every pure submodule of the module C is a direct summand in C.*

Proof. Let A be a pure submodule of the module C. First, we prove that the closure $\bar{A}$ is a pure submodule in C. Let $c \in \bar{A}$ and let $c = p^k x$, where $x \in C$ and k is a positive integer. We have $c = \lim c_i$ and $c_i \in A$, $i \geq 1$. Then for a sufficiently large integer i, we have $c - c_i \in p^k M$. Consequently, $c_i \in p^k M$. We can assume that this holds for all i. In addition, the sequence $\{c_i\}_{i \geq 1}$ can be replaced by a subsequence of it such that $c_{i+1} - c_i \in p^{k+i} M$ for all i. In this case, it follows from the purity of A that $c_{i+1} - c_i = p^{k+i} a_i$ for all i, where $a_i \in A$. Let also $c_1 = p^k b_1$, where $b_1 \in A$. We set $b_i = b_1 + p a_1 + \cdots + p^{i-1} a_{i-1}$ for all $i \geq 1$. Then $\{b_i\}_{i \geq 1}$ is a Cauchy sequence in A. Let b be the limit of the sequence in $\bar{A}$. We have

$$c_i = c_1 + (c_2 - c_1) + \cdots + (c_i - c_{i-1}) = p^k b_1 + p^{k+1} a_1 + \cdots + p^{k+i-1} a_{i-1} = p^k b_i.$$

Therefore $c = p^k b$ and the submodule $\bar{A}$ is pure in C. The submodule $\bar{A}$ is a complete submodule as a pure closed submodule of the complete module C. By Corollary 11.5, $\bar{A}$ is a direct summand in C. □

Corollary 11.11. (1) *Let C be a complete module. Then $C = \overline{t(C)} \oplus G$, where the module $\overline{t(C)}$ does not contain torsion-free direct summands.*

(2) *If B is some basic submodule of the complete module C and $B = X \oplus Y$, then $C = \bar{X} \oplus \bar{Y}$.*

Proof. (1) Since $t(C)$ is a pure submodule in C, it follows from Theorem 11.10 that $C = \overline{t(C)} \oplus G$ for some torsion-free module G. We assume that $\overline{t(C)} = A \oplus H$, where H is some torsion-free module. Then $t(C) \subseteq A$ and $\overline{t(C)} \subseteq A$; this is impossible.

(2) Since B is a pure dense submodule in C (see Corollary 11.3), we have $C = \bar{B}$. It can be easy verified that $\bar{B} = \bar{X} \oplus \bar{Y}$. □

We consider homomorphisms into complete modules.

Proposition 11.12. *Let C be a complete module, A be a pure dense submodule of some module M, and let φ be some homomorphism from A into C. Then φ can be uniquely extended to a homomorphism from M in C.*

Proof. Let $m \in M$ and let $m = \lim a_i$, where $a_i \in A$ for all $i \geq 1$. Then $\{\varphi(a_i)\}_{i \geq 1}$ is a Cauchy sequence in C. We assume that $\psi(m) = \lim \varphi(a_i)$. This definition does not depend on the choice of the sequence $\{a_i\}$. We obtain that ψ is a homomorphism from M in C and ψ extends φ. □

Theorem 11.13. *Let M be a module without elements of infinite height. Then M is isomorphic to a pure dense submodule of some complete module.*

Proof. It follows from Corollary 10.5, Corollary 10.6, and Theorem 11.4 that there exists a complete module C such that M is isomorphic to some pure submodule of the module C. We assume that M is a pure submodule of the module C. By Theorem 11.10, the closure $\bar{M}$ is a direct summand of the module C. Consequently, $\bar{M}$ is a complete module and M is a pure dense submodule of $\bar{M}$. □

Let M be a module without elements of infinite height. Every complete module with properties from Theorem 11.13 is called the *p-adic completion* or the *completion* of the module M. It is known that the completion can be constructed with the use of Cauchy sequences or the inverse limits (see the ring case in Section 3). It follows from Theorem 11.13 that the topology of the completion of the module M coincides with the p-adic topology. The completion is unique in the following sense.

Corollary 11.14. *Let M be a module without elements of infinite height and let C and E be two completions of the module M. Then the identity mapping of the module M can be extended to an isomorphism between C and E.*

Proof. By Proposition 11.12, the identity mapping of the module M can be extended to homomorphisms $\alpha\colon C \to E$ and $\beta\colon E \to C$. We consider the endomorphism $1 - \alpha\beta$ of the module C. By Corollary 11.3, C/M is a divisible module. Since $(1 - \alpha\beta)M = 0$, we have that $(1 - \alpha\beta)C$ is a divisible module. Consequently, $1 - \alpha\beta = 0$ and $\alpha\beta = 1_C$. Similarly, we have $\beta\alpha = 1_E$. We obtain that α and β are mutually inverse isomorphisms which do not move every element of the module M. □

The completion of the module M is denoted by $\widehat{M}$.

It is clear that the completion $\widehat{R}$ of the ring R coincides with the completion of the R-module R (see Section 3). We obtain that R is a pure dense submodule in $\widehat{R}$ (i.e., $\widehat{R}/R$ is a divisible torsion-free module). This can be also verified with the use of the construction of the ring $\widehat{R}$.

Corollary 11.15. (1) *Let M be a module without elements of infinite height and let B be some basic submodule of M. Then $\widehat{M} \cong \widehat{B}$. In particular, a complete module is the completion of every basic submodule of the module, i.e., a complete module is the completion of a direct sum of cyclic modules.*

(2) *Two complete modules are isomorphic to each other if and only if their basic submodules are isomorphic to each other.*

Proof. (1) The completion $\widehat{M}$ of the module M is also the completion for the basic submodule B of the module M. If $\widehat{B}$ is some completion of the module B, then $\widehat{B} \cong \widehat{M}$ by Corollary 11.14.

(2) Isomorphic modules have isomorphic basic submodules. Let C_1 and C_2 be two complete modules and let B_1 and B_2 be their basic submodules. We assume that $B_1 \cong B_2$. Then $\widehat{B}_1 \cong \widehat{B}_2$. It follows from (1) that $\widehat{B}_i \cong \widehat{C}_i = C_i$, $i = 1, 2$, and $C_1 \cong C_2$. □

Let C be a complete module and let B be some basic submodule of C. We have $B = \bigoplus_{n \geq 0} B_n$, where

$$B_0 = \bigoplus_{t_0} R \quad \text{and} \quad B_n = \bigoplus_{t_n} R(p^n) \quad \text{for} \quad n \geq 1.$$

A countable set of cardinal numbers t_0 and t_n, $n = 1, 2, \ldots$, is a complete, independent system of invariants of the module B (see Sections 6 and 8). By Corollary 11.15, we obtain that this set of numbers is also a complete, independent system of invariants of the complete module C. We briefly consider the property that this system is independent. Let $\{t_0, t_n(n \geq 1)\}$ be some set of cardinal numbers. We take a direct sum B of cyclic modules. A system of invariants of the module B is a given set of numbers. Furthermore, we form the completion $\widehat{B}$. The module B is a basic submodule for the module $\widehat{B}$. Consequently, the set $\{t_0, t_n(n \geq 1)\}$ is a system of invariants of the complete module $\widehat{B}$. If C is a complete torsion-free module, then t_0 coincides with p-rank $r_p(C)$ of the module C. Therefore, two complete torsion-free modules are isomorphic to each other if and only if the p-ranks of the modules are equal to each other.

Let Q be a pure-injective module and let $Q = D \oplus C$, where D is a divisible module and C is a reduced pure-injective module. By Corollary 10.5 and Theorem 11.4, C is a complete module. The invariants of the module D defined in Section 6 and the above-mentioned invariants of the module C form a complete, independent system of invariants of the module Q. As a result, we have a quite satisfactory structural theory of pure-injective modules and complete modules.

Proposition 11.16 (Rotman [262]). *Let C be a complete module and let A be a submodule of C such that C/A does not have elements of infinite height. Then C/A is a complete module.*

Proof. We take any pure Cauchy sequence $\{\bar{c}_i\}_{i \geq 1}$ of elements in C/A. Using the induction on i, we prove that there exists a representative c_i in the residue class $\bar{c}_i$ such that $c_{i+1} - c_i \in p^i C$ for all i. As c_1 we take every element from $\bar{c}_1$. We assume that we have obtained elements c_i for all $i \leq k$. We choose some element c'_{k+1} in $\bar{c}_{k+1}$. Then $c'_{k+1} - c_k = p^k b_k + a_k$, where $b_k \in C$ and $a_k \in A$. We set $c_{k+1} = c'_{k+1} - a_k$. Then $c_{k+1} \in \bar{c}_{k+1}$ and $c_{k+1} - c_k \in p^k C$. The constructed sequence $\{c_i\}_{i \geq 1}$ is a pure Cauchy sequence in C. Let $c = \lim c_i$. Then $c + A = \lim \bar{c}_i$ in C/A. Therefore C/A is complete. □

In Section 21 we will describe arbitrary homomorphic images of complete modules.

Let M be a primary module without elements of infinite height. If $M = t(\widehat{M})$, then M is called a *torsion complete* module. Torsion complete modules coincide with torsion submodules of complete modules (Corollary 11.11 (1)). Properties of torsion complete modules are similar to properties of complete modules. For example, two torsion complete modules are isomorphic to each other if and only if their basic submodules are isomorphic to each other. The theory of torsion complete Abelian p-groups is presented in the book of Fuchs [93, Sections 68–73].

Exercise 1. Let C be a complete module and let A be a closed submodule in C. Then A and C/A are complete modules. The converse assertion also holds.

Exercise 2. If α is an endomorphism of the complete module, then $\mathrm{Ker}(\alpha)$ and $\mathrm{Im}(\alpha)$ are complete modules.

Exercise 3. A complete torsion-free module is the completion of a free module.

Exercise 4. Let M be a module without elements of infinite height, B be a basic submodule of M, and let $B = X \oplus Y$. Then $\widehat{M} \cong \widehat{X} \oplus \widehat{Y}$.

Exercise 5. (1) In a torsion-free module, the closure of a pure submodule is a pure submodule.

(2) Construct an example of a module M and a pure submodule A in M such that the closure $\bar{A}$ is not a pure submodule.

Exercise 6. Let R be a complete discrete valuation domain and let M be an R-module. Then every pure submodule of finite rank is a direct summand in M.

Remarks. All main results of this chapter are classical; they were first proved for Abelian groups. In the case of Abelian groups, the results are contained in the books of Kaplansky [166] and Fuchs [92, 93]. In these books it is written that the mentioned results and their proofs with the corresponding changes can be extended to modules over commutative principal ideal domains. However, it has long been known that almost all these results also hold for modules over non-commutative discrete valuation domains (see the works [206, 207] of Liebert). It seems that the chapter contains the first systematical study of modules over not necessarily commutative discrete valuation domains. The authors tried to present the material in a compact form. The injectivity and the existence of injective hulls were discovered by Baer [25].

A pure subgroup of the Abelian group was defined as "Servanzuntergruppe" in the epoch-making work [252] of Prüfer. Later Kaplansky [166] began to use the term "pure". The main results related to pure subgroups of Abelian groups, heights, and direct sums of cyclic groups, are given by Prüfer for countable groups (however, the countability did not play an important role). In base papers of Kulikov [187] and [188], the countability assumption was removed and many other important results were obtained (in particular, Theorem 8.4 and 8.5). Theorem 8.3 is the familiar Pontryagin criterion.

The notions of torsion, divisibility, and purity were generalized and developed in different directions. The axiomatic approach to these notions is presented in the book of Mishina and Skornyakov [236].

In the paper of Kulikov [188], basic subgroups of primary Abelian groups were first introduced and many different results on basic subgroups were proved. Kulikov [189, 190] also considered basic submodules of modules over two discrete valuation domains: the ring $\widehat{\mathbb{Z}}_p$ of p-adic integers and the ring $\mathbb{Z}_p$ of rational numbers whose denominators are coprime to a given prime integer p.

Torsion complete Abelian p-groups were defined and studied by Kulikov [188]. The theory of complete modules has been constructed in the book of Kaplansky [166].

Problem 1. A module M is called a module with the *summand intersection property* if the intersection of every finite set of direct summands of the module M is a direct summand of the module M. Describe modules with the summand intersection property in various classes of modules (cf. [316], [136]).

Chapter 3

Endomorphism rings of divisible and complete modules

In Chapter 3, we consider the following topics:

- examples of endomorphism rings (Section 12);
- the Harrison–Matlis equivalence (Section 13);
- the Jacobson radical (Section 14);
- the Galois correspondences (Section 15).

Divisible modules and complete modules form two quite simple, but very important classes of modules. The modules from the both classes have many remarkable properties; for example, they satisfy the universality property with respect to all modules (see Theorems 6.4 and 11.13). In this chapter, we study several problems on endomorphism rings of divisible primary modules and complete torsion-free modules. We note that the study of endomorphism rings of arbitrary divisible or complete modules can be reduced to the study of endomorphism rings of mentioned modules. We obtain solutions of the considered problems which are complete so far as possible. Results and arguments of the chapter can be samples in the study of similar or other problems for modules from other classes.

In Section 13 we obtain an equivalence between the category of divisible primary modules and the category of complete torsion-free modules. This gives the possibility to extend many results about endomorphism rings of modules in one of the these categories to endomorphism rings of modules in another category.

An important role of the Jacobson radical in the structural ring theory is well known. By this reason, the study of this radical for endomorphism rings is of special interest. Section 14 contains characterizations of the Jacobson radical of the endomorphism ring of a divisible primary module and a complete torsion-free module.

In Section 15, we consider the interrelations between submodules of a divisible primary module or a complete torsion-free module and one-sided ideals of their endomorphism rings. More precisely, we obtain isomorphisms and anti-isomorphisms between lattices of some submodules and the lattices of some one-sided ideals. The results of the book of Baer [28] on the correspondence between subspaces of a vector space and one-sided ideals of the operator ring of the space are examples of such studies.

Instead of $\operatorname{Hom}_R(M, N)$ and $\operatorname{End}_R(M)$, we write $\operatorname{Hom}(M, N)$ and $\operatorname{End}(M)$, respectively. Unless otherwise stated, R denotes a discrete valuation domain (in Sections 13–15, R is complete). The considered notions and results of the lattice theory are presented in the book of Grätzer [124].

12 Examples of endomorphism rings

We describe endomorphism rings of some widely met modules.

Example 12.1. By Proposition 2.2 (b), we have $\operatorname{End}({}_RR) \cong R$. More precisely, every endomorphism of the module ${}_RR$ is the right multiplication of the ring R by some element of R. Therefore, endomorphisms of the module ${}_RR$ coincide with right module multiplications of the module R_R.

Example 12.2. We take the cyclic module $R(p^n)$ for some n. We recall that we denote by $R(p^n)$ the R-module R/p^nR. The factor ring R/p^nR is denoted by R_{p^n}. We have $R(p^n)$ is an R-R-bimodule and R_{p^n}-R_{p^n}-bimodule. In this case,

$$\operatorname{End}(R(p^n)) = \operatorname{End}_{R_{p^n}}(R(p^n)) \cong R_{p^n}.$$

In addition, every endomorphism of the module $R(p^n)$ is a right module multiplication $R(p^n)$ by some element from R or, equivalently, by an element of R_{p^n}. It is clear that an endomorphism α of the module $R(p^n)$ is divisible by p^k if and only if $\alpha(R(p^n)[p^k]) = 0$.

Example 12.3. After the proof of Theorem 6.3, it is noted that every endomorphism of a divisible torsion-free module is a linear operator of this module considered as a K-space. Therefore $\operatorname{End}(K) = \operatorname{End}_K(K) \cong K$. In addition, every endomorphism of the module K coincides with the right multiplication K by some element of K.

Example 12.4. Every R-endomorphism of a complete R-module is an $\widehat{R}$-endomorphism (see the paragraph after Example 11.9). Therefore $\operatorname{End}(\widehat{R}) \cong \widehat{R}$. We again obtain that endomorphisms of the module $\widehat{R}$ coincide with the right multiplication by elements of $\widehat{R}$.

Example 12.5. The quasicyclic module $R(p^\infty)$ is equal to K/R by definition. Consequently, $R(p^\infty)$ is an R-R-bimodule. It was noted in Section 4 that every primary R-module can be considered as an $\widehat{R}$-module. Thus, the right R-module $R(p^\infty)$ can be turned into a right $\widehat{R}$-module. We also note that the modules $R(p^\infty)$ and $\widehat{R}(p^\infty)$ are canonically isomorphic to each other, where $\widehat{R}(p^\infty) = \widehat{K}/\widehat{R}$. Thus, $R(p^\infty)$ is an $\widehat{R}$-$\widehat{R}$-bimodule.

We prove that $\operatorname{End}(R(p^\infty)) \cong \widehat{R}$. Let $r \in \widehat{R}$ and $r \neq 0$. Then $r = p^k v$, where k is a non-negative integer and v is an invertible element of the ring $\widehat{R}$. It is clear that $R(p^\infty)r \neq 0$. Therefore, the mapping $r \to \gamma_r$, $r \in \widehat{R}$, is an embedding of rings $\widehat{R} \to \operatorname{End}(R(p^\infty))$, where γ_r is an endomorphism of the module $R(p^\infty)$ such that $\gamma_r(a) = ar$ for all $a \in R(p^\infty)$. Let α be an endomorphism of the module $R(p^\infty)$. For every positive integer n, the restriction of α to $R(p^\infty)[p^n]$ is an endomorphism of this module which is isomorphic to $R(p^n)$. It follows from Example 12.2 that there exists an element $r_n \in R_{p^n}$ such that the effect of α on $R(p^\infty)[p^n]$ coincides with the right multiplication of $R(p^\infty)[p^n]$ by the element r_n. In the method of the construction of the ring $\widehat{R}$ considered in Section 3, the sequence of elements $\{r_n\}_{n\geq 1}$ defines some element $r \in \widehat{R}$. The right multiplication of the module $R(p^\infty)$ by the element r coincides on the submodule $R(p^\infty)[p^n]$ with the right multiplication by element r_n for every n; therefore, the multiplication coincides with α. Consequently, the embedding of rings $\widehat{R} \to \operatorname{End}(R(p^\infty))$ is an isomorphism. More precisely, every endomorphism of the module $R(p^\infty)$ is the right multiplication of $R(p^\infty)$ by some element from $\widehat{R}$. It is easy to prove that an endomorphism α of the module $R(p^\infty)$ is divisible by p^k if and only if $\alpha(R(p^\infty)[p^k]) = 0$.

Using Proposition 2.4, we can extend the list of examples of endomorphism rings by considering direct sums of various modules.

Example 12.6. We assume that M is a module which satisfies one of the following property.

(a) M is a free module of rank n.

(b) M is a divisible torsion-free module of rank n.

(c) M is a complete torsion-free module of rank n.

(d) M is a divisible primary module of rank n.

Then the ring $\operatorname{End}(M)$ is isomorphic to the ring of all $n \times n$ matrices over the ring R, K, $\widehat{R}$, and $\widehat{R}$, respectively.

Exercise 1. For every module M and each positive integer n, there exists an R-module isomorphism $\operatorname{Hom}(R(p^n), M) \cong M[p^n]$.

Exercise 2. Prove the isomorphism $\operatorname{Hom}(R(p^m), R(p^n)) \cong R(p^k)$, where $k = \min(m, n)$.

Exercise 3. Verify that there exists an isomorphism of modules

$$\operatorname{Hom}(K, R(p^\infty)) \cong \widehat{K},$$

where $\widehat{K}$ is the division ring of fractions of the domain $\widehat{R}$.

Exercise 4. Obtain the representation of endomorphisms by matrices for the following modules: $R(p^n) \oplus R$, $R(p^\infty) \oplus R(p^n)$, $K \oplus R$, and $R(p^\infty) \oplus K$.

Exercise 5. Describe the centers of endomorphism rings of modules from Exercise 4 under the additional assumption that R is a commutative domain.

13 Harrison–Matlis equivalence

The definition of a category and main corresponding notions are presented in Section 1. We recall that when we say about the category R-modules of certain form, then morphisms coincide with usual module homomorphisms, i.e., we mean a full subcategory of the category $R\text{-mod}$ (see the end of Section 2).

Until to the end of the chapter, R denotes some discrete valuation domain. Starting from the R-R-bimodule K/R (it was explicitly studied in Section 4), we define two functors in the category $R\text{-mod}$. We use the following property: since K/R is an R-R-bimodule, the groups $\operatorname{Hom}(K/R, M)$ and $K/R \otimes_R M$ are left R-modules for every left R-module M; the method to turn the group Hom and the tensor product into modules is presented in Section 2. Furthermore, we omit the subscript R in Hom and in the notation of the tensor product.

We denote by H the covariant functor $\operatorname{Hom}(K/R, -)\colon R\text{-mod} \to R\text{-mod}$, i.e., $H(M) = \operatorname{Hom}(K/R, M)$ and $H(\varphi) = \varphi_*\colon H(M) \to H(N)$ for every module M and every homomorphism of modules $\varphi\colon M \to N$, where $H(\varphi)(f) = f\varphi$ for $f \in H(M)$. On the other hand, there exists a covariant functor $K/R \otimes (-)\colon R\text{-mod} \to R\text{-mod}$; we denote this functor by T. We have

$$T(M) = K/R \otimes M \qquad \text{and} \qquad T(g) = 1 \otimes g\colon T(M) \to T(N)$$

for every module M and each module homomorphism $g\colon M \to N$. Here we have $(1 \otimes g)(a \otimes m) = a \otimes g(m)$ for any elements $a \in K/R$ and $m \in N$. The functor H is left exact; this means that for an arbitrary monomorphism $\varphi\colon M \to N$, the homomorphism $H(\varphi)$ is also a monomorphism. The functor T is right exact: if $g\colon M \to N$ is an epimorphism, then $T(g)$ is an epimorphism.

There exist natural transformations $\theta\colon TH \to 1$ and $\phi\colon 1 \to HT$, where 1 is the identity functor of the category $R\text{-mod}$. The natural homomorphism $\theta_M\colon K/R \otimes \operatorname{Hom}(K/R, M) \to M$ is defined by the relation $\theta_M(a \otimes f) = f(a)$ for the module M, an element $a \in K/R$ and $f \in \operatorname{Hom}(K/R, M)$. The naturalness of θ_M means that for every homomorphism $g\colon M \to N$, the relation $\theta_M g = TH(g)\theta_N$ holds. The natural homomorphism $\phi_M\colon M \to \operatorname{Hom}(K/R, K/R \otimes M)$ is defined by the relation $[\phi_M(m)](a) = a \otimes m$ for the module M and $a \in K/R$, $m \in M$. In addition, for every homomorphism $g\colon M \to N$, the relation $\phi_M HT(g) = g\phi_N$ holds.

Let $\mathcal{D}$ be the category of all divisible primary modules and let $\mathcal{E}$ be the category of all complete torsion-free modules.

Theorem 13.1 (Harrison [134], Matlis [220]). *The functors H and T define the equivalence of the categories $\mathcal{D}$ and $\mathcal{E}$.*

Proof. It is sufficient to prove the following properties.

(1) $H\colon \mathcal{D} \to \mathcal{E}$ (this means that the functor H maps modules in $\mathcal{D}$ into modules in $\mathcal{E}$) and $T\colon \mathcal{E} \to \mathcal{D}$.

(2) The transformations $\theta\colon TH \to 1$ and $\phi\colon 1 \to HT$ are equivalences. This means that θ_D and ϕ_C are isomorphisms for all $D \in \mathcal{D}$ and $C \in \mathcal{E}$.

Let $D \in \mathcal{D}$. We prove that $H(D) \in \mathcal{E}$, i.e., $\mathrm{Hom}(K/R, D)$ is a complete torsion-free module. We have $D = \bigoplus_{i\in I} Q_i$, where $Q_i \cong K/R$ for all $i \in I$ and $\bar{D} = \prod_{i\in I} Q_i$. We have $\mathrm{Hom}(K/R, K/R) \cong R$ (see Example 12.5 and consider the completeness of R) and $\mathrm{Hom}(K/R, \bar{D}) \cong \prod_{i\in I}(K/R, Q_i)$. Consequently, $\mathrm{Hom}(K/R, \bar{D})$ is a complete torsion-free module (Proposition 11.1).

Let $f \in H(D)$ and $f = p^n g$, where $n \geq 1$ and $g \in H(\bar{D})$. We have

$$(p^n g)K/R = g((K/R)p^n) = g(p^n(K/R)) = g(K/R).$$

Consequently, $g(K/R) \subseteq D$ and $g \in H(D)$. This implies that the submodule $H(D)$ is pure in $H(\bar{D})$.

We prove that $H(D)$ is a closed submodule of the module $H(\bar{D})$ in the p-adic topology. Let $f = \lim f_i$, where $f \in H(\bar{D})$ and $f_i \in H(D)$. We take some element $x \in K/R$ of order p^n. There exists an element $y \in K/R$ with $xp^n = p^n y$. We note that $o(x) = o(y)$ and $p^n y = 0$ (see property (6) from Section 3). Furthermore, there exists a subscript i such that $f - f_i = p^n g$ for some $g \in H(\bar{D})$. Now we have $(f - f_i)x = p^n g(x) = g(p^n y) = 0$, i.e., $f(x) = f_i(x)$. Consequently, $f \in H(D)$. Thus, $H(D)$ is a pure closed submodule in $H(\bar{D})$. Therefore $H(D)$ is a complete torsion-free module (see Theorem 11.10) and $H(D) \in \mathcal{E}$.

Let $C \in \mathcal{E}$. Since K/R is a divisible primary module, it is easy to verify that $K/R \otimes C$ is also a divisible primary module. Consequently, $T(C) \in \mathcal{D}$.

We prove that the transformation θ is an equivalence. For every $D \in \mathcal{D}$, we need to prove that θ_D is an isomorphism, where $\theta_D\colon K/R \otimes H(D) \to D$ with $\theta_D(a \otimes f) = f(a)$ for $a \in K/R$ and $f \in H(D)$.

Let $r(D) = 1$, i.e., $D \cong K/R$. It is directly verified that θ_D is an isomorphism. Assume that the rank of the module D is finite; $D \cong (K/R)^n$ for some positive integer n. Since H and T are permutable with finite direct sums, we obtain that θ_D is an isomorphism.

We assume that D has an infinite rank. Every element of the module D is contained in some direct summand which is isomorphic to K/R. It is clear that

θ_D is a surjective mapping. Let $\sum_{i=1}^{k}(a_i \otimes f_i) \in \mathrm{Ker}(\theta_D)$, where $a_i \in K/R$, $f_i \in H(D)$, and $f_i \neq 0$, $i = 1, \dots, k$. Since $\mathrm{Im}(f_1) \cong K/R$, we have $D = Q_1 \oplus E_1$, where $Q_1 = \mathrm{Im}(f_1)$. Let π be the projection $D \to E_1$ with kernel Q_1. If $f_2\pi \neq 0$, then we have $D = Q_1 \oplus Q_2 \oplus E_2$, where $Q_2 = \mathrm{Im}(f_2\pi) \cong K/R$. If $f_2\pi = 0$, then we pass to f_3 and repeat the argument. After some time, we obtain the decomposition $D = Q \oplus E$, where the module Q has finite rank and $\sum(a_i \otimes f_i) \in K/R \otimes H(Q)$. Since θ is natural, we have $\theta_D = \theta_Q + \theta_E$. Since $\theta_Q(\sum(a_i \otimes f_i)) = 0$, it follows from the proved property that $\sum(a_i \otimes f_i) = 0$. Consequently, θ_D is an isomorphism and θ is an equivalence.

We pass to the proof of the property that the transformation ϕ is an equivalence. This means that for every module $C \in \mathcal{E}$, the mapping $\phi_C \colon C \to H(K/R \otimes C)$ is a isomorphism, where $[\phi_c(x)](a) = a \otimes x$, $x \in C$, $a \in K/R$.

Let $C \cong R$. We note that $K/R \otimes R \cong K/R$ and endomorphisms of the module K/R coincide with right multiplications by elements of R (Example 12.5). Therefore ϕ_R is an isomorphism.

Let C be an arbitrary complete torsion-free module. We assume that $\phi_C(x) = 0$ for some element $x \in C$. By Corollary 11.7, there exists a decomposition $C = A \oplus B$, where $x \in A$ and $A \cong R$. Since ϕ is natural, we obtain $x = 0$. Consequently, ϕ_C is a monomorphism.

It remains to verify that ϕ_C is an epimorphism. Let $f \colon K/R \to K/R \otimes C$. We take some generator system $a_1, a_2, \dots, a_n, \dots$ of the module K/R such that $pa_1 = 0$ and $pa_{n+1} = a_n$, $n \geq 1$. Since $Ra_n \cong R(p^n)$, it is convenient to identify Ra_n with $R(p^n)$. The module C is flat, since it is a torsion-free module. Therefore, we have embeddings $R(p^n) \otimes C \to K/R \otimes C$ for every $n \geq 1$ (see the end of Section 2 on flat modules). The module $K/R \otimes C$ is equal to the union of modules $R(p^n) \otimes C$, $n \geq 1$. Therefore, we have the homomorphism $f \colon R(p^n) \to R(p^n) \otimes C$ for every $n \geq 1$. Let

$$f(a_n) = \sum(b_i \otimes y_i), \qquad b_i \in K/R, \qquad y_i \in C.$$

Since $b_i = a_n r_i$, where $r_i \in R$, we have that $f(a_n) = a_n \otimes y_n$ for some $y_n \in C$. We have the relation

$$a_n \otimes (y_{n+1} - y_n) = p(a_{n+1} \otimes y_{n+1}) - (a_n \otimes y_n) = f(a_n) - f(a_n) = 0.$$

We set $y = y_{n+1} - y_n$. Let $y = p^k z$, where $z \in C \setminus pC$. The submodule Rz is pure in C. Therefore, we can assume that the module $R(p^n) \otimes Rz$ is a submodule in $R(p^n) \otimes C$ (this result is mentioned in the end Section 2). Since $Rz \cong R$, we have $R(p^n) \otimes Rz \cong R(p^n)$, where $a_n \otimes rz \to a_n r$, $r \in R$. Now it follows from $a_n \otimes p^k z = 0$ that $a_n p^k = 0$. Therefore $k \geq n$. We have proved that $y_{n+1} - y_n \in p^n C$ for all $n \geq 1$. It is also possible to prove this inclusion with the use of the canonical isomorphism

$$R(p^n) \otimes C \cong C/p^n C, \qquad a_n \otimes x \to x + p^n C, \qquad x \in C.$$

Thus, $\{y_n\}_{n \geq 1}$ is a Cauchy sequence in C. Let x be the limit of the sequence. For every n, there exists an integer $i \geq n$ such that $x - y_i \in p^n C$. Therefore

$$a_n \otimes x = a_n \otimes y_i = a_n \otimes y_n = f(a_n).$$

Since $\phi_C(x)(a_n) = a_n \otimes x$, we have $\phi_C(x) = f$. Consequently, ϕ_C is an isomorphism and ϕ is an equivalence. □

Let D be a divisible primary module. We denote by $L(D)$ the ordered with respect to inclusion set of all direct summands of the module D. Then $L(D)$ is a lattice with the following lattice operations. If A and B are two direct summands of the module D, then $\inf(A, B)$ is a maximal divisible submodule in $A \cap B$ and $\sup(A, B) = A + B$.

Let C be a complete torsion-free module and let $L(C)$ be the ordered with respect to inclusion the set of all direct summands of the module C. Then $L(C)$ is a lattice, where for any two direct summands A and B of the module C, we have $\inf(A, B) = A \cap B$ and $\sup(A, B)$ is the closure in the p-adic topology of the pure submodule generated by $A + B$. So defined $\inf(A, B)$ and $\sup(A, B)$ are pure closed submodules. Consequently, they are direct summands in C.

For every positive integer n, we define ideals

$$A_n(D) = \{\alpha \in \mathrm{End}(D) \mid \alpha(D[p^n]) = 0\}$$

and

$$A_n(C) = \{\alpha \in \mathrm{End}(C) \mid \alpha C \subseteq p^n C\}$$

of the rings $\mathrm{End}(D)$ and $\mathrm{End}(C)$, respectively.

Corollary 13.2 (Liebert [206]). *Let D be a divisible primary module and let C be a complete torsion-free module. Then the following assertions hold.*

(1) *There exist canonical ring isomorphisms*

$$\mathrm{End}(D) \cong \mathrm{End}(H(D)) \quad \textit{and} \quad \mathrm{End}(C) \cong \mathrm{End}(T(C)).$$

(2) *The lattices $L(D)$ and $L(H(D))$ are isomorphic to each other and the lattices $L(C)$ and $L(T(C))$ are isomorphic to each other.*

(3) $r(D) = r_p(H(D))$ *and* $r_p(C) = r(T(C))$.

(4) *The isomorphisms of rings in* (1) *induce isomorphisms of the ideals*

$$A_n(D) \cong A_n(H(D)) \quad \textit{and} \quad A_n(C) \cong A_n(T(C))$$

for every $n \geq 1$.

Proof. The assertions (1) and (2) follow from the use of natural equivalences θ and ϕ. We present some explanations.

(1) For an endomorphism α of the module D, we take the induced endomorphism α_* of the module $H(D)$, where $\alpha_*(f) = f\alpha$ for all $f \in H(D)$. The mapping $\alpha \to \alpha_*$ with $\alpha \in \mathrm{End}(D)$ is the required isomorphism. The isomorphism $\mathrm{End}(C) \cong \mathrm{End}(T(C))$ is defined by the relation $\alpha \to 1 \otimes \alpha$ for all $\alpha \in \mathrm{End}(C)$.

(2) We obtain the lattice isomorphism between $L(D)$ and $L(H(D))$ as follows. We associate the direct summand $H(A)$ of the module $H(D)$ with the direct summand A of the module D. The converse mapping maps the direct summand F of the module $H(D)$ onto the direct summand $F(K/R)$ of the module D, where $F(K/R) = \sum_{f \in F} f(K/R)$.

We consider the lattices $L(C)$ and $L(T(C))$. The direct summand A of the module C is mapped into $T(A)$. Conversely, if B is a direct summand of the module $T(C)$, then it follows from Theorem 13.1 there exists a direct summand A of the module C such that $B = K/R \otimes A$. The converse lattice isomorphism is $B \to A$.

(3) We have $D = \bigoplus_{i \in I} Q_i$, where $Q_i \cong K/R$ for all i. Furthermore, we set

$$F = \bigoplus_{i \in I} \mathrm{Hom}(K/R, Q_i) = \bigoplus_{i \in I} F_i,$$

where $F_i \cong R$ for all i. We prove that F is a basic submodule of the module $H(D)$. Let $\bar{D} = \prod_{i \in I} Q_i$. Since F is a pure submodule in $H(\bar{D})$, we have that F is a pure submodule in $H(D)$. It remains to verify that the factor module $H(D)/F$ is divisible. We fix some element a of order p of the module K/R. Let $f \in H(D)$ and let π be the projection of the module D onto $Q_1 \oplus \cdots \oplus Q_k$, where $f(a) \in Q_1 \oplus \cdots \oplus Q_k$. Then $f(1-\pi)a = 0$ and, consequently, $f(1-\pi) \in pH(D)$ (Example 12.5). Now we have $f = f\pi + f(1-\pi)$, where $f\pi \in F$. We obtain that $F + pH(D) = H(D)$ and $H(D)/F$ is a divisible module. Consequently, F is a basic submodule of the module $H(D)$. Therefore $r_p(H(D)) = r(F) = r(D)$.

The relation $r_p(C) = r(T(C))$ follows from the above and Theorem 13.1. The relation can be also proved directly. Let B be a basic submodule of the module C. We have the induced exact sequence R-modules

$$0 \to K/R \otimes B \to K/R \otimes C \to K/R \otimes C/B \to 0$$

(we need the purity of B, see the end of Section 2). Since K/R is a primary module and C/B is a divisible module, we have $K/R \otimes C/B = 0$. Consequently,

$K/R \otimes C \cong K/R \otimes B$. Let $B = \bigoplus_{i \in I} R_i$, where $R_i \cong R$, $i \in I$. Then $K/R \otimes C \cong \bigoplus_{i \in I}(K/R \otimes R_i)$, where $K/R \otimes R_i \cong K/R$ for all i. Therefore $r_p(C) = r(B) = |I| = r(T(C))$.

(4) It is sufficient to prove the isomorphism $A_n(C) \cong A_n(T(C))$. Let $\alpha \in A_n(C)$, i.e., $\alpha C \subseteq p^n C$. Then $(1 \otimes \alpha)T(C) = K/R \otimes \alpha C$. It follows from the proof of Theorem 13.1 that $T(C)[p^n] = (K/R)[p^n] \otimes C$. Now it is clear that

$$(1 \otimes \alpha)(T(C)[p^n]) = 0 \qquad \text{and} \qquad 1 \otimes \alpha \in A_n(T(C)).$$

Every endomorphism of the module $T(C)$ is equal to $1 \otimes \alpha$, where $\alpha \in \mathrm{End}(C)$. Therefore, assume that $(1 \otimes \alpha)(T(C)[p^n]) = 0$. Consequently, $(K/R)[p^n] \otimes \alpha C = 0$. It follows from the proof of Theorem 13.1 that $\alpha C \subseteq p^n C$, i.e., $\alpha \in A_n(C)$. Therefore, the restriction of the isomorphism $\mathrm{End}(C) \cong \mathrm{End}(T(C))$ to the ideal $A_n(C)$ provides the required isomorphism. □

The following exercises are results of the work [206] of Liebert.

Exercise 1. If A and B are divisible primary modules or complete torsion-free modules, then the lattice isomorphism between $L(A)$ and $L(B)$ implies the isomorphism of the modules A and B.

Exercise 2. Let D be a divisible primary module and let C be a complete torsion-free module. Prove that the following conditions are equivalent.

(1) $r(D) = r_p(C)$.

(2) The lattices $L(D)$ and $L(C)$ are isomorphic to each other.

(3) $\mathrm{End}(D) \cong \mathrm{End}(C)$.

14 Jacobson radical

For a ring S, the *Jacobson radical* of S is called the *radical*; it is denoted by $J(S)$. It is well known that left-side characterizations of the radical are equivalent to right-side characterizations of the radical. By this reason, we present only the right-side definition and several properties of the radical. The radical of the ring S, can be defined as the intersection of all maximal right ideals of S. The element x of the ring S is contained in $J(S)$ if and only if the element $1 - xy$ is right invertible for every element $y \in S$. The ideal L of the ring S is contained in $J(S)$ if and only if the element $1 - x$ is right invertible for every $x \in L$. The factor ring $S/J(S)$ has the zero radical. Under ring homomorphisms, the radical is mapped into the radical of the image.

We recall that invertible elements of the ring $\mathrm{End}(M)$ coincide with automorphisms of the module M.

We will obtain a quite complete information about the radical of the endomorphism ring of a divisible primary module or a complete torsion-free module over a complete domain R. It is sufficient to consider only one of these two classes of modules. Using the equivalences from the previous section, we can to extend the results to another class of modules. We also touch on the radical of the endomorphism ring of an arbitrary torsion-free module and a primary module. First, we consider complete torsion-free modules.

Let C be a complete torsion-free module and let B be some basic submodule of C. Then $B = \bigoplus_{i \in I} R_i$, where $R_i \cong R$ for all i. Using the isomorphism $R_i \cong R$, we can consider R_i as a right R-module for every $i \in I$. This naturally turns B into a right R-module. Therefore B is an R-R-bimodule. We also have $p^n B = Bp^n$ and $(p^n B)r \subseteq p^n B$ for every positive integer n and $r \in R$. Let $x \in C$ and $x = \lim x_i$ with $x_i \in B$. For $r \in R$, we assume that $xr = \lim x_i r$. We obtain the right R-module C; more precisely, we obtain an R-R-bimodule. We note that C is a right torsion-free R-module. The right module multiplication by C can be defined with the use of some additive automorphism of the module C. For the construction of the multiplication, we use automorphisms of the form σ_y of the ring R from Section 3 (see Liebert [206]). Unfortunately, both methods to turn the module C into a right R-module are not canonical. For example, they depend on the choice of the isomorphism $R_i \cong R$.

We present several properties of the R-R-bimodule C.

(1) For an element $r \in R$, the mapping $x \to xr$, $x \in C$, is an endomorphism of the left R-module C. We denote it by λ_r. The mapping $R \to \operatorname{End}(C)$ such that $r \to \lambda_r$ for all $r \in R$ is an embedding of rings. We can assume that R is a subring in $\operatorname{End}(C)$ (we identify r with λ_r). Now it is clear that $(p^n C)r \subseteq p^n C$ for every positive integer n and $r \in R$. If R is a commutative ring, then λ_r is a module multiplication $x \to rx$, $x \in C$.

(2) $p^n C = Cp^n$ for every positive integer n. We have $p^n B = Bp^n$. Let

$$x \in C, \quad x = \lim x_i, \quad x_i \in B, \quad p^n x_i = y_i p^n, \quad y_i \in B.$$

Then

$$p^n x = p^n \lim x_i = \lim p^n x_i = \lim y_i p^n = (\lim y_i) p^n.$$

Consequently, $p^n C \subseteq Cp^n$. The converse inclusion can be proved similarly.

(3) Let E be one more complete torsion-free module. Since C is an R-R-bimodule, we have that $\operatorname{Hom}(C, E)$ is a left R-module (see Section 2). For an element $r \in R$ and a homomorphism $f \in \operatorname{Hom}(C, E)$, we denote by rf the homomorphism such that $(rf)(x) = f(xr)$ for every $x \in C$.

We prove that the relation $p^n \operatorname{Hom}(C, E) = \{f : C \to E \mid f(C) \subseteq p^n E\}$ holds for every positive integer n. If $g \in \operatorname{Hom}(C, E)$, then

$$(p^n g)C = g(Cp^n) = g(p^n C) = p^n(gC) \subseteq p^n E,$$

which provides one inclusion.

Let $f \in \operatorname{Hom}(C,E)$, $f(C) \subseteq p^n E$, and let f_i be the restriction f to R_i for every $i \in I$. Then $f_i = p^n g_i$ for some $g_i\colon R_i \to E$. Consequently, $f_{|B} = p^n g'$, where g' is the sum of all homomorphisms g_i, $i \in I$. By Proposition 11.12, g' can be extended to a homomorphism $g\colon C \to E$. Let $x \in C$ and $x = \lim x_i$ with $x_i \in B$. Then

$$f(x) = f(\lim x_i) = \lim f(x_i) = \lim(p^n g)x_i = p^n g(\lim x_i) = (p^n g)x$$

and $f = p^n g$. Therefore $f \in p^n \operatorname{Hom}(C,E)$.

On the module $\operatorname{Hom}(C,E)$, we define the p-adic topology which has a basis of neighborhoods of zero formed by the submodules $p^n \operatorname{Hom}(C,E)$ $(n \geq 1)$. It follows from (3) that this topology does not depend on the method to turn the module C into a right R-module.

(4) The R-module $\operatorname{Hom}(C,E)$ is a complete in the p-adic topology torsion-free module. Indeed, the exact sequence of R-R-bimodules $0 \to B \to C \to C/B \to 0$ induces the exact sequence of left R-modules

$$\operatorname{Hom}(C/B,E) \to \operatorname{Hom}(C,E) \to \operatorname{Hom}(B,E) \to 0.$$

We have 0 in the right part of the sequence, since C is a pure-injective module by Theorem 11.4. In addition, $\operatorname{Hom}(C/B,E) = 0$, since C/B is divisible and E is reduced. We have the module isomorphism

$$\operatorname{Hom}(C,E) \cong \operatorname{Hom}(B,E) \cong \prod_{i \in I} \operatorname{Hom}(R_i,E) \cong \prod_{|I|} E.$$

Therefore $\operatorname{Hom}(C,E)$ is a complete module by Proposition 11.1.

(5) Let A be a module. Since C is an R-R-bimodule, $\operatorname{Hom}(A,C)$ is a right R-module, where $(fr)(a) = f(a)r$ for all $f \in \operatorname{Hom}(A,C)$, $r \in R$ and $a \in A$. Therefore $\operatorname{End}(C)$ is a R-R-bimodule. Let $\alpha \in \operatorname{End}(C)$, $r \in R$, and let $x \in C$. Then

$$(r\alpha)x = \alpha(xr) \quad \text{and} \quad (\lambda_r\alpha)x = \alpha(\lambda_r(x)) = \alpha(xr).$$

Therefore $r\alpha = \lambda_r\alpha$. In addition,

$$(\alpha r)x = \alpha(x)r, \qquad (\alpha\lambda_r)x = \lambda_r(\alpha(x)) = \alpha(x)r, \quad \text{and} \quad \alpha r = \alpha\lambda_r.$$

Consequently, the left and right module multiplication on the R-module $\operatorname{End}(C)$ by the element r coincide with multiplications by λ_r in the ring $\operatorname{End}(C)$.

We prove that for every positive integer n, the relation $A_n(C) = p^n \operatorname{End}(C) = \operatorname{End}(C)p^n$ holds (the ideal $A_n(C)$ is defined in Section 13). The first relation is contained in (3). Let $\alpha \in \operatorname{End}(C)$ and let α_i be the restriction of α to R_i, $i \in I$. We have $p^n\alpha_i = \beta_i p^n$ for some $\beta_i\colon R_i \to C$ (it should be used Proposition 2.2 (b) and properties (5) or (6) from Section 3). Therefore $p^n\alpha_{|B} = \beta' p^n$, where $\beta'\colon B \to C$ and β' is the sum of β_i for all $i \in I$. We denote by $\beta\colon C \to C$ an extension of β'. If $x \in C$ and $x = \lim x_i$, $x_i \in B$, then, similar to (3), we can obtain $(p^n\alpha)x = (\beta p^n)x$. Consequently,

$$p^n\alpha = \beta p^n \quad \text{and} \quad p^n \operatorname{End}(C) \subseteq \operatorname{End}(C)p^n.$$

In addition,

$$(\alpha p^n)C = (\alpha C)p^n \subseteq Cp^n = p^nC, \quad \alpha p^n \in A_n(C), \quad \text{and} \quad \operatorname{End}(C)p^n \subseteq A_n(C),$$

which completes the proof.

(6) If $r \neq 0$, then λ_r is not a zero-divisor in the ring $\operatorname{End}(C)$. Let $\alpha \in \operatorname{End}(C)$ and $\alpha \neq 0$. Then

$$(\lambda_r\alpha)C = (r\alpha)C = \alpha(Cr) = \alpha(rC) = r(\alpha C) \neq 0,$$

since C is a torsion-free module. In addition,

$$(\alpha\lambda_r)C = (\alpha r)C = (\alpha C)r \neq 0,$$

since C is a right torsion-free module. Therefore $\lambda_r\alpha, \alpha\lambda_r \neq 0$.

We define so-called *J-adic topology* on the ring. A basis of neighborhoods of zero of this topology consists of all powers $J(\operatorname{End}(C))^n$ $(n \geq 1)$ of the radical of the ring $\operatorname{End}(C)$. The ring $\operatorname{End}(C)$ is a topological ring with respect to the J-adic topology. If we consider the ring $\operatorname{End}(C)$ as a left or right R-module, then it follows from the theorem presented below that the J-adic topology coincides with the p-adic topology (see also (5)).

Theorem 14.1 (Liebert [206]). *For a complete torsion-free module C, the following assertions hold.*

(a) *The ring* $\operatorname{End}(C)$ *is complete in the J-adic topology and* $\operatorname{End}(C)$ *has an element π such that π is not a zero-divisor and*

$$J(\operatorname{End}(C)) = \operatorname{Hom}(C, pC) = \pi \operatorname{End}(C) = \operatorname{End}(C)\pi.$$

(b) *The factor ring* $\operatorname{End}(C)/J(\operatorname{End}(C))$ *is isomorphic to the operator ring of the space C/pC over the residue division ring R_p, where $R_p = R/pR$.*

Proof. We take λ_p as π. (If R is a commutative ring, then we denote by π the multiplication by p.) We define the mapping

$$f\colon \operatorname{End}(C) \to \operatorname{End}_{R_p}(C/pC), \qquad \alpha \to f(\alpha), \qquad \alpha \in \operatorname{End}(C),$$

where $f(\alpha)(x+pC) = \alpha(x)+pC$ for all $x \in C$. It is easy to verify that f is a ring homomorphism with kernel $\mathrm{Hom}(C, pC)$ (this is the ideal $A_1(C)$ in the notation of Section 13). Let β be some linear operator of the R_p-space C/pC. We fix a basis $\{a_i + pC\}_{i\in I}$ of the space C/pC. Then $B = \bigoplus_{i\in I} Ra_i$ is a basic submodule of the module C (property (6) from Section 9). There exists a homomorphism $\alpha\colon B \to C$, where $\alpha(a_i) + pC = \beta(a_i + pC)$ for every $i \in I$. Since the module C is complete, α can be extended to an endomorphism of the module C which is also denoted by α. It follows from properties of basic submodules that the relation $C = B + pC$ holds. Therefore, for $x \in C$ we have $x = a + py$, where $a \in B$, $y \in C$. This implies

$$f(\alpha)(x + pC) = f(\alpha)(a + pC) = \alpha(a) + pC = \beta(a + pC) = \beta(x + pC).$$

Therefore $f(\alpha) = \beta$ and f is a surjective homomorphism. Consequently,

$$\mathrm{End}(C)/\,\mathrm{Hom}(C, pC) \cong \mathrm{End}_{R_p}(C/pC).$$

The ring of all linear operators of the space has the zero radical. Therefore $J(\mathrm{End}(C)) \subseteq \mathrm{Hom}(C, pC)$. Let $\alpha \in \mathrm{Hom}(C, pC)$. It can be verified that $1 - \alpha$ is a monomorphism and $(1 - \alpha)C$ is a pure submodule in C; therefore $(1 - \alpha)C$ is a complete module. Consequently, $(1 - \alpha)C$ is a closed submodule in C. For every $x \in C$ we have $x = (1 - \alpha)x + \alpha(x)$. Consequently, $(1 - \alpha)C + pC = C$ and $(1 - \alpha)C$ is a dense submodule in C. Therefore $(1 - \alpha)C = C$ and $1 - \alpha$ is an automorphism of the module C. We have proved that $1 - \alpha$ is an invertible element in $\mathrm{End}(C)$ for every $\alpha \in \mathrm{Hom}(C, pC)$. Since $\mathrm{Hom}(C, pC)$ is an ideal, $\mathrm{Hom}(C, pC) \subseteq J(\mathrm{End}(C))$. Consequently, $J(\mathrm{End}(C)) = \mathrm{Hom}(C, pC)$ which implies (b) and the first relation in (a). The remaining equalities in (a) are contained in (5). It follows from (4) that the ring $\mathrm{End}(C)$ is complete in the J-adic topology. □

We pass to divisible primary modules. By applying Theorem 13.1 and Corollary 13.2, all results obtained for complete torsion-free modules and their endomorphism rings can be formulated for divisible primary modules with some changes. Clearly, these results can be proved independently (see Liebert [206]). We present only a brief presentation.

Let D be a divisible primary module. Then $D \cong K/R \otimes C$ for some complete torsion-free module C. We identify D with $K/R \otimes C$. The rings $\mathrm{End}(D)$ and $\mathrm{End}(C)$ are canonically isomorphic to each other, as it is known from Section 13. Since C is an R-R-bimodule, the module D can be considered as an R-R-bimodule (see the beginning of Section 13). We can turn D into an R-R-bimodule as follows. Let $D = \bigoplus_{i\in I} Q_i$, where $Q_i \cong K/R$ for all i. The module

K/R is an R-R-bimodule. Therefore D can be naturally considered as an R-R-bimodule. It is also possible to consider some additive automorphism of the module D based on the automorphisms σ_y from Section 4. We note that D is a primary divisible right R-module.

If G is some divisible primary module, then $\mathrm{Hom}(D, G)$ is a left R-module (details are given in (3)). It follows from (3) and Corollary 13.2 that $p^n \mathrm{Hom}(D, G) = \{f : D \to G \mid f(D[p^n]) = 0\}$ for every $n \geq 1$. We obtain that the p-adic topology on $\mathrm{Hom}(D, G)$ does not depend on the method to turn the module D into a right R-module. For the module G, there exists a complete torsion-free module C with $G \cong K/R \otimes C$. Considering (4) and Theorem 13.1, we can state that $\mathrm{Hom}(D, G)$ is a complete torsion-free module. Similar to (1) and (5), R we can assume a subring in $\mathrm{End}(C)$. In Section 13, the ideals

$$A_n(D) = \{\alpha \in \mathrm{End}(D) \mid \alpha(D[p^n]) = 0\} \quad \text{and} \quad A_n(C) = \mathrm{Hom}(C, p^n C)$$

are defined for every positive integer n; it is also proved that they are isomorphic to each other. Now it follows from (5) that $A_n(D) = p^n \mathrm{End}(C) = \mathrm{End}(C)p^n$ $(n \geq 1)$.

Theorem 14.2 (Liebert [206]). *For a divisible primary module D, the following assertions hold.*

(a) *The ring* $\mathrm{End}(D)$ *is complete in the* J*-adic topology and there exists a non-zero-divisor* π *in* $\mathrm{End}(D)$ *such that*

$$J(\mathrm{End}(D)) = A_1(D) = \pi \,\mathrm{End}(D) = \mathrm{End}(D)\pi.$$

(b) *The factor ring* $\mathrm{End}(D)/J(\mathrm{End}(D))$ *is isomorphic to the linear operator ring of the* R_p*-space* $D[p]$.

Proof. (a) The assertion follows from Theorem 14.1, Corollary 13.2, and results proved before.

(b) We define a ring homomorphism $f : \mathrm{End}(D) \to \mathrm{End}_{R_p}(D[p])$, where $f(\alpha)$ is the restriction α to $D[p]$ for every $\alpha \in \mathrm{End}(D)$. We recall that endomorphisms of the R-module $D[p]$ coincide with linear operators of the R_p-space $D[p]$ (this was mentioned in Section 4). Every endomorphism of the submodule $D[p]$ can be extended to an endomorphism of the divisible module D. Since $\mathrm{Ker}(f) = A_1(D)$, we have isomorphism $\mathrm{End}(D)/A_1(D) \cong \mathrm{End}_{R_p}(D[p])$. However, $A_1(D) = J(\mathrm{End}(D))$ and we obtain (b). □

In the remaining part of the section, we present several basic results about the radical of the endomorphism ring of an arbitrary torsion-free module or a primary module over the domain R (we recall that R is complete).

Let M be some torsion-free module. An endomorphism α of the module M is said to *finite* if the submodule $\mathrm{Im}(\alpha)$ has finite rank. All finite endomorphisms of the module M form an ideal in $\mathrm{End}(M)$; we denote it by $\mathrm{Fin}(M)$. This ideal is studied in Section 17.

Proposition 14.3. *Let M be a reduced torsion-free module.*

(a) $J(\mathrm{End}(M)) \subseteq \mathrm{Hom}(M, pM)$.

(b) *Let L be some ideal of the ring* $\mathrm{End}(M)$ *contained in* $\mathrm{Hom}(M, pM)$. *The inclusion $L \subseteq J(\mathrm{End}(M))$ holds if and only if for the ideal L, the following condition $(*)$ holds*:

$(*)$ *if $x \in M$, $\alpha \in L$, and $y_n = x + \alpha(x) + \cdots + \alpha^{n-1}(x)$ for every positive integer n, then the sequence $\{y_n\}_{n \geq 1}$ converges in the module M.*

(c) $J(\mathrm{End}(M)) = \mathrm{Hom}(M, pM)$ *if and only if for the ideal* $\mathrm{Hom}(M, pM)$, *the condition $(*)$ from* (b) *holds.*

(d) $\mathrm{Hom}(M, pM) \cap \mathrm{Fin}(M) \subseteq J(\mathrm{End}(M))$.

Proof. (a) We assume that there exist $x \in M$ and $\alpha \in J(\mathrm{End}(M))$ such that $\alpha(x) \notin pM$. The submodules Rx and $R(\alpha x)$ are direct summands in M (Corollary 11.5). We can obtain $\beta \in \mathrm{End}(M)$ with $\beta(\alpha x) = x$. Then $(1 - \alpha\beta)x = 0$. However, $\alpha\beta \in J(\mathrm{End}(M))$. Therefore $1 - \alpha\beta$ is an automorphism of the module M. This is impossible, since $x \neq 0$. Consequently, $\alpha \in \mathrm{Hom}(M, pM)$.

(b) We assume that the inclusion holds, $x \in M$, and $\alpha \in L$. Then $1 - \alpha$ is an automorphism of the module M and $(1 - \alpha)y = x$ for some y. We have

$$\begin{aligned} y_n &= x + \alpha(x) + \cdots + \alpha^{n-1}(x) \\ &= (1-\alpha)y + (\alpha - \alpha^2)y + \cdots + (\alpha^{n-1} - \alpha^n)y \\ &= (1 - \alpha^n)y. \end{aligned}$$

Therefore $y - y_n = \alpha^n(y)$. It follows from $\alpha \in \mathrm{Hom}(M, pM)$ that $\alpha^n(y) \in p^n M$ for all n. Therefore $y = \lim y_n$. Therefore $(*)$ holds for the ideal L.

Conversely, assume that condition $(*)$ holds. We prove that $1 - \alpha$ is an automorphism of the module M for every $\alpha \in L$. Since $\alpha M \subseteq pM$, we have $\mathrm{Ker}(1 - \alpha) = 0$. For an element $x \in M$, we set $y = \lim y_n$, where $y_n = x + \alpha(x) + \cdots + \alpha^{n-1}(x)$. We have the relation

$$\begin{aligned} (1-\alpha)y = (1-\alpha)(\lim y_n) &= \lim(y_n - \alpha(y_n)) = \lim(x - \alpha^n(x)) \\ &= x - \lim \alpha^n(x) \\ &= x, \end{aligned}$$

since $\alpha^n(x) \in p^n M$ for every n. Consequently, $(1-\alpha)y = x$ and $1-\alpha$ is an automorphism of the module M, i.e., an invertible element of the ring $\operatorname{End}(M)$. Therefore $L \subseteq J(\operatorname{End}(M))$.

(c) We obtain from (a) and (b).

(d) It follows from (b) that it is sufficient to verify that condition $(*)$ holds for the ideal $\operatorname{Hom}(M, pM) \cap \operatorname{Fin}(M)$. We take

$$x \in M, \quad \alpha \in \operatorname{Hom}(M, pM) \cap \operatorname{Fin}(M), \quad y_n = x + \alpha(x) + \cdots + \alpha^{n-1}(x)$$

for $n \geq 1$. Let C be the pure submodule generated in M by elements x and y_n for all n. By Corollary 11.7, the free module C of finite rank is a complete module. Consequently, the Cauchy sequence $\{y_n\}_{n \geq 1}$ converges in C. Therefore, condition $(*)$ holds. □

For a complete torsion-free module M, we have the relation

$$J(\operatorname{End}(M)) = \operatorname{Hom}(M, pM)$$

proved in Theorem 14.1. If M has an infinite rank, then the inclusion in (d) is proper. The inclusion in (a) is proper for a free module of infinite rank by Proposition 14.4.

We note the following result. We assume that $M = A \oplus B$, $\varkappa\colon A \to M$, and $\pi\colon M \to A$. Then for every $\alpha \in J(\operatorname{End}(M))$, we have $\varkappa\alpha\pi \in J(\operatorname{End}(A))$. We take some $\beta \in \operatorname{End}(A)$. We can assume that β is an endomorphism of the module M with $\beta B = 0$. If γ is a right inverse endomorphism for $1 - \alpha\beta$, then $\varkappa\gamma\pi$ is right inverse for $1 - (\varkappa\alpha\pi)\beta$ in $\operatorname{End}(A)$. Consequently, $\varkappa\alpha\pi \in J(\operatorname{End}(A))$.

Proposition 14.4. *For a free module M, we have $J(\operatorname{End}(M)) = \operatorname{Hom}(M, pM) \cap \operatorname{Fin}(M)$.*

Proof. It is sufficient to verify the inclusion $J(\operatorname{End}(M)) \subseteq \operatorname{Fin}(M)$. We assume that $\alpha \in J(\operatorname{End}(M))$, but $\alpha \notin \operatorname{Fin}(M)$, i.e., the rank of the module αM is infinite. Let $M = \bigoplus_{i \in I} A_i$, where $A_i \cong R$, $i \in I$. We can choose pairwise different summands $A_{i_1}, A_{i_2}, \dots$ such that $\alpha N \subseteq N$, where $N = \bigoplus_{n \geq 1} A_{i_n}$, and the rank of the submodule αN is infinite. By the remark before the proposition, $\alpha_{|N} \in J(\operatorname{End}(N))$. Therefore, we assume that $M = \bigoplus_{i \geq 1} A_i$.

We set $n_0 = 1$. We assume that the subscript k_1 satisfies the property $\alpha A_1 \subseteq A_1 \oplus \cdots \oplus A_{k_1}$ and k_1 is the minimal number with such a property. Since the rank of the module αM is infinite, there exist two subscripts n_1 and k_2 such that

$$n_1, k_2 > k_1, \qquad \alpha A_{n_1} \subseteq A_1 \oplus \cdots \oplus A_{k_2},$$

and k_2 is the minimal number with such a property. Furthermore, there exist two subscripts n_2 and k_3 with analogous properties. Continuing this procedure, we obtain the sequence of the submodules $A_{n_0}, A_{k_1}, A_{n_1}, A_{k_2}, \dots$. Now we define an endomorphism $\beta \in \mathrm{End}(M)$ as follows. We assume that $\beta\colon A_{k_i} \to A_{n_i}$ be some nonzero homomorphism for all $i \geq 1$ and $\beta A_j = 0$ for all other j. We set $N = \bigoplus_{i \geq 1} A_{k_i}$. Let $\varkappa\colon N \to M$ and $\pi\colon M \to N$ be the embedding and the projection, respectively. Then $\varkappa\beta\alpha\pi \in J(\mathrm{End}(N))$. However, the endomorphism $1 - \varkappa\beta\alpha\pi$ is not an automorphism, since A_{k_1} is not contained in the image of the endomorphism. This is a contradiction. Therefore $J(\mathrm{End}(M)) \subseteq \mathrm{Fin}(M)$. □

We need some lemma about endomorphisms of primary modules.

Lemma 14.5. *For a primary module M, an endomorphism α of M is an automorphism if and only if the relations*

$$\mathrm{Ker}(\alpha) \cap M[p] = 0 \quad \textit{and} \quad \alpha((p^n M)[p]) = (p^n M)[p]$$

hold for $n \geq 0$.

Proof. The necessity of the condition of the lemma is obvious. We verify the sufficiency. If $\alpha x = 0$ and $o(x) = p^k$, then $o(p^{k-1}x) = p$ and $\alpha(p^{k-1}x) = 0$. This is impossible, since $\mathrm{Ker}(\alpha) \cap M[p] = 0$. Therefore $\mathrm{Ker}(\alpha) = 0$. We prove that α is a surjective mapping. For every $y \in M[p]$, there exists an element $x \in M[p]$ such that $y = \alpha x$, and we can assume $h(x) = h(y)$ for $h(y) < \infty$. If $h(y) = \infty$, then we can choose as x an element of any large height. Now we assume that $y \in M$, $o(y) = p^k$, $k > 1$, and all elements of order $< p^k$ are contained in the image of the endomorphism α. The element $p^{k-1}y$ is contained in $M[p]$, thus $p^{k-1}y = \alpha(x_1)$ for some x_1, and $h(x_1) \geq k-1$. Therefore $x_1 = p^{k-1}x$ for some x, and so $p^{k-1}(y - \alpha x) = 0$. By assumption, there exists $z \in M$ with $y - \alpha x = \alpha z$. Consequently, $y = \alpha(x + z) \in \mathrm{Im}(\alpha)$. Therefore $\mathrm{Im}(\alpha) = M$ and α is an automorphism of the module M. □

In the primary case, we can define ideals which are similar to $\mathrm{Hom}(M, pM)$ and $\mathrm{Fin}(M)$.

Let M be a reduced primary module. We set $P(M) = \{\alpha \in \mathrm{End}(M) \mid x \in M[p],\ h(x) < \infty \Longrightarrow h(x) < h(\alpha x)\}$. It is clear that

$$P(M) = \{\alpha \in \mathrm{End}(M) \mid \alpha((p^n M)[p]) \subseteq (p^{n+1} M)[p] \text{ for every } n \geq 0\}$$

and $P(M)$ is an ideal of the ring $\mathrm{End}(M)$. It is called the *Pierce ideal* of the module M. Furthermore, we set

$$F(M) = \{\alpha \in \mathrm{End}(M) \mid \alpha((p^k M)[p]) = 0 \text{ for some } k \geq 0\}.$$

The following assertions are similar to Proposition 14.3. By this reason, our proofs are not very detailed.

Proposition 14.6 (Pierce [248]). *Let M be a reduced primary module.*

(a) $J(\mathrm{End}(M)) \subseteq P(M)$.

(b) *Let L be some ideal of the ring* $\mathrm{End}(M)$ *contained in $P(M)$. The inclusion $L \subseteq J(\mathrm{End}(M))$ holds if and only if for the ideal L, the following condition $(*)$ holds:*

$(*)$ *if $x \in M[p]$, $\alpha \in L$, and $y_n = x + \alpha(x) + \cdots + \alpha^{n-1}(x)$ for every positive integer n, then the sequence $\{y\}_{n \geq 1}$ converges in the module M.*

(c) $J(\mathrm{End}(M)) = P(M)$ *if and only if for the ideal $P(M)$, the condition $(*)$ from* (b) *holds.*

(d) $P(M) \cap F(M) \subseteq J(\mathrm{End}(M))$.

Proof. (a) Let $\alpha \in J(\mathrm{End}(M))$. We assume that $h(x) = h(\alpha x)$ for some element x of finite height whose order is p. By Corollary 7.5, there exist decompositions $M = Ra \oplus A = Rb \oplus B$ for some elements a, b and submodules A, B, where $\alpha x \in Ra$ and $x \in Rb$. It follows from $h(\alpha x) = h(x)$ that $o(a) = o(b)$ and $Ra \cong Rb$. We can define an endomorphism β such that

$$\beta\alpha \in Rb, \qquad \beta(\alpha x) = x, \quad \text{and} \quad \beta A = 0.$$

Then $\alpha\beta \in J(\mathrm{End}(M))$. Therefore $1 - \alpha\beta$ is an automorphism of the module M. However, $(1-\alpha\beta)x = x - x = 0$ which contradicts to this property. Consequently, $h(x) < h(\alpha x)$ for every element x of order p and of finite height and $\alpha \in P(M)$.

The proof of (b) is similar to the proof of Proposition 14.3 (b). It is also necessary to use Corollary 11.5.

(c) The assertion follows from (a) and (b).

(d) It is sufficient to verify that condition $(*)$ holds for the ideal $P(M) \cap F(M)$. We take

$$x \in M[p], \quad \alpha \in P(M) \cap F(M), \quad \text{and} \quad y_n = x + \alpha(x) + \cdots + \alpha^{n-1}(x)$$

for all $n \geq 1$. There exists an integer $k \geq 0$ such that $\alpha^n(x) = 0$ for all $n > k$. Therefore the sequence $\{y_n\}_{n \geq 1}$ converges. □

Let M be a torsion complete module (these modules are defined before exercises in Section 11). Every sequence such that the orders of elements of the sequence are bounded by the element p^k for some k converges in the module M. Using Proposition 14.6, we can prove the following result (see Pierce [248]).

Corollary 14.7. *If M is a torsion complete module, then* $J(\mathrm{End}(M)) = P(M)$. *In particular, this relation holds for the bounded module M.*

We present a description of the radical $J(\mathrm{End}(M))$ for one more class of primary modules.

Proposition 14.8 (Liebert [209]). *Let M be a direct sum of primary cyclic modules. Then $J(\mathrm{End}(M)) = P(M) \cap F(M)$.*

Proof. By considering Proposition 14.6, we need to only verify the inclusion

$$J(\mathrm{End}(M)) \subseteq F(M).$$

We assume that $\alpha \in J(\mathrm{End}(M))$ and $\alpha \notin F(M)$. In this case, there exist two infinite sequences $\{x_i\}_{i\geq 1}$ and $\{z_i\}_{i\geq 1}$ of elements from $M[p]$ such that $\alpha(x_i) = z_i$ for every i and $h(x_1) < h(z_1) < h(x_2) < h(z_2) < \cdots$. We apply Corollary 7.5 to all the elements x_i and z_i, $i \geq 1$. We obtain the decomposition $M = \bigoplus_{k\geq 1} Ra_k$ such that every of the elements $x_1, x_2, \dots$ and $z_1, z_2, \dots$ is contained in some cyclic summand Ra_k. It follows from the conditions for the height of elements that distinct elements are contained in distinct summands. Since $h(z_i) < h(x_{i+1})$, there exists an endomorphism β of the module M which maps z_i into x_{i+1} for every $i \geq 1$. We have $(\beta\alpha)^{n-1}(z_1) = z_n$, $n \geq 1$. Now we set

$$y_n = z_1 + (\beta\alpha)(z_1) + \cdots + (\beta\alpha)^{n-1}(z_1)$$

for every n. Then $y_n = z_1 + z_2 + \cdots + z_n$ and it is clear that the sequence $\{y_n\}_{n\geq 1}$ does not have the limit in the module M. This contradicts to the inclusion $\beta\alpha \in J(\mathrm{End}(M))$, since by Proposition 14.6, condition $(*)$ holds for every endomorphism in $J(\mathrm{End}(M))$. Consequently, $J(\mathrm{End}(M)) \subseteq F(M)$. □

For a primary module M or a torsion-free module M, the description problem for the radical $J(\mathrm{End}(M))$ seems to be very difficult. The case of mixed modules was not practically studied.

Exercise 1. Let $M = A \oplus B$, where B is a fully invariant submodule in M. Prove that

$$J(\mathrm{End}(M)) = \begin{pmatrix} J(\mathrm{End}(A)) & \mathrm{Hom}(A, B) \\ 0 & J(\mathrm{End}(B)) \end{pmatrix}.$$

Using Exercise 1, solve Exercise 2.

Exercise 2. (a) Reduce the study of the radical $J(\mathrm{End}(M))$ to the case of the reduced module M.

(b) Describe the radicals $J(\mathrm{End}(D))$ and $J(\mathrm{End}(C))$, where D is a divisible module and C is a complete module.

Hint. Represent the module D as the direct sum of a primary module and the torsion-free module. Represent the module C in the form of Corollary 11.11 (1) and also use Corollary 14.7.

Exercise 3. For a ring S, the additive group of S is generated by invertible elements of S if and only if the factor ring $S/J(S)$ is generated by invertible elements of $S/J(S)$.

Using Exercise 3 and Theorem 14.1, solve the following exercise.

Exercise 4. Let C be a complete or torsion complete module. Obtain the conditions, under which every endomorphism of the module C is the sum of a finite number of automorphisms (or is a sum of two automorphisms).

Exercise 5. Let $M = \bigoplus_{i=1}^{k} A_i$. If every endomorphism of each of the module A_i is the sum of n automorphisms of A_i ($n \geq 2$), then the same property has the module M.

Exercise 6. Let $M = \bigoplus_{i \in I} A_i$. We assume that every endomorphism of each summand A_i is the sum of n automorphisms of A_i ($n \geq 3$). Then the same property holds for every endomorphism α of the module M such that the image of αA_j is contained in the sum of a finite number of summands A_i for all $j \in I$.

Exercise 7. If $M = \bigoplus_{i \in I} A_i$, where all summands A_i are isomorphic to each other, then every endomorphism of the module M is a sum of four (three) automorphisms of M.

Exercise 8. Let S be a local ring. Under what conditions, every element of the ring S is a sum of two invertible elements?

15 Galois correspondences

Let M be a module. Every submodule X of the module M provides the left ideal $\{\alpha \in \mathrm{End}(M) \mid \alpha M \subseteq X\}$ and the right ideal $\{\alpha \in \mathrm{End}(M) \mid \alpha X = 0\}$ of the endomorphism ring of the module M. We use both these methods to obtain characterizations of one-sided ideals of the endomorphism ring of a complete torsion-free module and of the divisible primary module over a complete discrete valuation domain R. We have chosen results which are not most difficult. However, the results well demonstrate our approach to the characterization of one-sided ideals.

We follow to the work [210] of Liebert. We use without citations the material of Sections 6 and 11. Everywhere, C is a complete torsion-free module and D is a divisible primary module. We recall that $r_p(M)$ is the p-rank and $r(M)$ is the rank of the module M.

Proposition 15.1. *For a submodule X of the module C, the following conditions are equivalent.*

(1) *There exists an endomorphism $\alpha \in \operatorname{End}(C)$ with $X = \alpha C$.*

(2) *X is isomorphic to some direct summand of the module C.*

(3) *X is a complete module and $r_p(X) \leq r_p(C)$.*

Proof. (1) $\Longrightarrow$ (2). The submodule $\operatorname{Ker}(\alpha)$ is a pure closed submodule in C. Consequently, $\operatorname{Ker}(\alpha)$ is a direct summand in C and $C = \operatorname{Ker}(\alpha) \oplus Y$ for some submodule Y. Therefore $X = \alpha C \cong C/\operatorname{Ker}(\alpha) \cong Y$.

(2) $\Longrightarrow$ (3). Let $C = A \oplus Y$ and $X \cong Y$. Then X is a complete module. It follows from $r_p(C) = r_p(A) + r_p(Y)$ that $r_p(X) = r_p(Y) \leq r_p(C)$.

(3) $\Longrightarrow$ (1). Let B be some basic submodule of the module C. Since $r(B) = r_p(C)$, there exists a direct decomposition of B into a direct sum $V \oplus W$, where $r(V) = r_p(X)$. We have

$$C = \bar{V} \oplus \overline{W} \quad \text{and} \quad r_p(\bar{V}) = r(V) = r_p(X).$$

Consequently, $\bar{V} \cong X$ (consider, that $\bar{V}$ and X are complete modules). We take an endomorphism α of the module C such that $\alpha\colon \bar{V} \to X$ is some isomorphism and $\alpha\overline{W} = 0$. Then $\alpha C = X$ and we obtain (1). □

To the following lemma, Exercise 1 is related.

Lemma 15.2. *Let X be a submodule in C. Then $r_p(X) > r_p(C)$ if and only if $r((C/X)[p]) > r_p(C)$.*

Proof. There is a submodule Y of X such that

$$pX \subseteq Y \qquad \text{and} \qquad X/pX = (X \cap pC)/pX \oplus Y/pX.$$

Here we have

$$Y/pX \cong X/(X \cap pC) \cong (X + pC)/pC,$$

i.e., Y/pX is isomorphic to some submodule in C/pC. There exists an isomorphism from $(C/X)[p]$ onto $(X \cap pC)/pX$ such that $a + X \to pa + pX$ for every $a \in C$ with $pa \in X$. Now the assertion of the lemma is obvious. □

We denote by $\mathcal{L}(C)$ the ordered with respect to inclusion set of all submodules X in C such that X is a complete module with $r_p(X) \leq r_p(C)$. We prove that $X + Y, X \cap Y \in \mathcal{L}(C)$ for any two submodules X and Y of $\mathcal{L}(C)$. First, we note that if $r_p(C)$ is a finite cardinal number, then C is a free module of finite rank and $\mathcal{L}(C)$ is the set of all submodules of the module C. We assume that $r_p(C)$ is an infinite number. Using a basic submodule, we repeat the proof of

Proposition 15.1 and obtain the decomposition $C = E \oplus G \oplus H$, where $E \cong X$ and $G \cong Y$. We define the homomorphism $\beta\colon C \to X \oplus Y$ such that β coincides on E (resp., G) with some isomorphism $E \cong X$ (resp., $G \cong Y$) and $\beta H = 0$. Let $\gamma\colon X \oplus Y \to X + Y$ be the homomorphism such that $(x, y) \to x + y$ for all $x \in X$ and $y \in Y$. Now we set $\alpha = \beta\gamma$. Then α is an endomorphism of the module C and $\alpha C = X + Y$. By Proposition 15.1, $X + Y \in \mathcal{L}(C)$.

The intersection of every number of complete submodules of the module C is a complete submodule. We have a module embedding $C/(X \cap Y) \to C/X \oplus C/Y$, where $a + (X \cap Y) \to (a + X, a + Y)$ for every $a \in C$. By Lemma 15.2, we have $r_p(X \cap Y) \le r_p(C)$. Consequently, $X \cap Y \in \mathcal{L}(C)$. We obtain, that the ordered set $\mathcal{L}(C)$ is a lattice, where

$$\inf(X, Y) = X \cap Y \qquad \text{and} \qquad \sup(X, Y) = X + Y$$

for any two elements $X, Y \in \mathcal{L}(C)$.

For a left ideal L of the ring $\operatorname{End}(C)$, we define the submodule LC of the module C consisting of all finite sums $\sum \alpha_i(a_i)$, where $\alpha_i \in L$ and $a_i \in C$. If L is a principal left ideal with generator α, then $LC = \alpha C$.

If X is a submodule of the module C, then we denote by $\operatorname{Hom}(C, X)$ the left ideal $\{\alpha \in \operatorname{End}(C) \mid \alpha C \subseteq X\}$ of the ring $\operatorname{End}(C)$. We consider various properties of the submodule LC and the left ideal $\operatorname{Hom}(C, X)$.

(1) If L is a principal left ideal of the ring $\operatorname{End}(C)$, then $LC \in \mathcal{L}(C)$. This follows from Proposition 15.1.

(2) If $X \in \mathcal{L}(C)$, then $\operatorname{Hom}(C, X)$ is a principal left ideal of the ring $\operatorname{End}(C)$.

By Proposition 15.1, there exists an endomorphism $\alpha \in \operatorname{End}(C)$ with $X = \alpha C$. We prove that $\operatorname{Hom}(C, X) = \operatorname{End}(C)\alpha$. Let $\beta \in \operatorname{Hom}(C, X)$. Similar to Proposition 15.1, we have $C = \operatorname{Ker}(\alpha) \oplus Y$, where $\alpha_{|Y}$ is a isomorphism $Y \to X$. Let δ be the converse mapping to $\alpha_{|Y}$. We set $\gamma = \beta\delta$. Then $\beta = \gamma\alpha$ and $\beta \in \operatorname{End}(C)\alpha$. Consequently, $\operatorname{Hom}(C, X) \subseteq \operatorname{End}(C)\alpha$. The converse inclusion is directly verified.

(3) If L is a principal left ideal of the ring $\operatorname{End}(C)$, then $\operatorname{Hom}(C, LC) = L$. Let $L = \operatorname{End}(C)\alpha$. Then $LC = \alpha C$ and $\operatorname{Hom}(C, \alpha C) = \operatorname{End}(C)\alpha$ by property (2). Consequently, $\operatorname{Hom}(C, LC) = L$.

(4) If $X \in \mathcal{L}(C)$, then $\operatorname{Hom}(C, X)C = X$. We choose $\alpha \in \operatorname{End}(C)$ with $X = \alpha C$. By (2), $\operatorname{Hom}(C, X) = \operatorname{End}(C)\alpha$. Therefore $\operatorname{Hom}(C, X)C = \alpha C = X$.

(5) Let L and N be two principal left ideals of the ring $\operatorname{End}(C)$. By (3) and Proposition 15.1, there exist submodules $X, Y \in \mathcal{L}(C)$ such that $L = \operatorname{Hom}(C, X)$ and $N = \operatorname{Hom}(C, Y)$. We have $L \cap N = \operatorname{Hom}(C, X \cap Y)$, where $X \cap Y \in \mathcal{L}(C)$. By (2), $L \cap N$ is a principal left ideal of the ring $\operatorname{End}(C)$.

We prove that $L + N = \operatorname{Hom}(C, X + Y)$. We choose some basic submodule

$B = \bigoplus_{i \in I} Rb_i$ of the module C. Let $\alpha \in \operatorname{Hom}(C, X+Y)$. We have $\alpha(b_i) = x_i + y_i$, where $x_i \in X$ and $y_i \in Y$. We assume that $\beta(rb_i) = rx_i$ and $\gamma(rb_i) = ry_i$ for every $i \in I$ and $r \in R$. Thus, we obtain homomorphisms $\beta\colon B \to X$ and $\gamma\colon B \to Y$, and $\alpha = \beta + \gamma$ on B. Since X and Y are complete modules, β and γ have unique extensions to homomorphisms $C \to X$ and $C \to Y$, respectively. We also denote these extensions by β and γ, respectively. Then $\alpha = \beta + \gamma$. Thus, $\operatorname{Hom}(C, X+Y) \subseteq \operatorname{Hom}(C, X) + \operatorname{Hom}(C, Y)$. The second inclusion is obvious. Now $L + N$ is a principal left ideal of the ring $\operatorname{End}(C)$ by (2).

(6) For any two principal left ideals L and N of the ring $\operatorname{End}(C)$ we have the relations $(L+N)C = LC + NC$ and $(L \cap N)C = LC \cap NC$.

The first relation always holds. Using familiar properties, we have

$$L = \operatorname{Hom}(C, LC), \quad N = \operatorname{Hom}(C, NC), \quad \text{and} \quad L \cap N = \operatorname{Hom}(C, LC \cap NC).$$

Therefore, we have

$$(L \cap N)C = \operatorname{Hom}(C, LC \cap NC)C = LC \cap NC.$$

It follows from (5) that the intersection and the sum of every finite number of principal left ideals of the ring $\operatorname{End}(C)$ is a principal left ideal. We conclude that all principal left ideals of the ring $\operatorname{End}(C)$ with respect to inclusion form a lattice, where

$$\inf(L, N) = L \cap N \qquad \text{and} \qquad \sup(L, N) = L + N.$$

Theorem 15.3. *Let C be a complete torsion-free module. The mappings $X \to \operatorname{Hom}(C, X)$ and $L \to LC$ are mutually inverse isomorphisms between the lattice $\mathcal{L}(C)$ of all complete submodules X in C with $r_p(X) \le r_p(C)$, and the lattice of all principal left ideals of the ring $\operatorname{End}(C)$.*

Proof. It follows from (3) and (4) that the mentioned mappings are mutually inverse. We denote them Φ and Ψ, respectively. By (5) and (6), we have

$$\Phi(X \cap Y) = \Phi(X) \cap \Phi(Y), \qquad \Phi(X + Y) = \Phi(X) + \Phi(Y),$$
$$\Psi(L \cap N) = \Psi(L) \cap \Psi(N), \qquad \text{and} \qquad \Psi(L + N) = \Psi(L) + \Psi(N)$$

for any two elements $X, Y \in \mathcal{L}(C)$ and any two principal left ideals L and N of the ring $\operatorname{End}(C)$. Consequently, Φ and Ψ are lattice isomorphisms. □

We will obtain characterization of the lattice of all left ideals of the ring $\operatorname{End}(C)$. We consider the notion of the ideal of a lattice. A non-empty subset S of some lattice $\langle \mathcal{L}, \le \rangle$ is called an *ideal* in $\mathcal{L}$ if the following two conditions hold.

(1) $\sup(a, b) \in S$ for all $a, b \in S$.

(2) If $a \in S$ and $x \leq a$, then $x \in S$.

The ideals of the lattice $\mathcal{L}$ ordered with respect to inclusion form a lattice. We denote it by $\mathcal{JL}$; it is called the *ideal lattice* of the lattice $\mathcal{L}$. If S and T are two ideals of the lattice $\mathcal{L}$, then

$$\inf(S,T) = S \cap T \text{ and } \sup(S,T) = \{x \in \mathcal{L} \mid \exists a \in S,\, b \in T : x \leq \sup(a,b)\}.$$

Our aim is to prove that the lattice $\mathcal{JL}(C)$ of ideals of the lattice $\mathcal{L}(C)$ is isomorphic to the lattice $\mathcal{L}(\mathrm{End}(C))$ of all left ideals of the endomorphism ring of the module C. For these purposes, we use the following simple property: the left ideal L of the ring $\mathrm{End}(C)$ is equal to the union of principal left ideals $\mathrm{End}(C)\alpha$ for all $\alpha \in L$.

We denote by $\mathcal{PL}(\mathrm{End}(C))$ the lattice of all principal left ideals of the ring $\mathrm{End}(C)$. For the left ideal L of the ring $\mathrm{End}(C)$, let $\sigma(L)$ be the set of all principal left ideals of the ring $\mathrm{End}(C)$ contained in L. Then $\sigma(L)$ is an ideal of the lattice $\mathcal{PL}(\mathrm{End}(C))$. If T is an ideal of the lattice $\mathcal{PL}(\mathrm{End}(C))$, then we define $\tau(T)$ as the union of all ideals from T. In both cases, we consider that the sum of two principal left ideals is a principal left ideal. In addition, we have

$$\tau(\sigma(L)) = L \quad \text{and} \quad \sigma(\tau(T)) = T.$$

Consequently, the correspondences $T \to \tau(T)$ and $L \to \sigma(L)$ are mutually inverse lattice isomorphisms between the lattices $\mathcal{JPL}(\mathrm{End}(C))$ and $\mathcal{L}(\mathrm{End}(C))$. The isomorphism Φ between the lattices $\mathcal{L}(C)$ and $\mathcal{PL}(\mathrm{End}(C))$ in Theorem 15.3 induces isomorphism φ between the lattices $\mathcal{JL}(C)$ and $\mathcal{JPL}(\mathrm{End}(C))$. As a result, we obtain mutually inverse lattice isomorphisms between the lattices $\mathcal{JL}(C)$ and $\mathcal{L}(\mathrm{End}(C))$. We present the action of these isomorphisms.

For the ideal S of the lattice $\mathcal{L}(C)$, we set $\Phi(S) = \tau(\varphi(S))$. If L is a left ideal of the ring $\mathrm{End}(C)$, then let $\Psi(L) = \varphi^{-1}(\sigma(L))$. We can verify that the following relations hold:

$$\Phi(S) = \{\alpha \in \mathrm{End}(C) \mid \text{there exists a } Y \in S \text{ with } \alpha C \subseteq Y\}$$

and

$$\Psi(L) = \{X \in \mathcal{L}(C) \mid \text{there exists a } \alpha \in L \text{ with } \alpha C = X\}.$$

Theorem 15.4. *Let C be a complete torsion-free module. The mappings $S \to \Phi(S)$ and $L \to \Psi(L)$ are mutually inverse isomorphisms between the lattice $\mathcal{JL}(C)$ of ideals of the lattice $\mathcal{L}(C)$ and the lattice $\mathcal{L}(\mathrm{End}(C))$ of left ideals of the ring $\mathrm{End}(C)$.*

Again, let $\mathcal{L}$ be an arbitrary lattice and $a \in \mathcal{L}$. Then $\{x \in \mathcal{L} \mid x \leq a\}$ is an ideal of the lattice $\mathcal{L}$ which is called the *principal ideal* generated by the element a. We denote it by $(a]$. The mapping $a \to (a]$, $a \in \mathcal{L}$, is an embedding from the lattice $\mathcal{L}$ in the lattice of ideals $\mathcal{J}\mathcal{L}$. The image of this embedding coincides with the sublattice consisting of all principal ideals of the lattice $\mathcal{L}$. If we identify the lattice $\mathcal{L}(C)$ with the lattice of principal ideals of $\mathcal{L}(C)$, then it is clear that the isomorphisms Φ and Ψ are Theorem 15.4 induce the lattice isomorphisms from Theorem 15.3.

The work [210] of Liebert contains the description of left ideals of the ring $\operatorname{End}(D)$, where D is a divisible primary module. In addition, Theorem 13.1 on the equivalence of categories, Theorem 15.3, and Theorem 15.4 are used. In essence, the main idea of this description coincides with the idea used for the ring $\operatorname{End}(C)$, but the realization of the idea is much more complicated.

We pass to right ideals. In this case, a simple description is obtained for divisible primary modules. First, we prove that principal right ideals of the ring $\operatorname{End}(D)$ coincide with annihilators of submodules of the divisible primary module D.

Lemma 15.5. *Let X be a submodule of the module D. Then the following assertions hold.*

(1) $r(D/X) \leq r(D)$.

(2) *There exists an endomorphism $\alpha \in \operatorname{End}(D)$ with $X = \operatorname{Ker}(\alpha)$.*

Proof. (1) We have $D = \bigoplus_{i\in I} Q_i$, where $Q_i \cong K/R$ for all $i \in I$. Then

$$D/X = \sum_{i\in I} \bar{Q}_i \quad \text{and} \quad \bar{Q}_i = (Q_i + X)/X \cong Q_i/(Q_i \cap X),$$

where either $\bar{Q}_i = 0$ or $\bar{Q}_i \cong K/R$. We have the equality of R_p-spaces

$$(D/X)[p] = \sum_{i\in I} \bar{Q}_i[p],$$

where either $\bar{Q}_i[p] = 0$ or $\bar{Q}_i[p] \cong R_p$. We note that $r(D/X) = \dim_{R_p}(D/X)[p]$. Now it follows from properties of spaces that $r(D/X) \leq |I| = r(D)$.

(2) Let $\pi\colon D \to D/X$ be the canonical homomorphism. The module D/X is a divisible primary module and $r(D/X) \leq r(D)$ by (1). Consequently, there exists an isomorphism χ from the module D/X onto some direct summand of the module D. Then $\alpha = \pi\chi \in \operatorname{End}(D)$ and $\operatorname{Ker}(\alpha) = X$. □

For a subset X in D, we define $X^\perp = \{\alpha \in \operatorname{End}(D) \mid \alpha X = 0\}$. If P is a subset in $\operatorname{End}(D)$, then we set $P^\perp = \{a \in D \mid \gamma a = 0 \text{ for all } \gamma \in P\}$. We note that if $P = \alpha \operatorname{End}(D)$, then $P^\perp = \operatorname{Ker}(\alpha)$.

Lemma 15.6. *The relation* $\alpha\,\mathrm{End}(D) = (\mathrm{Ker}(\alpha))^{\perp}$ *holds for every* $\alpha \in \mathrm{End}(D)$.

Proof. It is clear that $\alpha\,\mathrm{End}(D) \subseteq (\mathrm{Ker}(\alpha))^{\perp}$. Let $\gamma \in (\mathrm{Ker}(\alpha))^{\perp}$. We can define the homomorphism $\beta\colon \alpha D \to \gamma D$ by the relation $\beta(\alpha x) = \gamma x$ for every $x \in D$. Since αD is a direct summand of the module D, we have that β can be extended to an endomorphism of the module D. Then $\gamma = \alpha\beta \in \alpha\,\mathrm{End}(D)$. Consequently, $(\mathrm{Ker}(\alpha))^{\perp} \subseteq \alpha\,\mathrm{End}(D)$, which completes the proof. □

We present some properties of the operation $\perp$.

(1) If P is a principal right ideal of the ring $\mathrm{End}(D)$, then $P^{\perp}$ is a submodule of the module D and $P^{\perp\perp} = P$.

Let $P = \alpha\,\mathrm{End}(D)$. Then

$$P^{\perp\perp} = (\mathrm{Ker}(\alpha))^{\perp} = \alpha\,\mathrm{End}(D) = P$$

(consider Lemma 15.6).

(2) Let X be a submodule in D. Then $X^{\perp}$ is a principal right ideal of the ring $\mathrm{End}(D)$ and $X^{\perp\perp} = X$.

By Lemma 15.5, there exists an endomorphism $\alpha \in \mathrm{End}(D)$ such that $X = \mathrm{Ker}(\alpha)$. We have

$$X^{\perp\perp} = (\alpha\,\mathrm{End}(D))^{\perp} = \mathrm{Ker}(\alpha) = X.$$

(3) For any two submodules X and Y of the module D, we have the relations

$$(X \cap Y)^{\perp} = X^{\perp} + Y^{\perp} \quad \text{and } (X + Y)^{\perp} = X^{\perp} \cap Y^{\perp}.$$

It is sufficient to prove that the left part of the first equality is contained in the right part. Let $\alpha \in (X \cap Y)^{\perp}$. We take the canonical homomorphism

$$\sigma\colon X \oplus Y \to X + Y, \qquad (x, y) \to x + y, \qquad x \in X, \qquad y \in Y.$$

We have $X + Y \cong (X \oplus Y)/\mathrm{Ker}(\sigma)$, where $\mathrm{Ker}(\sigma) = \{(x, y) \mid x + y = 0\}$. Let φ be the homomorphism from $X \oplus Y$ into D which coincides with α at X and annihilates Y. If $(x, y) \in \mathrm{Ker}(\sigma)$, then $x + y = 0$, $x = -y \in X \cap Y$, and $\alpha x = 0$. Therefore

$$\varphi(x, y) = \alpha x = 0 \qquad \text{and} \qquad \mathrm{Ker}(\sigma) \subseteq \mathrm{Ker}(\varphi).$$

Consequently, φ induces the homomorphism $\gamma\colon X \oplus Y \to D$ such that γ coincides with α on X and γ annihilates Y. Since D is injective, γ can be extended to an endomorphism of the module D which is also denoted by γ. We set $\beta = \alpha - \gamma$. Then $\alpha = \beta + \gamma$, where $\beta \in X^{\perp}$ and $\gamma \in Y^{\perp}$. Consequently, $\alpha \in X^{\perp} + Y^{\perp}$.

(4) The relations

$$(P \cap N)^{\perp} = P^{\perp} + N^{\perp} \quad \text{and} \quad (P + N)^{\perp} = P^{\perp} \cap N^{\perp}$$

hold for any two principal right ideals P and N of the ring $\mathrm{End}(D)$.

Considering (1)–(3), we have

$$(P \cap N)^{\perp} = (P^{\perp\perp} \cap N^{\perp\perp})^{\perp} = (P^{\perp} + N^{\perp})^{\perp\perp} = P^{\perp} + N^{\perp}.$$

The second relation is directly verified.

It follows from properties (1)–(3) that the intersection and the sum of every finite number of principal right ideals of the ring $\mathrm{End}(D)$ is a principal right ideal. Consequently, all principal right ideals of the ring $\mathrm{End}(D)$ form a lattice with respect to inclusion, where $\inf(P,N) = P \cap N$ and $\sup(P,N) = P + N$.

Let $\mathcal{S}(D)$ be the lattice of all submodules of the module D (naturally, the order with respect to inclusion is considered). We have

$$\inf(X,Y) = X \cap Y \qquad \text{and} \qquad \sup(X,Y) = X + Y$$

for any two submodules X and Y.

Theorem 15.7. *Let D be a divisible primary module. The mappings $X \to X^{\perp}$ and $P \to P^{\perp}$ are mutually inverse anti-isomorphisms between the lattice $\mathcal{S}(D)$ and the lattice of all principal right ideals of the ring* $\mathrm{End}(D)$.

Proof. It follows from properties (1) and (2) that both mappings $\perp$ are mutually inverse bijections. It follows from (3) and (4) that $\perp$ are lattice anti-isomorphisms. □

We describe all right ideals of the ring $\mathrm{End}(D)$ with the use of the method which was used for left ideals of the ring $\mathrm{End}(C)$.

For a lattice $\langle \mathcal{L}, \leq \rangle$, a non-empty subset T of $\langle \mathcal{L}, \leq \rangle$ is called a *coideal* in $\mathcal{L}$ if the following two conditions hold:

(1) $\inf(a,b) \in T$ for all $a, b \in T$;

(2) if $a \in T$ and $a \leq y$, then $y \in T$.

Coideals of the lattice $\mathcal{L}$, ordered with respect to inclusion, form a lattice. We denote this lattice by $\mathcal{DL}$; it is called the *coideal lattice* of the lattice $\mathcal{L}$. We prove, that the lattice $\mathcal{DS}(D)$ of all coideals of the lattice $\mathcal{S}(D)$ is isomorphic to the lattice $\mathcal{R}(\mathrm{End}(D))$ of all right ideals of the ring $\mathrm{End}(D)$.

We denote by $\mathcal{PR}(\mathrm{End}(D))$ the lattice of all principal right ideals of the ring $\mathrm{End}(D)$. Similar to the case of left ideals of the ring $\mathrm{End}(C)$, we can obtain mutually inverse isomorphisms between the lattices $\mathcal{JPR}(\mathrm{End}(D))$ and $\mathcal{R}(\mathrm{End}(D))$. An anti-isomorphism $\perp$ between $\mathcal{S}(D)$ and $\mathcal{PR}(\mathrm{End}(D))$ in Theorem 15.7 induces isomorphism between the lattice of coideals $\mathcal{DS}(D)$ of the lattice $\mathcal{S}(D)$ and the ideal lattice $\mathcal{JPR}(\mathrm{End}(D))$ of the lattice $\mathcal{PR}(\mathrm{End}(D))$. By combining these isomorphisms, we obtain mutually inverse isomorphisms between the lattices $\mathcal{DS}(D)$ and $\mathcal{R}(\mathrm{End}(D))$. Precisely, for coideal T of the lattice $\mathcal{S}(D)$ and a right ideal P of the ring $\mathrm{End}(D)$, we set

$$T^{\perp} = \{\alpha \in \mathrm{End}(D) \mid \text{there exists a } Y \in T \text{ with } \alpha Y = 0\}$$

and

$$P^{\perp} = \{X \in \mathcal{S}(D) \mid \text{there exists a } \alpha \in P \text{ with } X = \mathrm{Ker}(\alpha)\}.$$

The following theorem holds.

Theorem 15.8. *The mappings $T \to T^{\perp}$ and $P \to P^{\perp}$ are mutually inverse isomorphisms between the lattice $\mathcal{DS}(D)$ of coideals of the lattice $\mathcal{S}(D)$ and the lattice $\mathcal{R}(\mathrm{End}(D))$ of right ideals of the ring* $\mathrm{End}(D)$.

The lattice isomorphisms from Theorem 15.8 induce the lattice anti-isomorphisms from Theorem 15.7.

For a complete torsion-free module C, Liebert [210] determined interrelations between of right ideals of the endomorphism ring of C and submodules of C.

Exercise 1. Let C be a complete torsion-free module of countable p-rank over the ring of p-adic integers. Then C contains a complete submodule X such that the p-rank of X is equal to continuum.

Exercise 2. (a) Let C be a complete torsion-free module. Every left (resp., right) ideal of the ring $\mathrm{End}(C)$ is principal if and only if the p-rank $r_p(C)$ is finite.

(b) For a divisible primary module D, every left (resp., right) ideal of the ring $\mathrm{End}(D)$ is principal if and only if the rank $r(D)$ is finite.

Let S be a ring and let X be a subset of S. The set $\{r \in S \mid rX = 0\}$ is called the *left annihilator* of the set X. The left annihilator of some subset of the ring S is called the left annihilator in the ring S. The *right annihilators* are defined similarly. Any left (resp., right) annihilator is a left (resp., right) ideal.

Exercise 3 (Liebert [206]). Let C be a complete torsion-free module.

(1) The mappings $A \to \mathrm{Hom}(C, A)$ and $L \to LC$ are mutually inverse isomorphisms between the lattice of all direct summands A of the module C (see Section 13 in connection to this lattice) and the lattice of all left annihilators L of the ring $\mathrm{End}(C)$.

(2) The mappings $A \to A^{\perp}$ and $P \to P^{\perp}$ are mutually inverse anti-isomorphisms between the lattice of all direct summands A of the module C and the lattice of all right annihilators P of the ring $\mathrm{End}(C)$.

Baer [28] and Wolfson [318] have obtained analogous results for vector spaces.

Exercise 4 (Baer [28]). Let V be a left vector space over some division ring. Then the lattice $\mathcal{S}(V)$ of all subspaces of the space V is isomorphic to the lattice of all principal left ideals and $\mathcal{S}(V)$ is anti-isomorphic to the lattice of all principal right ideals of the linear operator ring of the space V.

Liebert [210] used the language of the lattice of all ideals (resp., coideals) of the lattice $\mathcal{S}(V)$ to obtain a characterization of all left ideals (resp., right ideals) of the linear operator ring of the space V.

Remarks. The source of the theory of endomorphism rings lies in the theory of linear operators of finite-dimensional vector spaces. Such operators form a simple ring which is isomorphic to the matrix ring over the initial division ring.

The existence of equivalences or dualities between some categories of modules always is a remarkable useful property. Let A be a left R-module and let S be the endomorphism ring of A. The additive group A can be turned into a right S-module if we assume that $a\alpha = \alpha(a)$ for any two $a \in A$ and $\alpha \in S$. In fact, we have an R-S-bimodule A. Consequently, we can consider the functors

$$\mathrm{Hom}_R(A, -) : R\text{-mod} \to S\text{-mod} \quad \text{and} \quad A \otimes_S (-) : S\text{-mod} \to R\text{-mod}.$$

The functors from Section 13 are examples of similar functors. The equivalences are often established by the functors $\mathrm{Hom}_R(A, -)$ and $A \otimes_S (-)$ or functors constructed from the functors. In the book of the authors and A. V. Mikhalev [183], this approach is developed for Abelian groups. As a result, the book contains various applications of obtained equivalences to endomorphism rings and some group problems. Some equivalence of categories of modules is constructed in Section 22. In Section 24, the standard duality for finite-dimensional vector spaces is generalized to torsion-free modules of finite rank. Equivalences and dualities will be also considered in the remarks to Chapter 5.

There exist quite many works about the radical of the endomorphism ring of a primary Abelian group inspired by the foundational work [248] of Pierce. Pierce has posed the description problem for elements of the radical of the endomorphism ring in terms of their action on the group or the module. This topic was developed in the books [133, 183, 238].

Several exercises in Section 14 touch on the problem on the representation of endomorphisms as a sum of automorphisms. This problem can be represented in an abstract ring-theoretical form (see Exercise 8 in Section 14, Problem 5, and the paper of Goldsmith–Pabst–Scott [119]). Deep results are obtained in the works [144] of Hill and [107] of Göbel–Opdenhövel. In the first paper, it is proved that every endomorphism of a totally projective Abelian p-group ($p \neq 2$) is the sum of two automorphisms; the second paper contains a similar theorem for Warfield modules of finite rank.

The problem considered in Section 15 is a partial case of a general problem of characterization of one-sided ideals of endomorphism rings of modules. The case of vector spaces is explicitly studied in the book of Baer [28] and the work [318] of Wolfson. Monk [239] has extended the results of Baer to primary Abelian groups. An explicit presentation of questions studied in Section 15 is contained in the works [206] and [210] of Liebert.

Problem 2 (Pierce). Describe the radical of the endomorphism ring of a primary module.

Problem 3. Describe the radical of the endomorphism ring of a torsion-free module over a complete discrete valuation domain.

The following problem is induced by Exercises 6 and 7 from Section 14.

Problem 4. (a) Does Exercise 6 hold for $n=2$ and an arbitrary endomorphism α?

(b) Under what conditions, every endomorphism of the module M from Exercise 7 is a sum of two automorphisms of M?

Problem 5. When is every element of the ring of all $n \times n$ matrices over some ring S the sum of two invertible elements of the ring?

Let S be a ring. We set $1 + J(S) = \{1 + x \mid x \in J(S)\}$. Then $1 + J(S)$ is a normal subgroup of the group $U(S)$ and

$$U(S)/(1 + J(S)) \cong U(S/J(S)).$$

(Here $U(T)$ is the group of invertible elements of the ring T.)

Problem 6. (a) Describe local rings S such that the group $U(S)$ is the semidirect product of the subgroup $1 + J(S)$ by some subgroup Γ (it is clear that $\Gamma \cong U(S/J(S))$). Semidirect products are defined in Exercise 1 from Section 3.

(b) Describe modules M such that the group $\mathrm{Aut}(M)$ is the semidirect product of the group $1 + J(\mathrm{End}(M))$ by some subgroup Γ.

Chapter 4

Representation of rings by endomorphism rings

In Chapter 4, we consider the following topics:

- the finite topology (Section 16);
- the ideal of finite endomorphisms (Section 17);
- characterization theorems for endomorphism rings of torsion-free modules (Section 18);
- realization theorems for endomorphism rings of torsion-free modules (Section 19);
- essentially indecomposable modules (Section 20);
- cotorsion modules and cotorsion hulls (Section 21);
- an embedding from the category of torsion-free modules in the category of mixed modules (Section 22).

In the previous chapter, we studied the inner structure of the endomorphism ring. We studied the radical and the interrelations between properties of the module and the endomorphism ring of the module (isomorphism of the submodule lattice and the lattice of one-sided ideals).

This chapter is devoted to one of the central problems of the theory of endomorphism rings: the search for criteria for an abstract ring to be isomorphic to the endomorphism ring of some module. In a general formulation, this problem is very difficult. The way to a complete solution of the problem is not clear. Nevertheless, at the present time there are many remarkable deep results for quite large classes of rings and modules.

When we solve the considered problem, we can consider some class of modules as the initial point. Here the problem is to give ring-theoretical description of endomorphism rings of modules from a given class. The corresponding theorems are naturally called *characterization theorems* (for endomorphism rings).

Conversely, we can consider some class of rings and prove that every ring from this class can be realized up to an isomorphism as the endomorphism ring of some module. Such results are called *realization theorems* (for endomorphism rings).

Among realization theorems, so-called *split realization theorems* are distinguished. In such theorems, the endomorphism ring is the direct sum of a subring isomorphic to the given ring and some special ideal of endomorphisms. As a rule, there is no a simple description of this ideal in standard ring-theoretical terms.

Split realization theorems provide examples of a partial representation of rings by endomorphism rings. In such theorems, the endomorphism ring is not given in a complete form, but a large part of the endomorphism ring is known. This part is sufficient for reduction some module-theoretical problems to solvable problems on certain classes of rings. In particular, this method allows us to prove the existence of module direct decompositions with unusual properties. Characterization theorems are of especial interest in themselves. However, they usually are far in nature from applications, since questions about the modules can be very difficult. Split realization theorems are more transparent and they partially overcome this weakness.

In this chapter, unless otherwise stated, R denotes a discrete valuation domain (and R is a commutative domain in Sections 19–22) and all modules are R-modules. If R is a complete domain, then this is specified.

16 Finite topology

The p-adic topology defined in Sections 3 and 11 is a very important topology for discrete valuation domains and their modules. The endomorphism ring (and every ring) has the J-adic topology. All powers of the radical form a basis of neighborhoods of zero of this topology (see Theorems 14.1 and 14.2). The finite topology on the endomorphism ring is more useful. The finite topology and the p-adic topology are examples of a linear topology on an Abelian group, a module, or a ring. The linear topology is the topology which has a basis of neighborhoods of zero consisting of subgroups (submodules or right ideals), and the residue classes with respect to these subgroups form a basis of open sets. More precisely, this definition can be formulated with the use of coideal in the lattice of all subgroups (submodules or right ideals) (see [92, Section 7]). It is also convenient to describe completeness and the completion in terms of this coideal (see [92, Section 13] and [2]).

For module M, the *finite topology* on the endomorphism ring $\mathrm{End}(M)$ of M can be defined with the use of the following subbasis of neighborhoods of zero:

$$U_x = \{\varphi \in \mathrm{End}(M) \mid \varphi x = 0\}$$

for all $x \in M$. This is obvious that U_x are right ideals of the ring $\mathrm{End}(M)$. A basis of neighborhoods of zero consists of right ideals $U_X = \{\varphi \in \mathrm{End}(M) \mid \varphi X = 0\}$ for all finite sets X of the module M. In the notation of Section 15, we have $U_x = x^{\perp}$ and $U_X = X^{\perp}$. It is clear that $U_X = \bigcap_{x \in X} U_x$ and a basis of neighborhoods of the element $\alpha \in \mathrm{End}(M)$ is formed by the residue classes $\alpha + U_X$, where X runs over all finite sets of the module M. The finite topology is always Hausdorff. The following main theorem about the finite topology holds.

Theorem 16.1. *For a module M, the endomorphism ring* $\mathrm{End}(M)$ *of M is a complete topological ring in the finite topology.*

Proof. Since U_x are right ideals, the continuity of addition and subtraction in $\mathrm{End}(M)$ is obvious. The continuity of multiplication is verified as follows. Let $\alpha, \beta \in \mathrm{End}(M)$ and let $\alpha\beta + U_x$ be a neighborhood of the element $\alpha\beta$. Since $\alpha U_{\alpha x} \subseteq U_x$, we have the inclusion

$$(\alpha + U_x)(\beta + U_{\alpha x}) \subseteq \alpha\beta + U_x,$$

which implies the required continuity.

Thus, $\mathrm{End}(M)$ is a topological ring. We prove that it is complete. We assume that $\{\alpha_i\}_{i \in I}$ is a Cauchy sequence in the ring $\mathrm{End}(M)$. It follows from the definitions of the finite topology and the Cauchy sequence that the subscript set I is partially ordered with the use of the order which is dual to the order on finite subsets of the module M. Furthermore, we have the following property: for the given $x \in M$ there exists a subscript $i_0 \in I$ such that $\alpha_i - \alpha_j \in U_x$ for all $i, j > i_0$. This means that $\alpha_i(x) = \alpha_j(x)$ for sufficiently large subscripts i and j. We define an endomorphism α of the module M assuming that αx is equal to the above value of all such $\alpha_i x$. It is easy to verify that $\alpha \in \mathrm{End}(M)$ and $\alpha - \alpha_i \in U_x$ for $i > i_0$. Therefore α is the limit of the considered Cauchy sequence $\{\alpha_i\}_{i \in I}$. Therefore, every Cauchy sequence converges in the ring $\mathrm{End}(M)$; therefore, the ring is complete. □

The completeness of the endomorphism ring in the finite topology and continuous isomorphisms between endomorphism rings are especially often used in various studies.

There are two important cases, when we can define the finite topology of the ring $\mathrm{End}(M)$ without use of the module M.

Proposition 16.2. (1) *For a reduced primary module M, the finite topology of the endomorphism ring of M is defined if we take as a subbasis of neighborhoods of zero the set of right annihilators of elements $\varepsilon\varphi$, where $\varphi \in \mathrm{End}(M)$ and ε is a primitive idempotent. If M does not have elements of infinite height, then it is sufficient to take only right annihilators of primitive idempotents.*

(2) *If M is a torsion-free module over a complete domain R, then the finite topology of the ring* $\mathrm{End}(M)$ *can be similarly defined if we take the set of right annihilators of primitive idempotents as a subbasis of neighborhoods of zero.*

Proof. (1) Let M be a reduced primary module. By using the property of basic submodules from Corollary 9.3, we obtain the following property. For every $x \in M$, there exists a cyclic direct summand Ry of the module M with $o(x) \leq o(y)$.

Consequently, there exists an endomorphism $\varphi \in \operatorname{End}(M)$ with $\varphi y = x$. Let $\varepsilon\colon M \to Ry$ be the projection. Then U_x coincides with the right annihilator of the element $\varepsilon\varphi$ and ε is a primitive idempotent. If M does not have elements of infinite height, then there is a subbasis of neighborhoods of zero for the finite topology which consists of all right ideals U_x, where x runs over only the elements such that Rx is a direct summand of the module M (this can be verified with the use of Corollary 7.6). As before, if $\varepsilon\colon M \to Rx$ is the projection, then $U_x = (1-\varepsilon)\operatorname{End}(M)$ is the right annihilator of the primitive idempotent ε.

The proof of (2) is similar to the proof of the second assertion in (1). We need only to use Corollary 11.7 instead of Corollary 7.6. □

Exercise 1. Prove that for a torsion-free module of finite rank, the finite topology of the endomorphism ring is always discrete.

Exercise 2. Let M and N be two modules over a complete discrete valuation domain. If either M and N are reduced primary modules or M and N are torsion-free modules, then every isomorphism between their endomorphism rings is continuous with respect to finite topologies on these rings.

Exercise 3. Let M be a primary module without elements of infinite height. We denote by E_0 the right ideal of the ring $\operatorname{End}(M)$ generated by primitive idempotents of the ring. Then in the finite topology E_0 is dense in $\operatorname{End}(M)$, and $\operatorname{End}(M)$ is the completion for E_0 (cf., Proposition 17.6).

17 Ideal of finite endomorphisms

In this section, R is a complete discrete valuation domain which is not necessarily commutative and M is a reduced torsion-free R-module. The main property of such modules is formulated in Corollary 11.7. Precisely, every finite set of elements of the module M can be embedded in a direct summand of the module M which is a free module of finite rank. If M has a finite or countable rank, then M is a free module. For one element, we have more strong properties (see the proofs of Corollaries 11.6 and 11.7). If $a \in M$ and $h(a) = 0$, then Ra is a direct summand of the module M which is isomorphic to R. If $h(a) = k$ and $k > 0$, then $a = p^k b$, where $h(b) = 0$. Then Rb is a direct summand of the module M, $Rb \cong R$, and $a \in Rb$. Therefore, every element of the module M can be embedded in a direct summand of the module M which is isomorphic to R. Every pure submodule A of finite rank of the module M is a direct summand in M. Indeed, since A is a free module of finite rank, then A is a complete module. Consequently, A is a direct summand in M as pure-injective module (Theorems 10.3 and 11.4).

Let $x, y \in M$ and let $h(x) \leq h(y)$. There exists an endomorphism φ of the module M with $\varphi x = y$. (A module with such a property for every pair x, y is called a *fully transitive* module. Fully transitive modules are studied in Chapter 8.) We have $M = X \oplus A = Y \oplus B$, where $X \cong Y \cong R$, $x \in X$, and $y \in Y$. We can always choose φ with $\varphi A = 0$. Therefore, we assume that x and y elements of the ring R. Then $x = p^m u$ and $y = p^n v$, where u and v are invertible elements of the ring R and m, n are non-negative integers, and $m \leq n$. The right multiplication of the ring R by the element $u^{-1}p^{n-m}v$ is an endomorphism φ of the module R with $\varphi x = y$.

Finite endomorphisms are defined in Section 14 (see Propositions 14.3 and 14.4). An endomorphism α of the module M is said to be *finite* if the submodule $\operatorname{Im}(\alpha)$ has finite rank. We note that the endomorphism φ constructed above is finite. All finite endomorphisms of the module M form the ideal $\operatorname{Fin}(M)$ of the ring $\operatorname{End}(M)$. The ideal $\operatorname{Fin}(M)$ determines many properties of the ring $\operatorname{End}(M)$ (for example, see Proposition 17.6). Our main aim is to characterize $\operatorname{Fin}(M)$ in ring-theoretical terms.

Lemma 17.1. *Let α be a finite endomorphism of the module M. Then $\operatorname{Ker}(\alpha)$ is a direct summand of the module M. In addition, there exists a decomposition $M = G \oplus N$, where G has finite rank, $\alpha M \subseteq G$, and $N \subseteq \operatorname{Ker}(\alpha)$.*

Proof. Since αM has finite rank, αM is a free module. It follows from the isomorphism $M/\operatorname{Ker}(\alpha) \cong \alpha M$ and Theorem 5.4 that $\operatorname{Ker}(\alpha)$ is a direct summand in M: $M = A \oplus \operatorname{Ker}(\alpha)$, where the module A has finite rank. Let $a_1+b_1, \dots, a_k+b_k$ be a maximal linearly independent system of elements of the module αM ($a_i \in A$, $b_i \in \operatorname{Ker}(\alpha)$) and let B be the pure submodule in $\operatorname{Ker}(\alpha)$ generated by the elements $b_1, \dots, b_k$. Then B is a free module and $\operatorname{Ker}(\alpha) = B \oplus N$ for some module N. We set $G = A \oplus B$. Then $M = G \oplus N$, where G and N have the required properties. □

A right (left) ideal I of some ring S is said to be *nonradical* if I is not contained in the radical $J(S)$; a right (resp., left) ideal I of S is said to be *minimal nonradical* if the ideal I is a minimal element of the set of all nonradical right (resp., left) ideals of the ring S.

Lemma 17.2. *Let e be a primitive idempotent of the ring S. The following conditions are equivalent.*

(1) *eS is a minimal nonradical right ideal in S.*

(2) *$eSe/J(eSe)$ is a division ring.*

(3) *Se is a minimal nonradical left ideal in S.*

Proof. (1) $\Longrightarrow$ (2). Let $a \in eSe$ and $a \notin J(eSe)$. Then $a \notin J(S)$, since $J(eSe) = eSe \cap J(S)$. However, $a \in eS$. Since eS is minimal, we obtain $aS = eS$ and $aeSe = eSe$. This implies that $eSe/J(eSe)$ does not have proper right ideals which is equivalent to (2).

(2) $\Longrightarrow$ (1). Let I be a right ideal of the ring S, $I \subseteq eS$, and let $I \neq eS$. Then eIe is a proper right ideal of the ring eSe, since it does not contain e. Consequently, $eIe \subseteq J(eSe) \subseteq J(S)$. In addition, we obtain

$$I^2 \subseteq eIeI \subseteq J(S) \quad \text{and} \quad I \subseteq J(S)$$

which implies (1).

The equivalence of (1) and (3) can be proved similarly. □

Let M be a reduced torsion-free module. It follows from Section 2 and properties of such modules M that the following properties of an endomorphism ε of the module M are equivalent.

(a) ε is a primitive idempotent.

(b) εM is an indecomposable direct summand of the module M.

(c) εM is a direct summand of the module M of rank 1.

We recall some notation of Section 15. Let X be a submodule of the module M. Then

$$\operatorname{Hom}(M, X) = \{\alpha \in \operatorname{End}(M) \mid \alpha M \subseteq X\}$$

and

$$X^{\perp} = \{\alpha \in \operatorname{End}(M) \mid \alpha X = 0\}.$$

The group $\operatorname{Hom}(M, X)$ is a left ideal of the ring $\operatorname{End}(M)$ and $X^{\perp}$ is a right ideal of $\operatorname{End}(M)$ (in the case of $X^{\perp}$, it is sufficient to require that X is a set). For a subset Γ from $\operatorname{End}(M)$, we set $\Gamma^{\perp} = \{a \in M \mid \gamma a = 0 \text{ for all } \gamma \in \Gamma\}$. The set $\Gamma^{\perp}$ is a submodule of the module M.

Lemma 17.3. *For a left (right) ideal L of the ring $\operatorname{End}(M)$, the following conditions are equivalent.*

(1) *There exists a direct summand A of rank 1 (such that M/A has rank 1) with $L = \operatorname{Hom}(M, A)$ (with $L = A^{\perp}$).*

(2) *L is a principal left (resp., right) ideal generated by a primitive idempotent.*

(3) *L is a minimal nonradical left (resp., right) ideal.*

Proof. The equivalence of (1) and (2) follows from the following results. Let $M = A \oplus B$ and let $\varepsilon\colon M \to A$ be the projection with kernel B. A submodule A

(resp., B) has rank 1 if and only if ε (resp., $1-\varepsilon$) is a primitive idempotent. We also have the relations

$$\operatorname{Hom}(M, A) = \operatorname{End}(M)\varepsilon \text{ and } A^{\perp} = (1-\varepsilon)\operatorname{End}(M).$$

(2) $\Longrightarrow$ (3). Let $L = \operatorname{End}(M)\varepsilon$ (resp., $L = \varepsilon\operatorname{End}(M)$), where ε is a primitive idempotent. It follows from property (b) from Section 2 and Proposition 2.2 that

$$\varepsilon\operatorname{End}(M)\varepsilon \cong \operatorname{End}(\varepsilon M) \cong \operatorname{End}(R) \cong R.$$

However, $R/J(R)$ is a division ring. Therefore, (3) holds by Lemma 17.2.

(3) $\Longrightarrow$ (2). Let L be a minimal nonradical right ideal of the ring $\operatorname{End}(M)$. Then $L^2 = L$. Otherwise, since L is minimal, $L^2 \subseteq J(\operatorname{End}(M))$. Therefore $L \subseteq J(\operatorname{End}(M))$ which contradicts to the definition of a nonradical ideal. We have $L = L^n \subseteq \operatorname{Hom}(M, p^n M)$ for every positive integer n. Therefore

$$L \subseteq \operatorname{Hom}(M, \bigcap_{n\geq 1} p^n M) = \operatorname{Hom}(M, 0) = 0;$$

this is impossible. Consequently, $L \nsubseteq \operatorname{Hom}(M, pM)$. There exist $x \in M$ and $\alpha \in L$ such that $x, \alpha x \notin pM$. We have $M = R(\alpha x) \oplus A$ for some submodule A. We define an endomorphism μ of the module M such that $\mu(\alpha x) = x$ and $\mu A = 0$. Then $(\alpha\mu)M = Rx$. By Lemma 17.1, we obtain $M = Ry \oplus (\alpha\mu)^{\perp}$ for some $y \in M$. Since $(\alpha\mu)x = x \notin pM$, it follows from Proposition 14.3 $\alpha\mu \notin J(\operatorname{End}(M))$. On the other hand, $\alpha\mu \in L \cap (\alpha\mu)^{\perp\perp}$. Since L is minimal, we have $L \subseteq (\alpha\mu)^{\perp\perp}$. Since the implications (1) $\Longrightarrow$ (2) and (2) $\Longrightarrow$ (3) have been proved, we have that $(\alpha\mu)^{\perp\perp}$ is a minimal nonradical right ideal (in (1) we need to take $(\alpha\mu)^{\perp}$ as A). Therefore $L = (\alpha\mu)^{\perp\perp}$ and (2) holds for L.

Now let L be a minimal nonradical left ideal. By using the above argument, we obtain elements $y \in M$ and $\beta \in L$ such that $y, \beta y \notin pM$. Therefore $M = Ry \oplus B$ for some submodule B. Let λ be the endomorphism of the module M such that $\lambda y = y$ and $\lambda B = 0$. It follows from $(\lambda\beta)y = \beta y \notin pM$ that $\lambda\beta \notin J(\operatorname{End}(M))$ (see Proposition 14.3). Furthermore, $\lambda\beta \in L \cap \operatorname{Hom}(M, R(\beta y))$. Since $R(\beta y)$ is a direct summand in M of rank 1, we have that $\operatorname{Hom}(M, R(\beta y))$ is a minimal nonradical left ideal (consider that the implications (1) $\Longrightarrow$ (2) and (2) $\Longrightarrow$ (3) have been proved). Consequently, $L = \operatorname{Hom}(M, R(\beta y))$ and (2) holds for L. □

The following theorem provides a characterization of the ideal of finite endomorphisms.

Theorem 17.4 (Liebert [207]). *Let M be a reduced torsion-free module. Then* $\operatorname{Fin}(M)$ *coincides with the sum of all minimal nonradical right ideals; it is also equal to the sum of all minimal nonradical left ideals of the ring* $\operatorname{End}(M)$.

Proof. Every minimal nonradical right ideal L is generated by some primitive idempotent ε (Lemma 17.3). Since $\varepsilon \in \mathrm{Fin}(M)$, we have $L \subseteq \mathrm{Fin}(M)$. Now we assume that $\alpha \in \mathrm{Fin}(M)$. By Lemma 17.1, we have $M = G \oplus N$, where G has finite rank, $\alpha M \subseteq G$ and $\alpha N = 0$. We have $G = A_1 \oplus \cdots \oplus A_n$, where all A_i have rank 1. Let $\varepsilon_i \colon G \to A_i$ be the projection ($i = 1, \dots, n$). Then $\alpha = \varepsilon_1\alpha + \cdots + \varepsilon_n\alpha$. Considering Lemma 17.3, we need only to note that $\varepsilon_i \,\mathrm{End}(M)$ is a minimal nonradical right ideal. The left-side case can be proved similarly. □

In the second theorem, we use the notion of a *ring without identity element*. Such rings satisfy all ring axioms except for the axiom about the existence of the identity element. For rings without identity element, we can similarly define ideals, factor rings, and so on. Under such an approach, left and right ideals of the ring are rings. We will consider the (Jacobson) radical of the ideal of finite endomorphisms as of the ring. For a ring without identity element, the radical of the ring can be defined either as the intersection of annihilators of all simple left modules over the ring or with the use of quasiregular elements (see the Jacobson's book [160]). Clearly, for a ring with identity element these definitions coincide with the definitions given in Sections 3 and 14. We need the following result. The radical of the ideal is equal to the intersection of the radical of the ring with this ideal.

A linear operator α of a vector space V over some division ring is said to be *finite* if the image of α has the finite dimension. All finite linear operators form an ideal in the operator ring of the space V.

Let M be a torsion-free module. Then M/pM is a left vector R_p-space, where R_p is the residue division ring of the ring R, $R_p = R/pR$. The outer multiplication is defined by the relation $(r+pR)(m+pM) = rm+pM$ for all $r \in R$ and $m \in M$ (see Section 4).

We note that several properties of the radical of the ring $\mathrm{End}(M)$ are contained in Proposition 14.3. It is also interesting to compare the following result with Theorem 14.1.

Theorem 17.5 (Liebert [207]). *Let M be a reduced torsion-free module. Then*

$$J(\mathrm{Fin}(M))^n = J(\mathrm{End}(M))^n \cap \mathrm{Fin}(M) = \mathrm{Hom}(M, p^n M) \cap \mathrm{Fin}(M)$$

for every positive integer n. In addition, the factor ring $\mathrm{Fin}(M)/J(\mathrm{Fin}(M))$ is isomorphic to the ring of all finite linear operators of the R_p-space M/pM.

Proof. We have $J(\mathrm{Fin}(M)) = J(\mathrm{End}(M)) \cap \mathrm{Fin}(M)$, since $\mathrm{Fin}(M)$ is an ideal in $\mathrm{End}(M)$. Consequently, $J(\mathrm{Fin}(M))^n \subseteq J(\mathrm{End}(M))^n \cap \mathrm{Fin}(M)$ for every positive integer n. It follows from Proposition 14.3 (1) that

$$J(\mathrm{End}(M))^n \cap \mathrm{Fin}(M) \subseteq \mathrm{Hom}(M, p^n M) \cap \mathrm{Fin}(M).$$

We prove that

$$\mathrm{Hom}(M, p^n M) \cap \mathrm{Fin}(M) \subseteq (\mathrm{Hom}(M, pM) \cap \mathrm{Fin}(M))^n.$$

By Proposition 14.3 (4), we have

$$\mathrm{Hom}(M, pM) \cap \mathrm{Fin}(M) \subseteq J(\mathrm{End}(M)) \cap \mathrm{Fin}(M) = J(\mathrm{Fin}(M)).$$

Therefore, we obtain the required property.

Let $\alpha \in \mathrm{Hom}(M, p^n M) \cap \mathrm{Fin}(M)$. There exists a decomposition $M = G \oplus N$ such that the rank of G is finite, $\alpha M \subseteq G$, and $\alpha N = 0$ (Lemma 17.1). In fact, $\alpha M \subseteq p^n G$. Therefore $\alpha_{|G} \in \mathrm{Hom}(G, p^n G)$. Since G is a complete module, it follows from property (5) from Section 14 that

$$\begin{aligned}\mathrm{Hom}(G, p^n G) &= p^n \mathrm{Hom}(G, G) = p\,\mathrm{Hom}(G, G) \cdots p\,\mathrm{Hom}(G, G) \\ &= \mathrm{Hom}(G, pG) \cdots \mathrm{Hom}(G, pG) = \mathrm{Hom}(G, pG)^n.\end{aligned}$$

Therefore $\alpha_{|G} \in \mathrm{Hom}(G, pG)^n$. This can be verified with the use of an other argument. Precisely, if $G = \bigoplus_k R$, then $\mathrm{End}(G)$ is ring of $k \times k$ matrices over R. Further, we need to use the properties of the ring R. For α, we obviously obtain $\alpha \in (\mathrm{Hom}(M, pM) \cap \mathrm{Fin}(M))^n$.

Every finite endomorphism α of the module M induces the finite linear operator $\bar{\alpha}$ of the R_p-space M/pM, where $\bar{\alpha}(x+pM) = \alpha(x)+pM$, $x \in M$. The mapping $\alpha \to \bar{\alpha}$, $\alpha \in \mathrm{Fin}(M)$, is a homomorphism from the ring $\mathrm{Fin}(M)$ into the ring of all finite linear operators of the space M/pM. The kernel of this homomorphism is equal to $\mathrm{Hom}(M, pM) \cap \mathrm{Fin}(M) = J(\mathrm{Fin}(M))$. Let β be an arbitrary finite linear operator of the space M/pM. It remains to prove that β is induced by some finite endomorphism α of the module M. We use a similar argument from the proof of Theorem 14.1. We fix the basis $\{a_i + pM\}_{i \in I}$ of the space M/pM. Let B be a basic submodule of the module M generated by elements a_i, $i \in I$. There exists a homomorphism $\alpha\colon B \to M$ such that $\alpha(a_i) + pM = \beta(a_i + pM)$ for every $i \in I$. The image of α has finite rank; consequently, the image is a complete module. Therefore α can be extended to an endomorphism α of the module M. Similar to Theorem 14.1, β is induced by α. □

At the end of the section, we obtain some interrelations between the ideal $\mathrm{Fin}(M)$ and the finite topology.

Proposition 17.6. *The ideal* $\mathrm{Fin}(M)$ *is dense in* $\mathrm{End}(M)$ *in the finite topology and* $\mathrm{End}(M)$ *is the completion for* $\mathrm{Fin}(M)$.

Proof. Let $\alpha \in \mathrm{End}(M)$ and let X be a finite subset in M. We need to prove that $\mathrm{Fin}(M) \cap (\alpha + X^{\perp})$ is a non-empty set. We embed X in a direct summand G

of the module M of finite rank. Let π be the projection from M onto G. Then $1-\pi \in X^{\perp}$; consequently, $(1-\pi)\alpha \in X^{\perp}$. Then $\pi\alpha = \alpha - (1-\pi)\alpha \in \mathrm{Fin}(M) \cap (\alpha + X^{\perp})$. □

With use of Proposition 16.2 (2), it is easy to prove that the finite topology on the ring $\mathrm{End}(M)$ is also defined if we take right annihilators of finite idempotent endomorphisms of the module M as a basis of neighborhoods of zero. For the possibility to define later an analogue of the finite topology for an abstract ring, we represent this result in slightly another form.

Proposition 17.7. *The finite topology on the ring* $\mathrm{End}(M)$ *coincides with the topology such that right annihilators of finite sets in* $\mathrm{Fin}(M)$ *form a basis of neighborhoods of zero.*

Proof. Let X be some finite subset of the module M and let X_* be the pure submodule generated by X (see Section 7). The module X_* has finite rank; consequently, X_* is a direct summand in M. Let π be the projection from M onto X_*. Then

$$\pi \in \mathrm{Fin}(M) \quad \text{and} \quad (X_*)^{\perp} = \{\alpha \in \mathrm{End}(M) \mid \pi\alpha = 0\}.$$

Conversely, let Γ be a finite subset in $\mathrm{Fin}(M)$ and let $\Gamma = \{\gamma_1, \dots, \gamma_k\}$. Then the submodule G is equal to $\gamma_1 M + \cdots + \gamma_k M$ and it has finite rank. Consequently, G is a free module. Therefore G is generated by some finite set X. We have

$$\{\alpha \in \mathrm{End}(M) \mid \Gamma\alpha = 0\} = G^{\perp} = X^{\perp}.$$ □

In relation to Theorem 17.4, we note that for a vector space over a division ring, the ideal of finite linear operators coincides with the sum of all minimal right (left) ideals of the operator ring. In addition, the ideal of finite linear operators is contained in every nonzero ideal of the operator ring (Baer [28], Wolfson [318]).

Exercise 1. Let e be an idempotent of a semiprime ring S. Prove that eS (resp., Se) is a minimal right (resp., left) ideal if and only if eSe is a division ring (cf., Lemma 17.2).

The remaining exercises are taken from the works [206] and [207] of Liebert.

Exercise 2. Let C be a complete torsion-free module. Then the ideal of finite endomorphisms $\mathrm{Fin}(C)$ is contained in every nonradical ideal of the ring $\mathrm{End}(C)$. This is not true for an arbitrary torsion-free module.

Exercise 3. Let M be a reduced torsion-free module and let I be an ideal of the endomorphism ring of M. Prove that if $I \cap \mathrm{Fin}(M)$ is a nonradical ideal, then $\mathrm{Fin}(M) \subseteq I$.

Exercise 4. Every right (resp., left) ideal of the ring $\mathrm{Fin}(M)$ is a right (resp., left) ideal of the ring $\mathrm{End}(M)$.

18 Characterization theorems for endomorphism rings of torsion-free modules

Similar to the previous section, M denotes some reduced torsion-free module over a complete discrete valuation domain R. In this section, main properties of such modules are also listed. We also use notations $\mathrm{Fin}(M)$, $\mathrm{Hom}(M, X)$, $X^{\perp}$, and $\Gamma^{\perp}$ from Section 17. We continue to use in the studies rings without identity element.

Our aim is to obtain for endomorphism rings of torsion-free modules over a complete domain several properties of the rings such that only the endomorphism rings of the mentioned modules satisfy all these properties. Clearly, these properties need to be ring-theoretical.

Lemma 18.1. *Let L be a minimal nonradical right (left) ideal of the ring $\mathrm{End}(M)$. Then L contains an element π such that π is not a left (resp., right) zero-divisor in L, $\pi \in L \cap J(\mathrm{End}(M))$ and $(L \cap J(\mathrm{End}(M)))^n = \pi^n L$ (resp., $(L \cap J(\mathrm{End}(M)))^n = L\pi^n$) for every positive integer n.*

Proof. First, assume that L is a minimal nonradical right ideal. By Lemma 17.3, there exist a submodule B of rank 1 and a submodule A in M such that $M = B \oplus A$ and $L = A^{\perp} = \varepsilon\,\mathrm{End}(M)$, where $\varepsilon\colon M \to B$ is the projection with kernel A. Considering Proposition 14.3 and Theorems 17.4, 17.5, we obtain

$$\begin{aligned} L \cap J(\mathrm{End}(M)) &\subseteq L \cap \mathrm{Hom}(M, pM) = L \cap \mathrm{Hom}(M, pM) \cap \mathrm{Fin}(M) \\ &= J(\mathrm{Fin}(M)) \subseteq L \cap J(\mathrm{End}(M)). \end{aligned}$$

Therefore $L \cap J(\mathrm{End}(M)) = L \cap \mathrm{Hom}(M, pM)$. We note that $B \cong R$. Now we assume that π is an endomorphism of the module M such that $\pi A = 0$ and the restriction of π to B is the endomorphism corresponding to the multiplication of R from the right by the prime element p under the isomorphism $\mathrm{End}(B) \cong \mathrm{End}(R)$ (see Proposition 2.2 (b); we can assume that $\pi_{|B}$ is an endomorphism λ_p from Theorem 14.1). Then $\pi \in L \cap \mathrm{Hom}(M, pM)$ by Theorem 14.1 (a) (consider that $J(\mathrm{End}(B)) = \mathrm{Hom}(B, pB)$). We verify the relation $(L \cap \mathrm{Hom}(M, pM))^n = \pi^n L$. Let $\alpha \in (L \cap \mathrm{Hom}(M, pM))^n$. There exists a homomorphism $\beta\colon B \to M$ such that $\alpha_{|B} = \pi^n_{|B}\beta$ (see Theorem 14.1 (a)). Then $\alpha = \pi^n\beta$, where we assume that $\beta A = 0$. Now we have

$$\alpha = \varepsilon\alpha = \varepsilon\pi^n\beta = \pi^n(\varepsilon\beta) \in \pi^n L.$$

One inclusion has been proved and the second inclusion is obvious. Let $\gamma \in L$ and $\pi\gamma = 0$. It follows from property (6) in Section 14 and Theorem 14.1 (a) that $\gamma = 0$. This means that π is not a left zero-divisor in L.

Now we assume that L is a minimal nonradical left ideal. By Lemma 17.3, we have $M = A \oplus B$, where $r(A) = 1$ and $L = \mathrm{Hom}(M, A) = \mathrm{End}(M)\varepsilon$, where ε is the projection $M \to A$ with kernel B. As above, we obtain $L \cap J(\mathrm{End}(M)) = L \cap \mathrm{Hom}(M, pM)$. There exists an endomorphism π of the module M such that $\pi B = 0$ and π coincides on A with the endomorphism λ_p from Theorem 14.1. It is clear that $L\pi^n \subseteq (L \cap \mathrm{Hom}(M, pM))^n$. If $\alpha \in (L \cap \mathrm{Hom}(M, pM))^n$, then $M = Rx \oplus \alpha^{\perp}$ for some $x \in M$ (see Lemma 17.1). Since $\alpha(Rx) \subseteq p^n A$, we can similarly define the homomorphism $\beta\colon Rx \to A$ such that $\alpha_{|Rx} = \beta\pi^n$. Then $\alpha = \beta\pi^n$, where we assume that $\beta\alpha^{\perp} = 0$ and $\alpha \in L\pi^n$. If $\gamma\pi = 0$ and $\gamma \in L$, then $\gamma = 0$. Consequently, π is not a right zero-divisor in L. □

For a ring S and a subset X of S, we denote by $\mathrm{Ann}_S(X)$ or $\mathrm{Ann}(X)$ the right annihilator $\{s \in S \mid Xs = 0\}$. The right annihilator is a right ideal of the ring S. Let Γ be some subgroup of the additive group of the ring $\mathrm{End}(M)$. We denote by ΓM the set of all finite sums of the form $\sum \gamma_i(m_i)$, where $\gamma_i \in \Gamma$ and $m_i \in M$. The set ΓM is a submodule in M (see also Section 15).

Lemma 18.2. *Let N be a minimal nonradical right ideal of the ring* $\mathrm{End}(M)$. *If L is a left ideal of the ring N such that $\bigcap_{i\geq 1}(\mathrm{Ann}(L))^i \subseteq J(\mathrm{End}(M))$, then L is not a proper direct summand in N.*

Proof. By Lemma 17.3, there exists a decomposition $M = Rm \oplus A$ with $N = A^{\perp}$. It is clear that $NM = M$. We assume that $N = L \oplus I$ for some left ideal I of the ring N. Then $M = NM = LM + IM$. We verify the relation $LM = \{\lambda(m) \mid \lambda \in L\}$. Let $a = \lambda_1(x)$ and $b = \lambda_2(y)$, where $\lambda_1, \lambda_2 \in L$ and $x, y \in M$. We define endomorphisms α and β of the module M by the relations $\alpha m = x$, $\beta m = y$, and $\alpha A = 0 = \beta A$. Then

$$\alpha, \beta \in N \qquad \text{and} \qquad a + b = \lambda_1(x) + \lambda_2(y) = (\alpha\lambda_1 + \beta\lambda_2)m,$$

where $\alpha\lambda_1 + \beta\lambda_2 \in L$. This proves the required relation. By repeating the argument, we obtain $IM = \{\mu(m) \mid \mu \in I\}$. Now we assume that $c \in LM \cap IM$. Then $c = \lambda(m) = \mu(m)$, where $\lambda \in L$ and $\mu \in I$. Since $\lambda A = \mu A = 0$, we have $\lambda = \mu = 0$. Consequently, we have the direct sum $M = LM \oplus IM$. Let $\varepsilon\colon M \to IM$ be the projection with kernel LM. Then $\varepsilon \in (\mathrm{Ann}(L))^i$ for every $i \geq 1$, since $\varepsilon = \varepsilon^i$. Therefore $\varepsilon = 0$, $IM = 0$, and $I = 0$. Therefore $L = N$. □

Lemma 18.3. *Let S be a ring such that the intersection of all powers of the radical of S is equal to zero and let N be a minimal nonradical right ideal of the ring S.*

We assume that there exists an element p which is not a left zero-divisor in N, such that $p \in N \cap J(S)$ and $(N \cap J(S))^n = p^n N$ for every $n \geq 1$. Then $N = eS$, where e is some idempotent of the ring S. In addition, if there exists an element $q \in Se \cap J(S)$ with $Se \cap J(S) = Seq$, then eSe is a discrete valuation domain with $J(eSe) = peSe = eSep$.

Proof. If $s^2 \in J(S)$ for all $s \in N$, then $(N + J(S))/J(S)$ is a right nil-ideal; consequently, $N \subseteq J(S)$. Therefore, there exists an element t in N with $t^2 \notin J(S)$. Since N is minimal, we have $N = tN$.

We set $T = \{x \in N \mid tx \in J(S)\}$. Then T is a right ideal in S. If T is not contained in $J(S)$, then $T = N$, since N is minimal. This implies $t^2 \in J(S)$ that contradicts to the choice of t. Consequently if for some $y \in N$ we have $ty \in J(S)$, then $y \in J(S)$. Therefore, it can be verified that $t(N \cap J(S))^n = (N \cap J(S))^n$ for all positive integers n. It follows from $\bigcap_{n \geq 1} J(S)^n = 0$ and $N \cap J(S) = pN$ that every nonzero element y from N is equal to $p^m y'$, where $y' \notin J(S)$ and $m \geq 0$.

We prove that $\mathrm{Ann}_N(t) = 0$, where $\mathrm{Ann}_N(t)$ is a right annihilator of the element t in N. We assume that $x \in \mathrm{Ann}_N(t)$ and $x \neq 0$. Then $x \in T$; consequently, $x \in J(S)$. We have $x = p^m x'$, where $x' \notin J(S)$ and $m \geq 1$. We have $x' \in \mathrm{Ann}_N(tp^m)$. If the right ideal $\mathrm{Ann}_N(tp^m)$ is nonradical, then it coincides with N. In this case,

$$0 = tp^m N = t(N \cap J(S))^m = (N \cap J(S))^m = p^m N.$$

This is impossible, since p is not a left zero-divisor in N. Therefore

$$\mathrm{Ann}_N(tp^m) \subseteq J(S) \quad \text{and} \quad x' \in J(S);$$

this is a contradiction. Therefore $\mathrm{Ann}_N(t) = 0$.

Now it is easy to obtain the required element e. Since $tN = N$, there exists a nonzero element $e \in N$ such that $te = t$. It follows from $e^2 - e \in \mathrm{Ann}_N(t)$ that $e^2 = e$. Since $e \notin J(S)$ and N is a minimal nonradical ideal, we have $N = eS$. It is convenient to replace the element p by the element pe which satisfies the same properties that and p. However, we have $p = epe$ after this replacement.

We consider eSe as ring with identity element e. By Lemma 17.2, $eSe/J(eSe)$ is a division ring and eSe is a local ring. In addition, $J(eSe) = eSe \cap J(S) = eS \cap Se \cap J(S)$. We assume that $Se \cap J(S) = Seq$. Then

$$(eS \cap J(S)) \cap Se = peS \cap Se = peSe$$

and

$$(Se \cap J(S)) \cap eS = Seq \cap eS = eSeq.$$

Therefore $J(eSe) = peSe = eSeq$. There exists an element $y \in eSe$ with $pe = yq$. We assume that $y \in J(eSe)$. Then $y = pz$, where $z \in eSe$. The relation $pe = pzq$ implies that $e = zq$. However, $zq \in J(S)$. This is impossible, since e is a nonzero idempotent. Therefore $y \notin J(eSe)$ and y is invertible in eSe (Proposition 3.1). Now we obtain $eSep = eSeyq = eSeq$. Therefore $J(eSe) = peSe = eSep$. This is obvious that p is a non-nilpotent element. It follows from $J(eSe) \subseteq J(S)$ and the assumption of the lemma that the intersection of all powers of the radical $J(eSe)$ is equal to zero. It follows from Definition 3.2 that eSe is a discrete valuation domain with prime element p. □

The assertions of Lemma 17.3, Theorem 17.4, and Proposition 17.7 lead to the following definition of the finite topology for an abstract ring. Let S be a ring and let S_0 be the sum of all minimal nonradical right ideals of S (if S does not have such ideals, then we set $S_0 = 0$). The set of right annihilators of all finite sets in S_0 forms a basis of neighborhoods of zero for some topology on S; we call it the finite topology. For the ring $\operatorname{End}(M)$, this finite topology coincides with the finite topology defined in Section 16.

We formulate our main characterization theorem. We recall that the J-adic topology is defined in Section 14.

Theorem 18.4 (Liebert [207]). *Let S be an arbitrary ring. Then there exists a reduced torsion-free module over some complete discrete valuation domain R such that S is isomorphic to the ring $\operatorname{End}(M)$ if and only if the following conditions hold.*

(1) *The intersection of all powers of the radical of the ring S is equal to zero.*

(2) *S has a minimal nonradical right ideal.*

(3) *If N is a minimal nonradical right (resp., left) ideal of S, then there exists an element $p \in N$ (resp., $q \in N$) such that p is not a left zero-divisor in N, $p \in N \cap J(S)$, and $(N \cap J(S))^n = p^n N$ for every $n \geq 1$ (resp., $q \in N \cap J(S)$ and $N \cap J(S) = Nq$).*

(4) *If N is a minimal nonradical right ideal in S and L is a left ideal of the ring N with $\bigcap_{i \geq 1} (\operatorname{Ann}_S(L))^i = 0$, then L is not a proper direct summand in N.*

(5) *If L is a left nonzero ideal of S and N is a right nonzero ideal of S, then $L \cap N \neq 0$.*

(6) *If e is a primitive idempotent in S, then the ring eSe is complete in the J-adic topology of eSe.*

(7) *The ring S is complete in the finite topology of S.*

Proof. Let M be a reduced torsion-free R-module, where R is a complete discrete valuation domain. We verify that the ring $\operatorname{End}(M)$ satisfies conditions (1)–(7). Condition (1) follows from Proposition 14.3 (a) and condition (2) follows from

Lemma 17.3. Condition (3) follows from Lemma 18.1 and condition (4) follows from Lemma 18.2.

Let L be a left nonzero ideal and let N be a right nonzero ideal in $\operatorname{End}(M)$. We choose endomorphisms $\alpha \in N$, $\beta \in L$, and elements $x, y \in M$ such that $\alpha x \neq 0$ and $\beta y \neq 0$. There exists a $\xi \in \operatorname{End}(M)$ such that $\xi(\alpha x) = p^n y$, where $n = h(\alpha x)$. We have

$$0 \neq p^n(\beta y) = \beta(p^n y) = \beta(\xi(\alpha x)) = (\alpha\xi\beta)x, \quad \alpha\xi\beta \neq 0, \text{ and } \alpha\xi\beta \in L \cap N.$$

This proves that (5) holds. Let e be a primitive idempotent of the ring $\operatorname{End}(M)$. Then

$$eM \cong R \qquad \text{and} \qquad e\operatorname{End}(M)e \cong \operatorname{End}(eM) \cong R$$

which implies (6). The assertion (7) follows from the assertions of Theorems 16.1, 17.4 and Proposition 17.7.

Now we assume that some ring S satisfies conditions (1)–(7). By (2), there exists a minimal nonradical right ideal M of the ring S. It follows from Lemma 18.3 and conditions (1) and (3) that $M = eS$, where $e^2 = e \in S$. By Lemma 17.2, Se is a minimal nonradical left ideal of the ring S. It follows from (3) and Lemma 18.3 that eSe is a discrete valuation domain with prime element p (it is equal to epe). Since eSe is a domain, e is a primitive idempotent. By (6), eSe is a complete domain. We take eSe as the ring R. We naturally consider M as a left R-module. Since p is not a left zero-divisor in the ring M, we have that M is a torsion-free R-module. It follows from $(M \cap J(S))^n = p^n M$, $n \geq 1$, and $\bigcap_{n\geq 1} J(S)^n = 0$ that $\bigcap_{n\geq 1} p^n M = 0$ and M is a reduced torsion-free R-module.

The right annihilator $\operatorname{Ann}(eS)$ is an ideal of the ring S. The relation $\operatorname{Ann}(eS) \cap eS$ and (5) imply $\operatorname{Ann}(eS) = 0$. The mapping of rings $f\colon S \to \operatorname{End}(M)$, where $f(s)(ex) = exs$, $x \in S$, is a homomorphism; consequently, f is an embedding. We identify the ring S with the image of this embedding.

We prove that S contains all finite endomorphisms R-module M. By Lemma 17.1, it is sufficient to prove that for every decomposition $M = Ra \oplus B$ and each $x \in M$, there exists an endomorphism α of the module M contained in S such that $\alpha a = x$ and $\alpha B = 0$. First, we note that the minimal nonradical right ideal eS of the ring S is a minimal nonradical right ideal of the ring $\operatorname{End}(M)$. This follows from Lemma 17.3. Indeed, we have the decomposition $M = eSe \oplus eS(1-e)$ of the module M. Then eS coincides with the set of all endomorphisms of the module M annihilating the summand $eS(1-e)$. It is clear that eS annihilates this summand. Conversely, let φ be an endomorphism of the module M annihilating $eS(1-e)$. Then φ coincides with the right multiplication of the module M by the element $e\varphi(e)$.

We have two decompositions of the module M:

$$M = Ra \oplus B \qquad \text{and} \qquad M = R \oplus eS(1-e).$$

On the other hand, M is a minimal nonradical right ideal of the ring $\mathrm{End}(M)$ and $M = (eS(1-e))^{\perp}$. Therefore M can be identified with $\mathrm{Hom}(R, M)$. In this case, $M = \mathrm{Hom}(R, Ra) \oplus \mathrm{Hom}(R, B)$ is a nontrivial direct sum of left ideals of the ring M. It follows from (4) that $\bigcap_{i\geq 1} A^i \neq 0$, where $A = \mathrm{Ann}_S(\mathrm{Hom}(R, B))$. Since $\mathrm{Hom}(R, B)M = B$, we have $A = \{x \in S \mid Bs = 0\}$. We have $MA \not\subseteq pM$, since otherwise,

$$M\Big(\bigcap_{i\geq 1} A^i\Big) \subseteq \bigcap_{i\geq 1} p^i M \qquad \text{and} \qquad \bigcap_{i\geq 1} A^i = 0.$$

This is impossible. Therefore

$$MA \not\subseteq M \cap J(S) \quad \text{and} \quad MA \not\subseteq J(S).$$

Consequently, MA is a nonradical right ideal of the ring S contained in M. Therefore

$$M = MA = aA \quad \text{and} \quad \mathrm{Fin}(M) \subseteq S.$$

By Proposition 17.7 and Theorem 17.4, the topology induced on S by the finite topology of the ring $\mathrm{End}(M)$ coincides with the finite topology of the ring S. It follow from Theorem 16.1 and (7) that $S = \mathrm{End}(M)$. □

We consider another approach to the characterization problem based on the use of basic submodules. It seems that the approach is more promising from the viewpoint of possible applications and the construction of examples.

We assume that all modules have the p-adic topology and all topological notions for modules are related to this topology.

As before, let M be a reduced torsion-free module over a complete domain R and let B be some basic submodule of M. Since Theorem 11.13 and Corollary 11.15 (1), we can assume that M is a pure dense submodule of the completion $\widehat{B}$ of the module B. Every endomorphism α of the module M can be uniquely extended to an endomorphism $\widehat{\alpha}$ of the module $\widehat{B}$ (see Proposition 11.12). We identify α with $\widehat{\alpha}$. Then we obtain that $\mathrm{End}(M)$ is a subring in $\mathrm{End}(\widehat{B})$. Similar to Section 14, we assume that the complete module $\widehat{B}$ is an R-R-bimodule; we also assume that the ring $\mathrm{End}(\widehat{B})$ is a left R-module.

Let S be a subring of some ring T and let T be a left R-module. The subring S is called a *pure subring* in T if $S \cap p^k T = p^k S$ for every positive integer k.

Now we assume that $\alpha \in \mathrm{End}(M)$ and $\alpha = p^k\beta$, where $\beta \in \mathrm{End}(\widehat{B})$. For every element a of M, we have $p^k(\beta a) \in M$ and $\beta a \in M$, since M is pure in $\widehat{B}$. Consequently, $\beta \in \mathrm{End}(M)$. We obtain that $\mathrm{End}(M)$ is a pure subring in $\mathrm{End}(\widehat{B})$.

We set

$$\mathrm{Ines}(B) = \{\alpha \in \mathrm{End}(\widehat{B}) \mid \alpha\widehat{B} \subseteq B\}.$$

Here, $\mathrm{Ines}(B)$ is a left ideal in $\mathrm{End}(\widehat{B})$ and $\mathrm{Ines}(B) \subseteq \mathrm{End}(M)$. This ideal plays a crucial role in our characterization of endomorphism rings. (The ideal will be defined in a more general form in the next section.)

Let B be a free R-module of infinite rank, S be a pure subring of the ring $\mathrm{End}(\widehat{B})$ containing $\mathrm{Ines}(B)$, and let SB be the set of all finite sums of elements of the form αb, where $\alpha \in S$ and $b \in B$.

Lemma 18.5. (1) *If $x \in SB$, then there exist $\beta \in S$ and $b \in B$ such that*

$$\widehat{B} = Rb \oplus C, \qquad \beta b = x, \quad \textit{and} \quad \beta C = 0.$$

(2) *SB is a pure submodule in $\widehat{B}$ containing B.*

Proof. (1) Let $\alpha, \beta \in S$ and $a, b \in B$. There exists an endomorphism λ of the module $\widehat{B}$ such that $\lambda a = b$ or $\lambda b = a$ and $\lambda \in \mathrm{Ines}(B)$. For definiteness, let $\lambda a = b$. Then $\lambda\beta \in S$ and $\alpha a + \beta b = (\alpha + \lambda\beta)a$. This proves that any element x in SB is equal to αa, where $\alpha \in S$ and $a \in B$. We choose every element b of the module B of zero height. Then $\widehat{B} = Rb \oplus C$ for some submodule C. Let μ be the endomorphism of the module $\widehat{B}$ such that $\mu b = a$ and $\mu C = 0$. It is clear that $\mu \in \mathrm{Ines}(B)$. We set $\beta = \mu\alpha$. Then $\beta \in S$ and β has the required properties.

(2) It is clear that SB is a submodule in $\widehat{B}$ and $B \subseteq SB$. We assume that $x = p^k y$, where $x \in SB$, $y \in \widehat{B}$, and k is a positive integer. Applying (1) to x, we obtain $x = \beta y = p^k y$ for some $\beta \in S$ and $b \in B$. We define an endomorphism λ of the module $\widehat{B}$ by the relations $\lambda b = y$ and $\lambda C = 0$. We have

$$\beta = p^k \lambda \in S \cap p^k \, \mathrm{End}(B) = p^k S.$$

Since $\mathrm{End}(\widehat{B})$ is an torsion-free R-module, we have $\lambda \in S$. We obtain $y \in SB$. Therefore SB is pure in $\widehat{B}$. □

We define the linear topology τ on the ring S such that right annihilators of primitive idempotents in S form a basis of neighborhoods of zero. A basis of neighborhoods of zero consists of all intersections of a finite number of such annihilators. These intersections coincide with right annihilators of finite sums of pairwise orthogonal primitive idempotents of the ring S. With respect to the topology τ, S is a topological ring (see Sections 3 and 16). If M is a reduced torsion-free module, then the topology τ on $\mathrm{End}(M)$ (the ring $\mathrm{End}(M)$ can be obtained as the ring S if we take some basic submodule of the module M as B), coincides with the finite topology (Proposition 16.2). By this reason, the topology τ is naturally called the finite topology.

Proposition 18.6 (Goldsmith [112]). *Let B be a free R-module of infinite rank and let S be a pure subring in $\mathrm{End}(\widehat{B})$ containing $\mathrm{Ines}(B)$. Then if S is a complete ring in the finite topology, then $S = \mathrm{End}(SB)$.*

Proof. By our assumption, $\operatorname{End}(SB) \subseteq \operatorname{End}(\widehat{B})$. Then it is obvious that the inclusion $S \subseteq \operatorname{End}(SB)$ holds. Let $\{\pi_i \mid i \in I\}$ be the set of all sums of a finite number of pairwise orthogonal primitive idempotents of the ring S. We assume that

$$i \leq j \qquad \Longleftrightarrow \qquad \operatorname{Im}(\pi_i) \subseteq \operatorname{Im}(\pi_j);$$

this is equivalent to the property that the right annihilator π_i contains the right annihilator π_j in S. Let $\alpha \in \operatorname{End}(SB)$. We construct a sequence $\{\alpha_i\}_{i \in I}$ in S such that $\pi_i(\alpha - \alpha_i) = 0$ for all $i \in I$. It follows from the definitions of a Cauchy sequence and the limit of the sequence in the linear topology that this means that $\{\alpha_i\}$ is a Cauchy sequence in the finite topology of the ring S and α is the limit of the sequence. Since S is complete, we have $\alpha \in S$.

The module $\pi_i \widehat{B}$ for every $i \in I$ is a direct summand of finite rank in $\widehat{B}$ contained in SB. Let $\pi_i \widehat{B} = Ra_1 \oplus \cdots \oplus Ra_k$, where $a_1, \ldots, a_k \in SB$. For every $j = 1, \ldots, k$, it follows from Lemma 18.5 that there exist $\beta_j \in S$ and $b_j \in B$ such that $\alpha(a_j) = \beta_j(b_j)$. We define endomorphisms ξ_j of the module $\widehat{B}$ assuming that $\xi_j(a_j) = b_j$ and ξ_j annihilates the complement summand for Ra_j. It is clear that $\xi_j \in \operatorname{Ines}(B)$ for all j. We set $\beta = \sum_{j=1}^{k} \xi_j \beta_j$. Then $\beta \in S$. We set $\alpha_i = \pi_i \beta$ for every i. Then $\alpha_i \in S$ and we obtain

$$(\pi_i(\alpha - \alpha_i))\widehat{B} = (\alpha - \beta)(\pi_i \widehat{B}) = 0.$$

Consequently, we get $\pi_i(\alpha - \alpha_i) = 0$ and the sequence $\{\alpha_i\}$ has the required properties. □

The following characterization theorem holds.

Theorem 18.7 (Goldsmith [112]). *An abstract ring S is the endomorphism ring of a reduced torsion-free R-module if and only if there exists a free R-module B such that S is isomorphic to some pure subring of the ring $\operatorname{End}(\widehat{B})$ containing $\operatorname{Ines}(B)$ and S is complete in the finite topology.*

Proof. If $S \cong \operatorname{End}(M)$, where M is some reduced torsion-free module, then every basic submodule of the module M is the required module B.

Now we assume that B is a free module and S is a pure complete subring in $\operatorname{End}(\widehat{B})$ containing $\operatorname{Ines}(B)$. If the rank B is finite, then $\widehat{B} = B$ and $S = \operatorname{End}(\widehat{B})$. If the rank B is infinite, then by Proposition 18.6, $S = \operatorname{End}(SB)$, where SB is a reduced torsion-free module. □

For the construction of examples, the relation $\operatorname{Ines}(B) \subseteq S \subseteq \operatorname{End}(\widehat{B})$ is acceptable, but the ideal $\operatorname{Ines}(B)$ is not natural in the ring theory. It is desirable to obtain ring-theoretical conditions which guarantee that we deal with subrings of the ring $\operatorname{End}(\widehat{B})$ containing $\operatorname{Ines}(B)$. This topic is developed in the paper of Goldsmith [112].

In the works [206, 207, 208] of Liebert, endomorphism rings of complete torsion-free modules are characterized. By Theorem 13.1, we immediately obtain characterization theorems for endomorphism rings of divisible primary modules. It is not surprising that the corresponding theorems of Liebert include the condition of the completeness of an abstract ring in the J-adic topology (see Theorem 14.1). In [208] the Wolfson's characterization [318] of operator rings of vector spaces is used. Precisely, it is required that for a ring S, the factor ring $S/J(S)$ is isomorphic to some operator ring (similar to Theorem 14.1).

Exercise 1 (reconstruction of a module by the endomorphism ring of the module). Let M be a reduced torsion-free module over a complete domain R and let ε be a primitive idempotent of the ring $\operatorname{End}(M)$. We identify R with $\varepsilon \operatorname{End}(M)\varepsilon$ (see the proof of Theorem 18.4). Then the module M is isomorphic to $\varepsilon \operatorname{End}(M)\varepsilon$-module $\varepsilon \operatorname{End}(M)$.

Exercise 2. If R is commutative, then the center of the endomorphism ring of a reduced torsion-free R-module is isomorphic to R.

Exercise 3. Let M be a reduced torsion-free R-module. Then M is a cyclic projective right $\operatorname{End}(M)$-module (see Section 2).

19 Realization theorems for endomorphism rings of torsion-free modules

In the four remaining sections of the chapter, we assume that R is a commutative discrete valuation domain. In addition to technical convenience, the commutativity condition sometimes gives large preferences. Let M be an R-module. For every element $r \in R$, the module multiplication $\lambda_r \colon x \to rx$, $x \in M$, is an endomorphism of the module M (see also the beginning of Section 14). The mapping $f \colon r \to \lambda_r$, $r \in R$, is a homomorphism of rings $R \to \operatorname{End}(M)$ such that the image of f is contained in the center of the ring $\operatorname{End}(M)$. Consequently, we can consider $\operatorname{End}(M)$ as an R-module if we set $(r\alpha)x = \alpha(rx)$ for all $r \in R$, $\alpha \in \operatorname{End}(M)$, and $x \in M$. More precisely, $\operatorname{End}(M)$ is an R-algebra. A ring A is called an *algebra* over a commutative ring S (or an S-algebra) if A is an S-module and $s(ab) = (sa)b = a(sb)$ for all $s \in S$ and $a, b \in A$. The above mapping f is an embedding if and only if M is not a bounded module (for example, M is a torsion-free module). In such a case, we identify r with λ_r and we assume that R is a subring of the center of the ring $\operatorname{End}(M)$. For a non-commutative domain R, there is not a unified method to turn the ring $\operatorname{End}(M)$ into an R-module. In Section 14 such a method is presented for a complete torsion-free module. However, it is not canonical. In the case of a commutative ring R, the ring $\operatorname{End}(M)$ is also called the *endomorphism algebra*.

The used topological notions imply the p-adic topology. In particular, every R-algebra A with respect to the p-adic topology is a topological algebra, i.e., A is a topological ring and a topological R-module. If the R-module A does not have elements of infinite height, then the completion $\widehat{A}$ is a complete topological R-algebra and $\widehat{R}$-algebra (see Section 11).

There are powerful results about the realization problem for endomorphism rings of torsion-free modules. As a rule, their proofs are technically difficult and extensive. For the presentation, we choose more weak and simple results. Nevertheless, they allow to feel the nature of more general theorems. Some theorems were formulated without proof. For more explicit acquaintance, the reader can use the book of Göbel–Trlifaj [109] and the original papers mentioned in the end of the chapter.

For endomorphism rings of modules over a complete domain, split realization theorems are more typical. Therefore, we first assume that R is a complete discrete valuation domain. It is difficult to use Theorem 18.4, since the formulation of the theorem is complicated. In addition, it follows from Theorem 33.5 that in this case, the structure of the endomorphism ring exactly reflects the quite unknown structure of the module. Split realization theorems are more useful for applications.

Let B be a free R-module of infinite rank and let $\widehat{B}$ be the completion of B. Then $\widehat{B}/B$ is a divisible torsion-free module (Corollary 11.3). A module M is called a *maximal pure submodule* of the complete module $\widehat{B}$ if M is a pure submodule in $\widehat{B}$ containing B and $\widehat{B}/M \cong K$, where K is the field of fractions of the domain R.

Let M be a maximal pure submodule in $\widehat{B}$. Then for every element $a \in \widehat{B} \setminus M$, the pure submodule generated by M and a coincides with $\widehat{B}$. It is clear that B is a basic submodule for M and $\widehat{B}$ is the completion of the module M. Every endomorphism α of the module M can be uniquely extended to an endomorphism $\widehat{\alpha}$ of the module $\widehat{B}$ (Proposition 11.12). As usual, we identify α with $\widehat{\alpha}$ and we assume that $\operatorname{End}(M)$ is a subring in $\operatorname{End}(\widehat{B})$. We note the following important property. The image of $\alpha\widehat{B}$ is a complete module for every endomorphism α of the module $\widehat{B}$ (Proposition 11.16).

An endomorphism α of the module M is said to *inessential* if $\alpha\widehat{B} \subseteq M$. All inessential endomorphisms of the module M form the ideal $\operatorname{Ines}(M)$ in $\operatorname{End}(M)$ and the left ideal in $\operatorname{End}(\widehat{B})$. If M is an arbitrary reduced torsion-free module over a domain which is not necessarily complete, then it is clear that an endomorphism α should be called inessential if $\alpha\widehat{M} \subseteq M$. Let α be a finite endomorphism of the module M. Then αM is a complete module as a free module of finite rank. Therefore $\alpha\widehat{B} \subseteq \alpha M$ and α is an inessential endomorphism. We obtain $\mathrm{Fin}(M) \subseteq \mathrm{Ines}(M)$. The relation $\mathrm{Fin}(M) = \mathrm{Ines}(M)$ holds if and only if M does not contain complete submodules of infinite rank. Indeed, assume that $\alpha \in \mathrm{Ines}(M)$ and $\alpha \notin \mathrm{Fin}(M)$. Then αM and $\alpha\widehat{B}$ have an infinite rank. Consequently, $\alpha\widehat{B}$ is a complete submodule in M of infinite rank. Now we assume that C is a complete submodule in M of infinite rank. There exists a homomorphism $\alpha' \colon B \to C$ such that the image of α' has an infinite rank. Since C is a complete module, α' can be extended to a homomorphism $\alpha \colon \widehat{B} \to C$. Then $\alpha \in \mathrm{Ines}(M)$ and $\alpha \notin \mathrm{Fin}(M)$.

By our definition, the p-rank of the module M coincides with the dimension of the R_p-space M/pM, where R_p coincides with the residue field R/pR of the domain R. The p-rank of the module M is equal to the rank of every basic submodule of M (see Sections 4 and 9).

Let S be some ring and let T and I be the subring of S and an ideal of S, respectively. If S as a group has a direct decomposition $S = T \oplus I$, then the ring S is called a *split extension* of the ring T with the use of the ideal I.

Theorem 19.1 (Goldsmith [114]). *For every infinite cardinal number λ, there exists a reduced torsion-free R-module M of p-rank λ such that* $\mathrm{End}(M)$ *is a split extension of the ring R with the use of the ideal* $\mathrm{Ines}(M)$, $\mathrm{End}(M) = R \oplus \mathrm{Ines}(M)$.

Proof. We take a free module B of rank λ. We prove that an arbitrary maximal pure submodule M of the module $\widehat{B}$ is the required module. It is clear that M is a reduced torsion-free module of p-rank λ.

We choose an element $a \in \widehat{B} \setminus M$. Let $\alpha \in \mathrm{End}(M)$. Since $\widehat{B}/M \cong K$, we have that $r(\alpha a) = sa + x$ for some $r, s \in R$ and $x \in M$. We can assume that $r = p^m$ and $s = p^n$, where m and n are non-negative integers. We assume that $m \leq n$. In this case $p^m(\alpha - p^{n-m}1_M)a = x$. Since M is pure in $\widehat{B}$, we have $(\alpha - p^{n-m}1_M)a \in M$. Therefore $(\alpha - p^{n-m}1_M)\widehat{B} \subseteq M$. Therefore $\alpha - p^{n-m}1_M \in \mathrm{Ines}(M)$. This proves the relation $\mathrm{End}(M) = R + \mathrm{Ines}(M)$. This is obvious that sum is a direct sum, and we have a split extension $\mathrm{End}(M) = R \oplus \mathrm{Ines}(M)$.

We assume that $m > n$. As above, we obtain that $1_M - p^{m-n}\alpha \in \mathrm{Ines}(M)$. By Theorem 14.1, $p^{m-n}\alpha \in J(\mathrm{End}(\widehat{B}))$. Consequently, $1_M - p^{m-n}\alpha$ is an automorphism of the module $\widehat{B}$. This is a contradiction, since $(1_M - p^{m-n}\alpha)\widehat{B} \subseteq M$ and $M \neq \widehat{B}$. Therefore, the case $m > n$ is impossible and the proof is completed. □

The are interesting situations, when it is possible to replace $\mathrm{Ines}(M)$ by $\mathrm{Fin}(M)$ in the proved theorem. We consider one corresponding example. First, we prove two lemmas. If σ is a cardinal number, then σ^+ denotes the cardinal number following σ.

Lemma 19.2 (Beaumont–Pierce [34]). *We assume that V is a vector space of infinite dimension σ over a field F. Let $\{W_i\}_{i<\tau}$ (where the cardinal number τ does not exceed σ) be the set of subspaces of the space V such that* $\dim W_i = \sigma$ *for all $i < \tau$. Then there exist σ^+ subspaces U in V with following properties*: $\dim(V/U) = 1$ *and every W_i is not contained in every U.*

Proof. First, we prove that there exists one such a subspace, say, U_1. Let e be a nonzero element in V. Using the transfinite induction, we prove that there exist sequences of elements $\{x_i\}_{i<\tau}$ such that $x_i \in W_i$ and the subspace A_τ generated by elements $e - x_i$, $i < \tau$, does not contain e. We choose an element $x_1 \in V$ such that the elements x_1 and e are linearly independent. Then the subspace A_1 generated by the element $e - x_1$ does not contain e. We assume that for all $\eta < i$ (where i is an ordinal number which is less than τ), the elements x_η have been constructed and the subspace generated by the elements $e - x_\eta$, $\eta \leq \zeta$, where $\zeta < i$, does not contain e. Then the subspace A_i generated by the elements $e - x_\eta$, $\eta < i$, also does not contain e. In addition, the space $A_i + Fe$ has the dimension $\leq |i| + 1$. Since τ and σ are cardinal numbers and $i < \tau$, we have

$$|i| + 1 < \tau + 1 \leq \sigma + 1 = \sigma = \dim W_i.$$

Consequently, $W_i \not\subset A_i + Fe$. We choose an element x_i in W_i such that $x_i \notin A_i + Fe$. Then $e \notin A_i + F(e - x_i)$. Let A_τ be the subspace, generated by the elements $e - x_i$, $i < \tau$. Then $e \notin A_\tau$, since $e - x_i \in A_\tau$ for every $i < \tau$. Let U_1 be a subspace in V which is maximal with respect to properties $A_\tau \subset U_1$ and $e \notin U_1$. Then V/U_1 is an one-dimensional space and $x_i \notin U_1$ for every i (otherwise, $e = x_i + (e - x_i) \in U_1$). Therefore $W_i \not\subseteq U_1$ for every $i < \tau$.

We assume that the subspace U_j have been constructed for all $j < \rho$, where $\rho < \sigma^+$. We take the set consisting of all these subspaces W_i and all subspaces U_j. This set consists of at most σ subspaces and each of the subspaces is of dimension σ. Applying the proved property to this set, we obtain the subspace U_ρ such that $\dim(V/U_\rho) = 1$ and every W_i is not contained in U_ρ. The required result now follows from the transfinite induction argument. □

We know that divisible torsion-free modules can be identified with K-spaces. If D is a divisible torsion-free module, then $r(D) = \dim_K D$ (see Section 6 and Lemma 4.5).

In the following lemma, we assume that $|R| = c$, where c is the continual cardinality $2^{\aleph_0}$ (for example, we can take the ring of p-adic integers $\widehat{\mathbb{Z}}_p$).

Lemma 19.3. *If B is a free module of countable rank, then there exist c^+ maximal pure submodules in $\widehat{B}$ which do not contain a submodule isomorphic to $\widehat{B}$.*

Proof. The module B is equal to the direct sum of $\aleph_0$ copies of the module R. Therefore $|B| = c$ and $|\widehat{B}| = c^{\aleph_0} = c$. This is clear, since the elements of $\widehat{B}$ are the limits of Cauchy sequences of elements of B. More precisely, we can use the following argument. We assume that the module $\widehat{B}$ is a submodule of the complete module $\prod_{\aleph_0} R$. Let x and y be two linearly independent elements in R. The set of vectors in $\widehat{B}$ of the form $(z, pz, p^2z, \ldots)$, where either $z = x$ or $z = y$, forms a linearly independent system. Now it is clear that $r(\widehat{B}) = c$. Since B is a pure dense submodule in $\widehat{B}$, we have that $\widehat{B}/B$ is a divisible torsion-free module, i.e., K-space. In addition, $r(\widehat{B}/B) = \dim_K(\widehat{B}/B) = c$.

Let $\{A_t\}_{t\in T}$ be the set of all submodules of the module $\widehat{B}$ which are isomorphic to $\widehat{B}$. Every submodule A_t is the image of some endomorphism of the module $\widehat{B}$. In its turn, every endomorphism of the module $\widehat{B}$ is completely determined by the action of it on B (it is even determined by the action of it on every free basis of the module B). Therefore $|\operatorname{End}(\widehat{B})| \le c^{\aleph_0} = c$. (Since $R \subseteq \operatorname{End}(\widehat{B})$, we have $|\operatorname{End}(\widehat{B})| = c$.) Consequently, the set $\{A_t\}_{t\in T}$ consists of at most c submodules in $\widehat{B}$. Let E_t be a pure submodule in $\widehat{B}$ generated by the submodule $A_t + B$ and $V_t = E_t/B$. Then V_t is a divisible module. Consequently, V_t is a subspace of the K-space $\widehat{B}/B$. We have also $\dim_K(V_t) = r(V_t) = r(A_t) = c$. Applying Lemma 19.2, we can obtain c^+ subspaces U in $\widehat{B}/B$ such that every V_t is not contained in every U and $\dim_K((\widehat{B}/B)/U) = 1$. If G is a submodule in $\widehat{B}$ with $G/B = U$, then G is a maximal pure submodule in $\widehat{B}$. It is also clear that every A_t is not contained in any G. We have constructed the required set of maximal pure submodules. □

Corollary 19.4. *Let R be a domain such that $|R| = 2^{\aleph_0}$. There exists a reduced torsion-free R-module M of countable p-rank such that* $\operatorname{End}(M) = R \oplus \operatorname{Fin}(M)$.

Proof. Let B be a free R-module of countable rank. By Lemma 19.3, there exists a maximal pure submodule M in $\widehat{B}$ which does not contain a copy of the module $\widehat{B}$. It follows from the proof of Theorem 19.1 that $\operatorname{End}(M) = R \oplus \operatorname{Ines}(M)$. Let $\alpha \in \operatorname{Ines}(M)$. Then $\alpha\widehat{B} \subseteq M$ and $\alpha\widehat{B}$ is a complete module. It follows from $\alpha\widehat{B} \cong \widehat{B}/\operatorname{Ker}(\alpha)$ that $r_p(\alpha\widehat{B}) \le r_p(\widehat{B}) = \aleph_0$. If $r_p(\alpha\widehat{B}) = r_p(\widehat{B})$, then $\alpha\widehat{B} \cong \widehat{B}$ by Corollary 11.15 (2). However, this contradicts to the choice of the module M. Therefore $r_p(\alpha\widehat{B}) < \aleph_0$ and $\alpha\widehat{B}$ is a free module of finite rank. Thus, $\alpha \in \operatorname{Fin}(M)$ and $\operatorname{Ines}(M) = \operatorname{Fin}(M)$. □

We present without proof the following result which is more general than Corollary 19.4. Module notions applied below to the R-algebra A are related to A

considered as an R-module. In this sense, we write "a reduced R-algebra", "a torsion-free R-algebra", "the rank of an R-algebra".

Theorem 19.5 (Göbel–Goldsmith [103]). *If A is a torsion-free R-algebra and the R-module A is generated by less than $2^{\aleph_0}$ elements, then there exist a free R-module B of countable rank and a pure submodule M in $\widehat{B}$ containing B such that* $\mathrm{End}(M) = A \oplus \mathrm{Fin}(M)$.

The most general result about split realization for torsion-free modules is given by the following theorem. Meanings of terms related to cardinal numbers are presented in the book of Jech [165].

Theorem 19.6 (Dugas–Göbel–Goldsmith [62]). *Let R be a complete discrete valuation domain and let A be an R-algebra. The following conditions are equivalent.*

(1) *A is a reduced torsion-free algebra.*

(2) *There exists a reduced torsion-free R-module M such that* $\mathrm{End}(M) \cong A \oplus \mathrm{Ines}(M)$.

(3) *There exists a reduced torsion-free R-module M with property* (2) *such that the cardinality $|M|$ of M is equal to any preassigned strongly limit cardinal of confinality $> |A|^{\aleph_0}$.*

There is a greatly stronger version of Corollary 19.4 and Theorem 19.5. A module M over a Dedekind domain R (Dedekind domains are defined in Exercise 3 of Section 3) is called a *cotorsion-free* module if M does not have submodules which are isomorphic to R/P, $\widehat{R}_P$, and K for all prime ideals P of the ring R, where $\widehat{R}_P$ is the completion in the P-adic topology of the localization R_P of the ring R with respect to the ideal P (see Section 3) and K is the field of fractions of the domain R. Different properties of cotorsion-free modules are contained in [57] and [58]. A complete discrete valuation domain R does not have cotorsion-free modules. A reduced torsion-free R-module is called a *$\aleph_0$-cotorsion-free* module if it does not contain complete submodules of infinite rank.

Theorem 19.7 (see [62]). *For an algebra A over a complete discrete valuation domain R, the following conditions are equivalent.*

(1) *A is a $\aleph_0$-cotorsion-free R-algebra.*

(2) *There exists a reduced torsion-free R-module M such that* $\mathrm{End}(M) \cong A \oplus \mathrm{Fin}(M)$.

(3) *There exists a reduced torsion-free R-module M with property* (2) *such that the cardinality $|M|$ is equal to any preassigned singular strongly limit cardinal of confinality $>|A|^{\aleph_0}$.*

In the work [102], Göbel and Goldsmith have obtained results which are similar to Theorems 19.6 and 19.7; the results are stronger in some sense than the theorems but they use the additional assumption that Gödel's constructibility axiom holds.

For torsion-free modules over an incomplete discrete valuation domain R, there also exist split realization theorems (see [49]). In these theorems, a torsion-free R-module M with $\mathrm{End}(M) \cong A \oplus \mathrm{Ines}(M)$ is determined for a given R-algebra A. Naturally, it is preferable to represent the algebra A by the whole endomorphism ring $\mathrm{End}(M)$. In this case, the condition that A is a cotorsion-free R-module is a natural restriction (see Exercise 3).

To orient the reader, we present some (complete) realization theorem for endomorphism rings of torsion-free modules over Dedekind domains. The theorem precisely specifies the place of complete discrete valuation domains among all Dedekind domains. A module is said to be *superdecomposable* if it does not have nonzero indecomposable direct summands.

Theorem 19.8 (Dugas–Göbel [58]). *Let R be a Dedekind domain which is not a field. The following conditions are equivalent.*

(1) *R is not a complete discrete valuation domain.*

(2) *There exists an indecomposable R-module of rank ≥ 2.*

(3) *There exists a superdecomposable R-module.*

(4) *There exists an R-module M of rank >1 with $\mathrm{End}(M) \cong R$.*

(5) *If A is of arbitrary cotorsion-free R-algebra, then there exists a cotorsion-free R-module M with $\mathrm{End}(M) \cong A$.*

Thus, if R is a incomplete discrete valuation domain, then assertion (5) holds for R. In fact, there always exists a set of modules M from (5) with some additional properties. The corresponding results are contained in the works [49, 87, 105] and other works. We have one theorem from [105] for the case, when an R-algebra A has finite rank. In two remaining assertions, we assume that R is a incomplete discrete valuation domain.

Lemma 19.9 (Göbel–May [106]). *Let A be a reduced torsion-free R-algebra of cardinality $<2^{\aleph_0}$. Then the completion $\widehat{R}$ contains a family $\{u_i\}_{i\in I}$ of algebraically independent over A of elements of continual cardinality. (The algebraic independence means that there does not exist a finite set of elements u_i such that every element e_i is a root of every polynomial with coefficients in A.)*

Lemma 19.9 plays the main role in the construction of modules X_i from the following theorem. In this theorem we denote by d the function on the set of positive integers such that $d(1) = 2$ and $d(s) = 2s + 1$ for $s > 1$.

Theorem 19.10 (Göbel–May [105]). *We assume that A is a reduced torsion-free R-algebra of finite rank n, the cardinality of A is less than $2^{\aleph_0}$, s is a positive integer, and d is a cardinal number such that $d(s) \leq d \leq 2^{\aleph_0}$. We set $B = \bigoplus_s A$. Then $\widehat{B}$ contains a set of free A-modules F_i, $i \in I$, $|I| = 2^{\aleph_0}$, isomorphic to $\bigoplus_d A$, such that if $X_i = (F_i)_*$ for every $i \in I$, then the following conditions hold*:

(1) $B \subseteq X_i \subseteq \widehat{B}$;

(2) $\mathrm{End}(X_i) \cong A$ *and* $\mathrm{Hom}(X_i, X_j) = 0$ *for all distinct* $i, j \in I$;

(3) *the module* X_i *has rank* $\leq d \cdot n$.

In Exercises 1 and 2, R is a complete discrete valuation domain and B is a free R-module of infinite rank. In the remaining exercises, the domain R is not necessarily complete.

Exercise 1. If M and N are two pure submodules in $\widehat{B}$ containing B, then $M \cong N$ if and only if there exists an automorphism θ of the module $\widehat{B}$ such that $\theta M = N$.

Exercise 2 (Goldsmith [114]). We assume that $|R| = 2^{\aleph_0}$ and B has a countable rank. Then there exists a pure submodule M in $\widehat{B}$ containing B such that $\widehat{B}/M \cong K \oplus K$ and $\mathrm{End}(M) = R \oplus \mathrm{Fin}(M)$.

Exercise 3. Prove that the endomorphism algebra of the reduced torsion-free R-module is a reduced torsion-free R-algebra. If the module has finite rank, then the endomorphism algebra also has finite rank.

Exercise 4. Verify that the R-algebras $K \times K$, $\widehat{R} \times \widehat{R}$, and $R_p \times R_p$ (where R_p is the residue field of the ring R) cannot be the endomorphism algebra of any R-module.

20 Essentially indecomposable modules

We know that torsion-free modules over a complete discrete valuation domain R have many remarkable properties which do not necessarily hold in modules over a incomplete domain. For example, every reduced indecomposable torsion-free R-module is isomorphic to R, every reduced torsion-free R-module of finite or countable rank is free (see Section 11 and the beginning of Section 17). It may appear that this class of modules does not have any pathological direct decompositions which exist for modules over incomplete discrete valuation domains and Abelian groups. (Many examples of such pathologies are contained in the papers mentioned in the end of the chapter.) We will see that this is not always true.

Unless otherwise stated, R is a commutative complete discrete valuation domain. As before, we consider reduced torsion-free R-modules.

A torsion-free module M is said to be *essentially indecomposable* if every nontrivial decomposition of M into a direct sum of two summands has at least one summand which is a free module of finite rank.

Corollary 20.1. *Let $|R| = 2^{\aleph_0}$. There exists an essentially indecomposable reduced torsion-free R-module of countable p-rank.*

Proof. Let M be an R-module which satisfies Corollary 19.4, i.e., $\operatorname{End}(M) = R \oplus \operatorname{Fin}(M)$. We consider an arbitrary nontrivial direct decomposition $M = M_1 \oplus M_2$ with corresponding projections π_1 and π_2. Their images $\bar{\pi}_1$ and $\bar{\pi}_2$ in the factor ring $\operatorname{End}(M)/\operatorname{Fin}(M)$ are idempotents and the sum of these idempotents is equal to the identity element. Since this factor ring is isomorphic to R, we have either $\bar{\pi}_1 = 0$ or $\bar{\pi}_2 = 0$. For definiteness, let $\bar{\pi}_1 = 0$. Then $\pi_1 \in \operatorname{Fin}(M)$. It follows from the proof of Corollary 19.4 that the summand M_1, which is equal to $\pi_1 M$, is a free module of finite rank. □

As a rule, it is useful to construct a set of essentially indecomposable modules with certain properties. We recall some general notion of the theory of modules. Let T be a subscript set. A set of modules $\{M_t\}_{t\in T}$ over some ring is called a *rigid system* if $\operatorname{Hom}(M_s, M_t) = 0$ for any two distinct subscripts $s, t \in T$. We specialize this notion as follows. Let R be a commutative ring and let A be an R-algebra. A set of R-modules $\{M_t\}_{t\in T}$ is called an *A-rigid system* if $\operatorname{End}(M_s) \cong A$ and $\operatorname{Hom}(M_s, M_t) = 0$ for any two distinct subscripts $s, t \in T$. Theorem 19.10 gives examples of A-rigid systems from $2^{\aleph_0}$ of torsion-free modules over a incomplete discrete valuation domain. In the works [49, 104], it is proved that there exist many A-rigid systems of torsion-free modules containing any large number of modules.

Over a complete discrete valuation domain R, there does not exist a rigid system from two and more of torsion-free modules. Therefore, we need to change the definition of a rigid system.

First, we define inessential homomorphisms and finite homomorphisms by analogy with inessential and finite endomorphisms. Let M and N be two reduced torsion-free modules. A homomorphism $\varphi\colon M \to N$ is said to be *inessential* (resp., *finite*) if $\widehat{\varphi}\widehat{M} \subseteq N$, where $\widehat{\varphi}$ is the unique extension of φ to a homomorphism from $\widehat{M}$ into $\widehat{N}$ (resp., if φM has finite rank). All inessential (resp., finite) homomorphisms from M into N form a group which is denoted by $\operatorname{Ines}(M, N)$ (resp., $\operatorname{Fin}(M, N)$). It is clear that we have the inclusion

$$\operatorname{Fin}(M, N) \subseteq \operatorname{Ines}(M, N).$$

Let T be a subscript set. A set $\{M_t\}_{t\in T}$ of torsion-free modules is called an *essentially rigid* system (or an *R-rigid* system) if $\operatorname{Hom}(M_s, M_t) = R \oplus \operatorname{Ines}(M_t)$ for $s = t$ and $\operatorname{Hom}(M_s, M_t) = \operatorname{Ines}(M_s, M_t)$ for $s \neq t$ for all $s, t \in T$.

Similar to Lemma 19.3, let $|R| = c$, where c is the continual cardinality $2^{\aleph_0}$.

Theorem 20.2 (Goldsmith [114]). *There exists an essentially rigid system of reduced torsion-free R-modules of countable p-rank which contains c^+ terms.*

First, we prove two lemmas close to Lemma 19.3. Similar to Section 19, let B be a free module of countable rank and let $\widehat{B}$ be the completion of B. We denote by $\mathcal{M}$ the set of all maximal pure submodules in $\widehat{B}$ which do not contain submodules isomorphic to $\widehat{B}$. By Lemma 19.3, the cardinality $\mathcal{M}$ is equal to c^+. It directly follows from the proof of Corollary 19.4 that $\operatorname{Ines}(M, N) = \operatorname{Fin}(M, N)$ for any two submodules $M, N \in \mathcal{M}$.

Lemma 20.3. *Let $\{M_s\}_{s\in S}$ be a subset of the set $\mathcal{M}$ of cardinality $\leq c$. Then there exist c^+ submodules M in $\mathcal{M}$ such that* $\operatorname{Hom}(M_s, M) = \operatorname{Fin}(M_s, M)$ *for all $s \in S$.*

Proof. For every s, we denote by $\{H_{si}\}$ the set of the images of the submodule M_s which have the p-rank $\aleph_0$ for all endomorphisms of the module $\widehat{B}$. Then $\{H_{si}\}$ has the cardinality which does not exceed c (see the proof of Lemma 19.3). Since $|S| \leq c$, the union of the sets $\{H_{si}\}$ for all $s \in S$ contains at most c submodules in $\widehat{B}$. We denote it by $\mathcal{H}$. Furthermore, we denote by $\mathcal{E}$ the set of the images of all endomorphisms of the module $\widehat{B}$ which are isomorphic to the module $\widehat{B}$. Then the set $\mathcal{H} \cup \mathcal{E}$ does not contain more c submodules; for example, let $\mathcal{H} \cup \mathcal{E} = \{X_j\}_{j\in J}$, where $|J| \leq c$. We note, that every X_j has rank c. Let G_j is a pure submodule in $\widehat{B}$ generated by the submodule $X_j + B$, and $W_j = G_j/B$. Then the set $\{W_j\}_{j\in J}$ consists of at most c subspaces of the K-space $\widehat{B}/B$ and the dimension of every space W_j is c. By Lemma 19.2, there exist c^+ subspaces U in $\widehat{B}/B$ such that every W_j is not contained in U and $\dim_K((\widehat{B}/B)/U) = 1$. Let M be a maximal pure submodule in $\widehat{B}$ with $M/B = U$. It is clear that $M \in \mathcal{M}$.

Now we take an arbitrary subscript $s \in S$ and the homomorphism $\varphi\colon M_s \to M$. It follows from properties of the module M that $r_p(\varphi M_s) < \aleph_0$. Consequently, φM_s is a free module of finite rank and φ is a finite homomorphism. The required set of submodules has been constructed. □

Lemma 20.4. *For a given maximal pure submodule M in $\widehat{B}$, there exist at most c maximal pure submodules M_s in $\widehat{B}$ with* $\operatorname{Hom}(M_s, M) \neq \operatorname{Fin}(M_s, M)$.

Proof. We assume that there exists a set $\{M_s\}_{s\in S}$ consisting of more than c submodules with the mentioned property. For every $s \in S$, we choose a homomorphism $\varphi_s\colon M_s \to M$ which is not finite. Then the set $\{\widehat{\varphi}_s\}_{s\in S}$ consists of more than c endomorphisms of the module $\widehat{B}$, where $\widehat{\varphi}_s$ is an extension of φ_s to an endomorphism of the module $\widehat{B}$. It follows from the proof of Lemma 19.3 that $|\operatorname{End}(\widehat{B})| = c$. Therefore, we have $\widehat{\varphi}_s = \widehat{\varphi}_t$ for some distinct s and t from S. However, then

$$\widehat{\varphi}_s \widehat{B} = \widehat{\varphi}_s(M_s + M_t) \subseteq \widehat{\varphi}_s M_s + \widehat{\varphi}_s M_t \subseteq M.$$

We obtain that φ_s is a finite homomorphism. This is a contradiction. □

Proof of Theorem 20.2. Every module in $\mathcal{M}$ forms an essentially rigid system. We assume that we have constructed an essentially rigid system $\{M_\sigma\}$ of modules in $\mathcal{M}$ for all $\sigma < \tau$, where σ and τ are some ordinal numbers and $|\tau| < c^+$. By Lemma 20.3, there exist c^+ submodules M in $\mathcal{M}$ such that $\operatorname{Hom}(M_\sigma, M) = \operatorname{Fin}(M_\sigma, M)$ for every $\sigma < \tau$. For such σ, there exist at most c submodules M with $\operatorname{Hom}(M, M_\sigma) \neq \operatorname{Fin}(M, M_\sigma)$. By removing all such submodules M, we remove at most c submodules from the initial system, since $|\tau| \leq c$. Thus, there exists a module $M \in \mathcal{M}$ such that

$$\operatorname{Hom}(M, M_\sigma) = \operatorname{Fin}(M, M_\sigma) \qquad \text{and} \qquad \operatorname{Hom}(M_\sigma, M) = \operatorname{Fin}(M_\sigma, M)$$

for all $\sigma < \tau$. We set $M_\tau = M$. Then the set $\{M_\sigma\}_{\sigma \leq \tau}$ is an essentially rigid system of modules from $\mathcal{M}$. The use of the transfinite induction completes the construction of the required essentially rigid system of modules. □

In the remaining part of the section, we prove some interesting property of reduced modules of infinite rank $< 2^{\aleph_0}$. A reduced torsion-free module of finite or countable rank is necessarily free. On the other hand, we have proved that there exists an essentially indecomposable module M of rank $2^{\aleph_0}$. This means that if $M = M_1 \oplus M_2$, then M_1 or M_2 has finite rank. If the module M has an infinite noncountable rank, then for every positive integer n, there exists a module A_n such that $M \cong \bigoplus_n R \oplus A_n$. These results lead us to the following question. If $\aleph_0 \leq r(M) < 2^{\aleph_0}$, then does the module M have a direct summand of countable rank, i.e., $M \cong \bigoplus_{\aleph_0} R \oplus A$ for some module A? It turns out that every such a module has an analogous decomposition.

Let M be a reduced torsion-free R-module and let $B = \bigoplus_{e \in E} Re$ be a basic submodule of M. In this case, E is a maximal purely independent system of elements of the module M. Purely independent systems of elements and their relations with basic submodules are studied in Section 9. Since M and B have the same completions with respect to the p-adic topology, we can assume that $\widehat{M} \subseteq \prod_{e \in E} Re$. Consequently, we can consider an element $a \in \widehat{M}$ as a sequence $(r_e)_{e \in E}$ of elements r_e in R. For the element a, we set $[a]_E = \{e \in E \mid r_e \neq 0\}$. It is clear that if $[a]_E$ is a finite set, then $a \in M$.

Lemma 20.5. *We assume that $\mathcal{X}_0 \leq r(M) < c = 2^{\mathcal{X}_0}$. If Y is some countable purely independent system in M, then there exists a countable subset X in Y such that the module $\widehat{\bigoplus_{x\in X} Rx} \cap M$ has a countable rank.*

Proof. We note that $\widehat{\quad\quad}$ denotes the completion of the corresponding module in the p-adic topology. The set Y can be extended to a maximal purely independent system E. Then $\bigoplus_{e\in E} Re$ is a basic submodule of the module M (Proposition 9.2). There exists a family $\mathcal{F}$ such that it consists of c countable sets in Y and for all $X_1, X_2 \in \mathcal{F}$, the intersection $X_1 \cap X_2$ is a finite set. For every $X \in \mathcal{F}$, we consider the submodule

$$A_X = \widehat{\bigoplus_{x\in X} Rx} \cap M.$$

If every submodule A_X has a noncountable rank (it is clear that it is infinite), then for every $X \in \mathcal{F}$, there exists an element $a_X \in A_X$ such that $[a_X]_E$ is an infinite subset in X. It follows from the definition of the family $\mathcal{F}$ that $\{a_X \mid X \in \mathcal{F}\}$ is a linearly independent subset in M of cardinality c. This implies that $r(M) \geq c$. This is a contradiction. Consequently, there exists a countable subset X in Y with $r(A_X) = \mathcal{X}_0$. □

If $\bigoplus_{e\in E} Re$ is a basic submodule of the module M and $I \subseteq E$, then $\widehat{\bigoplus_{e\in I} Re}$ is a direct summand in $\widehat{M}$ (Corollary 11.11 (2)). Let π_I be the projection from the module $\widehat{M}$ onto this summand.

Theorem 20.6 (Göbel–Paras [108]). *Assume that the module M and the set X are taken from Lemma 20.5. Then there exists a countable subset I of X with $\pi_I(M) \subseteq M$. Therefore $M \cong \bigoplus_{\mathcal{X}_0} R \oplus A$ for some module A.*

Proof. We take a family $\mathcal{G}$ consisting of c countable subsets of the set X such that for all $I_1, I_2 \in \mathcal{G}$, the intersection $I_1 \cap I_2$ is a finite set. We assume that $\pi_I(M) \nsubseteq M$ for all $I \in \mathcal{G}$. Then there exists an element $a_I \in M$ with $\pi_I(a_I) \notin M$. We note that $[\pi_I(a_I)]_E$ is an infinite subset in I; otherwise, $\pi_I(a_I)$ is an element of M. (Here E is the maximal purely independent system considered in Lemma 20.5.) The intersections of distinct sets $[\pi_I(a_I)]_E$ are finite sets. Consequently, $\{a_I \mid I \in \mathcal{G}\}$ is a linearly independent subset in M of cardinality c. This contradicts to the assumption about the rank of the module M. Therefore, there exists a set $I \in \mathcal{G}$ such that $\pi_I(M) \subseteq M$ and π_I is an idempotent of the ring $\operatorname{End}(M)$. Since $\pi_I(M) \subseteq \widehat{\bigoplus_{e\in I} Re}$ and $\widehat{\bigoplus_{e\in I} Re} \cap M$ has a countable rank by Lemma 20.5, we have that $\pi_I(M)$ is a free module of countable rank, $\pi_I(M) \cong \bigoplus_{\mathcal{X}_0} R$. It follows from $M = \pi_I(M) \oplus (1 - \pi_I)M$ that $M \cong \bigoplus_{\mathcal{X}_0} R \oplus A$ for some module A. □

According to the beginning of Section 19, we consider the ring $\operatorname{End}(M)$ as an R-algebra. In Corollary 19.4 and Theorem 20.2, the R-module rank of the factor algebra $\mathrm{End}(M)/\mathrm{Fin}(M)$ is equal to 1. For the modules from Theorem 20.6, we have the directly opposed situation.

Corollary 20.7. *Let M be a module with $\aleph_0 \leq r(M) < c$. Then the following assertions hold.*

(1) *The rank of the factor algebra* $\mathrm{End}(M)/\mathrm{Fin}(M)$ *is not less than* c.

(2) $M \cong M \oplus \bigoplus_\sigma R$ *for every* $\sigma \leq \aleph_0$.

Proof. By Theorem 20.6, there exists a decomposition $M = \bigoplus_{i \geq 1} Rb_i \oplus A$.

(1) For every infinite set I of positive integers, assume that ε_I denotes the projection from the module M onto the direct summand $\bigoplus_{i \in I} Rb_i$. Then the set of the residue classes $\varepsilon_I + \mathrm{Fin}(M)$ for all such sets I forms the set consisting of c linearly independent elements in $\mathrm{End}(M)/\mathrm{Fin}(M)$.

(2) If σ is a cardinal number with $\sigma \leq \aleph_0$, then

$$\bigoplus_{i \geq 1} Rb_i \cong \bigoplus_{i \geq 1} Rb_i \oplus \bigoplus_{\sigma} R.$$

Therefore, we obtain $M \cong M \oplus \bigoplus_\sigma R$. □

Further information about the topic of this section is contained in the papers [62, 102, 103, 115]. For example, it is proved that there exist essentially indecomposable modules of any large ranks and essentially rigid systems containing any large number of modules. Remarks to the chapter contain the corresponding bibliography on primary modules, mixed modules, and some modules over other commutative rings.

Exercise 1 (see [62]). Let R be a complete discrete valuation domain and let t be a positive integer. There exists a torsion-free R-module X such that

$$X^m \cong X^n \qquad \Longleftrightarrow \qquad m \equiv n \pmod t$$

for any two positive integers m and n.

21 Cotorsion modules and cotorsion hulls

We assume that R is a commutative discrete valuation domain which is not necessarily complete. All modules are R-modules.

We considered earlier two forms of "hulls". Precisely, every module M can be embedded in the minimal divisible module D. The module D is called the *divisible hull* of the module M (see Section 6). If M does not have elements of infinite height, then M can be embedded in some complete module as a pure dense submodule. This complete module is denoted by $\widehat{M}$ and is called the *completion* of the module M (completions were considered in Section 11). Sometimes it is not sufficient to use only these two "hulls". In this section, we define and study cotorsion modules and cotorsion hulls. These remarkable notions are successively used in some studies. They appeared in the works of Harrison [134], Nunke [245], and Fuchs [91].

For the presentation of the material, it should be used the *extension group*. We neither define this group nor prove the properties of the group. Different aspects of the theory of extensions of modules are contained in the books of Cartan–Eilenberg [38] and MacLane [216]. In the case of Abelian groups, the theory is presented in the book of Fuchs [92]. This presentation is completely adequate for us, since it can be extended to modules over commutative principal ideal domains with obvious changes. We only list main used results.

Let M and N be two R-modules. We write the extension group $\mathrm{Ext}_R(M,N)$ of the module N by the module M without the subscript R. Some general properties of the group Ext are similar to the corresponding properties of the group Hom. For example, the module $\mathrm{Ext}(M,N)$ is defined provided M and N are bimodules (see the corresponding situation for Hom in Section 2). Since the ring R is commutative, $\mathrm{Ext}(M,N)$ is an R-module. Similar to Hom, the group Ext satisfies isomorphisms related to direct sums and products (see also Section 2).

We recall some information about induced homomorphisms and exact sequences for Ext (similar notions for Hom are defined in Section 2). Let L be an R-module and let $f\colon L \to N$ be a homomorphism. We have the induced homomorphism of R-modules $f_*\colon \mathrm{Ext}(M,L) \to \mathrm{Ext}(M,N)$. For homomorphism $g\colon L \to M$, we obtain the induced homomorphism $g^*\colon \mathrm{Ext}(M,N) \to \mathrm{Ext}(L,N)$. If we have an exact sequence of modules

$$0 \to A \xrightarrow{f} B \xrightarrow{g} C \to 0,$$

then we can construct short exact sequences for Ext. It is very useful that exact sequences for Hom and Ext can be combined in long exact sequences. As a result, we have the following two exact sequences:

$$0 \to \mathrm{Hom}(M,A) \to \mathrm{Hom}(M,B) \to \mathrm{Hom}(M,C) \xrightarrow{E_*}$$
$$\xrightarrow{E_*} \mathrm{Ext}(M,A) \xrightarrow{f_*} \mathrm{Ext}(M,B) \xrightarrow{g_*} \mathrm{Ext}(M,C) \to 0$$

and

$$0 \to \mathrm{Hom}(C,M,) \to \mathrm{Hom}(B,M) \to \mathrm{Hom}(A,M) \xrightarrow{E^*}$$
$$\xrightarrow{E^*} \mathrm{Ext}(C,M) \xrightarrow{g^*} \mathrm{Ext}(B,M) \xrightarrow{f^*} \mathrm{Ext}(A,M) \to 0.$$

To the above homomorphism groups, it is possible to apply induced homomorphisms defined in Section 2 and the so-called connecting homomorphisms E_* and E^*.

We note that it is not necessary to know the exact action of the induced homomorphisms and the connecting homomorphisms. In the most cases, only the existence of the above exact sequences is used.

The extension module $\operatorname{Ext}(B, A)$ is the zero module if and only if every exact sequence $0 \to A \to M \to B \to 0$ is split (split sequences are defined in Section 1). It follows from this property and the results of Sections 5 and 6 that $\operatorname{Ext}(B, A) = 0$ for every module A (resp., B) if and only if B is a free module (resp., A is a divisible).

We recall some necessary results about pure-injective and complete modules (the theories of such modules are presented in Sections 10, 11). A module Q is pure-injective if and only if every exact sequence $0 \to Q \xrightarrow{i} M \to B \to 0$, where $\operatorname{Im}(i)$ is a pure submodule in M, is split. The pure-injective module Q is equal to $C \oplus D$, where C is a complete module and D is a divisible module. Complete modules coincide with pure-injective modules without elements of infinite height.

We often use the exact sequence

$$0 \to R \to K \to K/R \to 0,$$

where K is the field of fractions of the ring R and K/R is the quasicyclic module $R(p^\infty)$ (see Section 4). We also note that the module M is reduced if and only if $\operatorname{Hom}(K, M) = 0$.

We define a central notion of the section. A module M is called a *cotorsion* module if $\operatorname{Ext}(G, M) = 0$ for every torsion-free module G.

We begin with several simple results about cotorsion modules.

(1) If $\operatorname{Ext}(K, M) = 0$, then M is a cotorsion module.

Indeed, assume that $\operatorname{Ext}(K, M) = 0$ and G is an arbitrary torsion-free module. There exists an exact sequence $0 \to G \to D$, where D is a divisible torsion-free module. It is a direct sum of some number of copies of the module K. This provides an exact sequence $\operatorname{Ext}(D, M) \to \operatorname{Ext}(G, M) \to 0$. However,

$$\operatorname{Ext}(D, M) = \operatorname{Ext}(\bigoplus K, M) \cong \prod \operatorname{Ext}(K, M) = 0.$$

Therefore $\operatorname{Ext}(G, M) = 0$ and M is a cotorsion module.

(2) Any epimorphic image of a cotorsion module is a cotorsion module.

Let M be a cotorsion module and let $M \to N$ be an epimorphism. We have the exact sequence

$$0 = \operatorname{Ext}(K, M) \to \operatorname{Ext}(K, N) \to 0.$$

Therefore $\mathrm{Ext}(K, N) = 0$ and N is a cotorsion module by (1).

(3) If A is a submodule of the module M and A and M/A are cotorsion modules, then M is a cotorsion module.

The exact sequence $0 \to A \to M \to M/A \to 0$ provides an exact sequence

$$0 = \mathrm{Ext}(K, A) \to \mathrm{Ext}(K, M) \to \mathrm{Ext}(K, M/A) = 0.$$

Therefore, we obtain the required property.

(4) Let M be a reduced cotorsion module. A submodule A of the module M is a cotorsion module if and only if M/A is a reduced module.

It follows from the exact sequence in (3) that there exists the exact sequence

$$0 = \mathrm{Hom}(K, M) \to \mathrm{Hom}(K, M/A) \to \mathrm{Ext}(K, A) \to \mathrm{Ext}(K, M) = 0.$$

We obtain $\mathrm{Ext}(K, A) \cong \mathrm{Hom}(K, M/A)$. The module $\mathrm{Hom}(K, M/A)$ is equal to zero if and only if the module M/A is reduced.

The following properties follow from (2) and (4).

(5) If M is a reduced cotorsion module, then $\mathrm{Ker}(\alpha)$ and $\mathrm{Im}(\alpha)$ are cotorsion modules for every endomorphism α of the module M.

(6) If M is a reduced cotorsion module, then there is the canonical isomorphism $M \cong \mathrm{Ext}(K/R, M)$.

The exact sequence $0 \to R \to K \to K/R \to 0$ induces the exact sequence

$$0 = \mathrm{Hom}(K, M) \overset{i_M}{\to} \mathrm{Ext}(K/R, M) \to \mathrm{Ext}(K, M) = 0,$$

where i_M is the connecting homomorphism. We identify $\mathrm{Hom}(R, M)$ with M, similar to Proposition 2.2 (a). Then i_M provides the required isomorphism.

(7) A reduced cotorsion R-module M is an $\widehat{R}$-module, where $\widehat{R}$ is the completion of the ring R in the p-adic topology defined in Section 3.

By (6), we have $M \cong \mathrm{Ext}(K/R, M)$. The module K/R is an $\widehat{R}$-module (Section 4). According the beginning of the section, $\mathrm{Ext}(K/R, M)$ is also an $\widehat{R}$-module. We can use a somewhat different argument. Module multiplications by elements of $\widehat{R}$ are endomorphisms of K/R inducing endomorphisms of $\mathrm{Ext}(K/R, M)$ which are multiplications by elements of $\widehat{R}$.

We determine exact interrelations between cotorsion modules and pure-injective or complete modules.

Proposition 21.1. *A module is a cotorsion module if and only if it is an epimorphic image of a pure-injective module. A reduced module is a cotorsion module if and only if it is an epimorphic image of the complete torsion-free module. If a cotorsion module does not have elements of infinite height, then it is complete.*

Proof. Let Q be a pure-injective module. In every exact sequence $0 \to Q \to G \to K \to 0$, the image of the module Q is pure in G. By Theorem 10.3, the sequence is split. We can conclude that $\operatorname{Ext}(K, Q) = 0$ and Q is a cotorsion module. It follows from (2) that any epimorphic image of a pure-injective module is a cotorsion module.

Let M be a cotorsion module. We can assume that M is a reduced module. We have the exact sequence $0 \to M \to D \to D/M \to 0$, where D is the divisible hull of the module M. In this case, D/M is a primary module (Proposition 6.6 and 4.2). We have the exact sequence

$$\operatorname{Hom}(K/R, D/M) \to \operatorname{Ext}(K/R, M) \to \operatorname{Ext}(K/R, D) = 0.$$

By Theorem 13.1, $\operatorname{Hom}(K/R, D/M)$ is a complete torsion-free module. It remains to note that $\operatorname{Ext}(K/R, M) \cong M$ by (6). In addition, we also have proved the second assertion. Considering Proposition 11.16, we obtain also that if M is a module without elements of infinite height, then M is a complete module. □

Proposition 21.2. *For every module M, the module $\operatorname{Ext}(K/R, M)$ is a reduced cotorsion module and $\operatorname{Ext}(K/R, M) \cong \operatorname{Ext}(K/R, M/t(M)) \oplus \operatorname{Ext}(K/R, t(M))$.*

Proof. We can assume that M is a reduced. Similar to the previous proposition, we can obtain that $\operatorname{Ext}(K/R, M)$ is an epimorphic image of the complete torsion-free module $\operatorname{Hom}(K/R, D/M)$. By the same proposition, $\operatorname{Ext}(K/R, M)$ is a cotorsion module. We verify that it is a reduced module. By Lemma 10.1 (1), there exists an exact sequence $0 \to A \to B \to K/R \to 0$, where B is a direct sum of primary cyclic modules. Therefore A is a direct sum of primary cyclic modules. We have the induced exact sequence

$$0 \to \operatorname{Hom}(B, M) \to \operatorname{Hom}(A, M) \to \operatorname{Ext}(K/R, M) \to \operatorname{Ext}(B, M).$$

We describe the structure of the module $\operatorname{Ext}(B, M)$. Since B is a direct sum of primary cyclic modules, we have that $\operatorname{Ext}(B, M)$ is isomorphic to the product of modules of the form $\operatorname{Ext}(R(p^n), M)$. The sequence $0 \to R \overset{p^n}{\to} R \to R(p^n) \to 0$ is exact, where p^n is a multiplication by element p^n. We have that the sequence

$$\operatorname{Hom}(R, M) \overset{p^n}{\to} \operatorname{Hom}(R, M) \to \operatorname{Ext}(R(p^n), M) \to 0$$

is exact. Since $\operatorname{Hom}(R, M) \cong M$, we obtain $\operatorname{Ext}(R(p^n), M) \cong M/p^n M$. Therefore $\operatorname{Ext}(B, M)$ is the product of modules of the form $M/p^n M$. In particular, $\operatorname{Ext}(B, M)$ is a reduced module. Let X be the image of the homomorphism $\operatorname{Hom}(A, M) \to \operatorname{Ext}(K/R, M)$. It is sufficient to prove that X is a reduced module. We have an exact sequence

$$0 \to \operatorname{Hom}(B, M) \to \operatorname{Hom}(A, M) \to X \to 0.$$

The module $\operatorname{Hom}(B, M)$ is isomorphic to the product of modules of the form $\operatorname{Hom}(R(p^n), M)$ (see Section 2). Furthermore, we have $\operatorname{Hom}(R(p^n), M) \cong M[p^n]$ (this follows from Proposition 2.2 (a); $M[p^n]$ is an R_{p^n}-module). Therefore $\operatorname{Hom}(B, M)$ and $\operatorname{Hom}(A, M)$ are the products of modules of the form $M[p^n]$. Consequently, this reduced complete modules (Proposition 11.1). We consider the induced exact sequence

$$0 = \operatorname{Hom}(K, \operatorname{Hom}(A, M)) \to \operatorname{Hom}(K, X) \to \operatorname{Ext}(K, \operatorname{Hom}(B, M)).$$

Since $\operatorname{Hom}(B, M)$ is a complete module, the group Ext is equal to zero (Proposition 21.1). Thus, $\operatorname{Hom}(K, X) = 0$ and X is a reduced the module. Consequently, $\operatorname{Ext}(K/R, M)$ is a reduced module.

Now we take an exact sequence $0 \to t(M) \to M \to M/t(M) \to 0$ and the induced sequence

$$0 \to \operatorname{Ext}(K/R, t(M)) \to \operatorname{Ext}(K/R, M) \to \operatorname{Ext}(K/R, M/t(M)) \to 0.$$

We denote the torsion-free module $M/t(M)$ by the symbol G. We embed G in an exact sequence $0 \to G \to D \to D/G \to 0$, where D is a divisible hull of the module G. This provides the exact sequence

$$\begin{aligned} 0 &= \operatorname{Hom}(K/R, D) \to \operatorname{Hom}(K/R, D/G) \to \operatorname{Ext}(K/R, G) \to \\ &\to \operatorname{Ext}(K/R, D) = 0. \end{aligned}$$

Consequently, $\operatorname{Ext}(K/R, G) \cong \operatorname{Hom}(K/R, D/G)$. Since D/G is a divisible primary module, $\operatorname{Hom}(K/R, D/G)$ is a complete torsion-free module (see the proof of Proposition 21.1). By the proved property, $\operatorname{Ext}(K/R, t(M))$ is a cotorsion module. Therefore, the induced sequence is split which is required. □

We specialize the structure of the module $\operatorname{Ext}(K/R, M)$. A reduced cotorsion module without torsion-free nonzero direct summands is called an *adjusted* module (see Corollary 11.11 (1)).

Proposition 21.3. *If G is a torsion-free module, then* $\operatorname{Ext}(K/R, G)$ *is a complete torsion-free module. If T is a reduced primary module, then* $\operatorname{Ext}(K/R, T)$ *is an adjusted cotorsion module, the torsion submodule of the module* $\operatorname{Ext}(K/R, T)$ *is isomorphic to the module T, and the factor module with respect to the torsion submodule is a divisible torsion-free module.*

Proof. In fact, the case of the torsion-free module G is completely studied in Proposition 21.2. By the same proposition, $\operatorname{Ext}(K/R, T)$ is a reduced cotorsion module. We consider the induced exact sequence

$$0 \to \operatorname{Hom}(R, T) \to \operatorname{Ext}(K/R, T) \to \operatorname{Ext}(K, T) \to \operatorname{Ext}(R, T) = 0$$

(see (6)). If T is a bounded module, then $\mathrm{Ext}(K,T) = 0$ by Theorem 7.2. Therefore $\mathrm{Ext}(K/R,T) \cong \mathrm{Hom}(R,T) \cong T$. Let T be a nonbounded module. Since K is an R-K-bimodule, $\mathrm{Ext}(K,T)$ is a K-module, i.e., $\mathrm{Ext}(K,T)$ is a divisible torsion-free module. More explicitly, the multiplications of K by elements of K induce endomorphisms of the module $\mathrm{Ext}(K,T)$ which are multiplications by elements of K (see also (7)). It is also easy to prove that if B is a basic submodule of the module T, then $\mathrm{Ext}(K,T)$ is a direct summand of the module $\mathrm{Ext}(K,B)$, and $\mathrm{Ext}(K,B)$ is isomorphic to $\mathrm{Hom}(K,\widehat{B}/B)$. Therefore if T is not bounded, then $\mathrm{Ext}(K,T)$ is a divisible torsion-free module. Since $\mathrm{Hom}(R,T) \cong T$, we obtain the required property. Why is $\mathrm{Ext}(K/R,T)$ adjusted? We obtain that every torsion-free direct summand of the module $\mathrm{Ext}(K/R,T)$ is a divisible module. However, this is impossible, since $\mathrm{Ext}(K/R,T)$ is a reduced module. □

Now we can prove the main result about cotorsion modules.

Theorem 21.4 (Harrison [134]). (1) *Let M be a reduced cotorsion module. Then there exists a direct decomposition $M = C \oplus A$, where C is a complete torsion-free module and A is an adjusted cotorsion module which is uniquely determined by the module M.*

(2) *Two adjusted cotorsion modules are isomorphic to each other if and only if their torsion submodules are isomorphic to each other.*

Proof. (1) We set

$$C = \mathrm{Ext}(K/R, M/t(M)) \quad \text{and} \quad A = \mathrm{Ext}(K/R, t(M)).$$

The most part of the assertion follows from Propositions 21.2 and 21.3. The uniqueness of the submodule A follows from the following properties. Since $t(M) = t(A)$ and $A/t(M)$ is a divisible torsion-free module, $A/t(M)$ coincides with the maximal divisible submodule of the module $M/t(M)$.

(2) Let M and N be two adjusted cotorsion modules with $t(M) \cong t(N)$. Since $M/t(M)$ is a divisible module, it follows from Proposition 21.2 that $\mathrm{Ext}(K/R,M) \cong \mathrm{Ext}(K/R,t(M))$, and the same property holds for the module N. Using (6), we obtain that the relation

$$\begin{aligned} M &\cong \mathrm{Ext}(K/R,M) \cong \mathrm{Ext}(K/R,t(M)) \\ &\cong \mathrm{Ext}(K/R,t(N)) \cong \mathrm{Ext}(K/R,N) \\ &\cong N \end{aligned}$$

holds. □

It is possible to prove (1) without direct use of extension groups. By Proposition 21.1, there exist a complete torsion-free module G and a submodule H of G with $M \cong G/H$. Let E be the closure of the pure submodule H_* generated by the submodule H. Then $G = C \oplus E$ for some submodule C (Theorem 11.10). In addition, $G/H \cong C \oplus E/H$. We have

$$t(E/H) = H_*/H \qquad \text{and} \qquad (E/H)/t(E/H) \cong E/H_*,$$

where the last module is a divisible torsion-free module. We set $A = E/H$. Here A is an adjusted cotorsion module. Thus, $M \cong C \oplus A$, where C and A satisfy the conditions of the theorem. It is easy to verify that

$$C \cong \operatorname{Ext}(K/R, M/t(M)) \quad \text{and} \quad A \cong \operatorname{Ext}(K/R, t(M)).$$

We can give a category nature to (2). Precisely, we have that the category of reduced primary modules is equivalent to the category of adjusted cotorsion modules. The equivalence is realized by the functors $\operatorname{Ext}(K/R, -)$ and $t(-)$, where $t(-)$ is the functor of passage to the torsion submodule. It follows from Theorem 13.1 that the category of divisible primary modules and the category of complete torsion-free modules are equivalent. Considering Theorem 21.4, we can say about the equivalence of the category of primary modules and the category of reduced cotorsion modules. (This explains the name of these modules.)

For a reduced module M, we set $M^\bullet = \operatorname{Ext}(K/R, M)$. The module $M^\bullet$ is called the *cotorsion hull* of the module M. We obtain main properties of cotorsion hulls.

Again, we take exact sequences $0 \to R \to K \to K/R \to 0$ and

$$0 \to \operatorname{Hom}(R, M) \overset{i_M}{\to} \operatorname{Ext}(K/R, M) \to \operatorname{Ext}(K, M) \to 0,$$

where i_M is the connecting homomorphism. We will identify M with $\operatorname{Hom}(R, M)$ (see (6)). The module $\operatorname{Ext}(K, M)$ is a divisible torsion-free module (see the proof of Proposition 21.3). When it is required, the module M will be identified with the image of the monomorphism i_M. It directly follows from (6) and Proposition 21.2 that we have the following three properties:

(a) $M^\bullet$ is a reduced cotorsion module, $t(M) = t(M^\bullet)$, and $M^\bullet/M$ is a divisible torsion-free module.

(b) $M^{\bullet\bullet} = M^\bullet$.

(c) The relation $M = M^\bullet$ holds if and only if M is a cotorsion module.

Let $f\colon M \to N$ be a homomorphism of reduced modules. We denote by $f^\bullet$ the homomorphism $\operatorname{Ext}(K/R, M) \to \operatorname{Ext}(K/R, N)$ induced by f.

(d) The relation $i_M f^\bullet = f i_N$ holds and $f^\bullet$ is the unique homomorphism from $M^\bullet$ into $N^\bullet$ which has a similar property.

The relation is a familiar property of group homomorphisms and group extensions. If $g\colon M^\bullet \to N^\bullet$ is a homomorphism with $i_M g = f i_N$, then the difference $g - f^\bullet$ annihilates M. Since $M^\bullet/M$ is a divisible module, $g - f^\bullet = 0$ and $g = f^\bullet$.

The cotorsion hull is unique in some sense.

(e) If G is a cotorsion module such that $M \subseteq G$ and G/M is a divisible torsion-free module, then the identity mapping of the module M can be uniquely extended to an isomorphism from $M^\bullet$ onto G.

It follows from (d) that the embedding $j\colon M \to G$ can be uniquely extended to a homomorphism $j^\bullet\colon M^\bullet \to G^\bullet = G$. It follows from the exact sequence

$$0 = \operatorname{Hom}(K/R, G/M) \to M^\bullet \to G^\bullet \to \operatorname{Ext}(K/R, G/M) = 0$$

that $j^\bullet$ is an isomorphism.

(f) Let $M \subseteq N$, where N is a reduced module and N/M is a divisible torsion-free module. Then the identity mapping of the module M can be uniquely extended to an isomorphism $M^\bullet \cong N^\bullet$.

We obtain (f) from (e) if we take $G = N^\bullet$.

In the next section, we will need a result about the first Ulm submodule of the cotorsion hull (they were defined in Section 11). Ulm submodules and Ulm factors of cotorsion modules are explicitly studied in the paper of Harrison [135] and the book of Fuchs ([92, Section 56]). In particular, the following proposition shows interesting interrelations between the cotorsion hull and the completion.

Proposition 21.5. *Let M be a module without elements of infinite height. Then $(M^\bullet)^1$ is a complete torsion-free module and there exists an isomorphism $(M^\bullet)^1 \cong \operatorname{Hom}(K/R, \widehat{M}/M)$. In addition, $M^\bullet/(M^\bullet)^1 \cong \widehat{M}$.*

Proof. We have

$$\operatorname{Ext}(K/R, M) = M^\bullet \qquad \text{and} \qquad \operatorname{Ext}(K/R, \widehat{M}) \cong \widehat{M}^\bullet.$$

Since $\widehat{M}$ is a cotorsion module, $\widehat{M}^\bullet = \widehat{M}$. The exact sequence $0 \to M \to \widehat{M} \to \widehat{M}/M \to 0$ induces the exact sequence

$$0 \to \operatorname{Hom}(K/R, \widehat{M}/M) \xrightarrow{j} M^\bullet \to \widehat{M} \to 0, \tag{4.1}$$

where j is the connecting homomorphism. We prove that $\operatorname{Im}(j) = (M^\bullet)^1$. Since the module $\widehat{M}$ does not have elements of infinite height, $(M^\bullet)^1 \subseteq \operatorname{Im}(j)$. The proof of the converse inclusion is more difficult. For every positive integer n, we have one more induced exact sequence

$$\mathrm{Hom}(R(p^n), \widehat{M}) \to \mathrm{Hom}(R(p^n), \widehat{M}/M) \xrightarrow{i} \mathrm{Ext}(R(p^n), M),$$

where i is the connecting homomorphism. Any cyclic module is pure-projective by Theorem 10.2. Therefore, the first mapping is a surjection and $i = 0$. There exists an exact sequence $R(p^n) \to K/R \to K/R$, where the second mapping is the multiplication by p^n. We obtain the induced homomorphism $f\colon \mathrm{Hom}(K/R, \widehat{M}/M) \to \mathrm{Hom}(R(p^n), \widehat{M}/M)$ and the induced exact sequence $\mathrm{Ext}(K/R, M) \to M^\bullet \xrightarrow{g} \mathrm{Ext}(R(p^n), M)$ where the first mapping is the multiplication by p^n. Consequently, $\mathrm{Ker}(g) = p^n M^\bullet$. In addition, $jg = fi = 0$. Therefore $\mathrm{Im}(j) \subseteq p^n M^\bullet$ for every n and $\mathrm{Im}(j) \subseteq (M^\bullet)^1$. Therefore $\mathrm{Im}(j) = (M^\bullet)^1$. We obtain $(M^\bullet)^1 \cong \mathrm{Hom}(K/R, \widehat{M}/M)$. Since $\widehat{M}/M$ is a divisible module, $\mathrm{Hom}(K/R, \widehat{M}/M)$ is a complete torsion-free module. Considering the exactness of (4.1), we obtain $\widehat{M} \cong M^\bullet/(M^\bullet)^1$. □

Exercise 1. (a) Prove that if M is a pure-injective (resp., cotorsion) module, then $\mathrm{Hom}(A, M)$ is a pure-injective module (resp., a cotorsion module) for every module A.

(b) If T is a primary module, then $\mathrm{Hom}(T, A)$ is a complete module for every module A.

Exercise 2. Prove that the first Ulm submodule of a cotorsion module is a cotorsion module.

Exercise 3. A primary module is a cotorsion module if and only if it is a direct sum of a divisible module and a bounded module.

Exercise 4. For any two modules M and N, the module $\mathrm{Ext}(M, N)$ is a cotorsion module.

Exercise 5. Let M and N be two adjusted cotorsion modules. Every homomorphism from $t(M)$ into $t(N)$ can be uniquely extended to a homomorphism from M in N. This leads to the canonical isomorphism $\mathrm{Hom}(M, N) \cong \mathrm{Hom}(t(M), t(N))$.

Exercise 6. A reduced cotorsion module M is an adjusted module if and only if all basic submodules of M are primary.

22 Embedding of category of torsion-free modules in category of mixed modules

It is difficult to characterize the endomorphism ring of a mixed module "as a whole". It is not accidental that existing realization theorems for endomorphism rings of mixed modules are split realization theorems. They are very similar to some assertion from Section 19 (e.g., see Theorems 19.5–19.7). For a given torsion-free R-algebra A, it is constructed a mixed module M with the preassigned torsion submodule $t(M)$ such that $M/t(M)$ is a divisible module and $\mathrm{End}(M)$ is a split extension of the ring A with the use of the ideal $\mathrm{Hom}(M, t(M))$. Therefore $\mathrm{End}(M) = A \oplus \mathrm{Hom}(M, t(M))$ (the direct sum of groups). There are similar realizations in the work [55] of Dugas.

Another promising approach to the considered problem is to obtain and use some functors from the category of torsion-free modules into the category of mixed modules. There are such functors in the works [105] of Göbel–May and [88] of Franzen–Goldsmith. They allow to extend various results from the category of torsion-free modules to the category of mixed modules. For example, we can automatically obtain split realizations for endomorphism rings of mixed modules. The existence of these functors demonstrates a close relation to the "torsion-free case".

As before, R is a commutative discrete valuation domain which is not necessarily complete. The symbols X and Y denote some reduced torsion-free R-modules.

For every module X, we set $X_0 = \bigoplus_{i\geq 1} X/p^iX$. Every homomorphism $\varphi\colon X \to Y$ induces the homomorphism $\varphi_0\colon X_0 \to Y_0$ which acts on the summand X/p^iX by the relation $\varphi_0(a + p^iX) = \varphi(a) + p^iY$, $a \in X$. The module $\prod_{i\geq 1} X/p^iX$ is complete, since it is the product of bounded modules. Since the module X_0 does not have elements of infinite height, there exists the completion $\widehat{X}_0$. The submodule X_0 is pure in the product $\prod_{i\geq 1} X/p^iX$. Therefore, we can assume that the completion $\widehat{X}_0$ is contained in this product.

For every element $a \in X$ and any positive integer k, we define the sequence

$$v_a^k = (0, \ldots, 0, a + p^kX, pa + p^{k+1}X, \ldots)$$

with zeros in the first $k-1$ places. It is clear that $v_a^k \in \widehat{X}_0$ and $pv_a^{k+1} - v_a^k \in X_0$. Consequently, the elements $v_a^k + X_0$ with $k \geq 1$ generate in $\widehat{X}/X_0$ the submodule which is isomorphic to the quasicyclic module K/R (see Section 4). Therefore, for the given element a there exists the unique homomorphism $f_a\colon K/R \to \widehat{X}_0/X_0$ such that $f_a(1/p^k + R) = v_a^k + X_0$ for all $k \geq 1$. Now we can define the mapping $f\colon X \to \mathrm{Hom}(K/R, \widehat{X}_0/X_0)$ by the relation $f(a) = f_a$, $a \in X$. It is a homomorphism.

Using Proposition 21.5, we identify $\operatorname{Hom}(K/R, \widehat{X}_0/X_0)$ with $(X_0^\bullet)^1$, where $(X_0^\bullet)^1$ is the first Ulm submodule of the cotorsion hull $X_0^\bullet$ of the module X_0. Using this identification, we obtain the homomorphism $f\colon X \to (X_0^\bullet)^1$. We write f_X instead of f if it is not clear which module is associated with f.

Lemma 22.1. *The mapping f is a monomorphism and the image of f is a pure submodule in $(X_0^\bullet)^1$.*

Proof. Let a be a nonzero element of the module X. We have $a \notin p^k X$ for some k. Consequently, $v_a^k \notin X_0$ and $f_a(1/p^k + R) \neq 0$. We obtain that f is a monomorphism. We assume that $f_a = p^n g$ for some positive integer n and homomorphism $g \in \operatorname{Hom}(K/R, \widehat{X}_0/X)$. Then $f_a(1/p^n + R) = 0$ (see Example 12.5). It follows from the first part of the proof that $a \in p^n X$ and $a = p^n c$, where $c \in X$. Therefore $f_a = p^n f_c$ and we obtain the required purity. □

We set $G(X) = \{\alpha \in X_0^\bullet \mid p^n\alpha \in f(X) \text{ for some positive integer } n\}$. Here $G(X)$ is a pure submodule in $X_0^\bullet$ containing X_0.

We assume that there is an arbitrary homomorphism $\varphi\colon X \to Y$. Starting from φ, we define several compatible homomorphisms. First, the homomorphism $\varphi_0\colon X_0 \to Y_0$ induces the homomorphism $\varphi_0^\bullet\colon X_0^\bullet \to Y_0^\bullet$ of cotorsion hulls (see the previous section). We also have the induced homomorphism of completions $\widehat{\varphi}_0\colon \widehat{X}_0 \to \widehat{Y}_0$. It is important that $\widehat{\varphi}_0$ acts coordinate-wise on every sequence in $\widehat{X}_0$. Indeed, $\widehat{\varphi}_0$ coincides on $\widehat{X}_0$ with the homomorphism $\bar{\varphi}\colon \prod_{i\geq 1} X/p^iX \to \prod_{i\geq 1} Y/p^iY$, where

$$\bar{\varphi}\colon (a_1 + pX, a_2 + pX, \dots) \to (\varphi a_1 + pX, \varphi a_2 + pX, \dots)$$

for all $a_i \in X$, $i \geq 1$. Therefore, it is clear that $\widehat{\varphi}_0$ satisfies the relation $\widehat{\varphi}_0(v_a^k) = v_{\varphi(a)}^k$ for all a and k. We also note that $\widehat{\varphi}_0$ induces the homomorphism $\widehat{\varphi}_0\colon \widehat{X}_0/X_0 \to \widehat{Y}/Y_0$. Now it follows from the proof of Proposition 21.5 that the homomorphism $\operatorname{Hom}(K/R, \widehat{X}_0/X_0) \to \operatorname{Hom}(K/R, \widehat{Y}/Y_0)$ induced by $\widehat{\varphi}_0$ coincides with $\varphi_0^\bullet$. Using all above arguments, we can prove the relation $\varphi f_Y = f_X \varphi_0^\bullet$. Therefore, we obtain that if we take the restriction homomorphism $\varphi_0^\bullet$ to $G(X)$, then we obtain the homomorphism $\varphi_0^\bullet\colon G(X) \to G(Y)$.

We define a covariant functor G from the category of reduced torsion-free modules into the category of mixed modules. We assume that $G\colon X \to G(X)$ and $G\colon \varphi \to \varphi_0^\bullet$ for every module X and each homomorphism $\varphi\colon X \to Y$. It is easy to verify that G is an additive functor (categories and functors are defined in Section 1).

We present main properties of the functor G.

Theorem 22.2 (Göbel–May [105]). *For every reduced torsion-free module X, $G(X)$ is a reduced mixed module. The torsion submodule of X is equal to X_0, and the factor module $G(X)/X_0$ is isomorphic to the divisible hull of the module X. For any two reduced torsion-free modules X and Y, the following assertions hold.*

(1) *If we identify X with $f(X)$, then $G(X)^1 = X$.*

(2) *The natural mapping $\mathrm{Hom}(X,Y)$ in $\mathrm{Hom}(G(X),G(Y))$ is a monomorphism. If we identify $\mathrm{Hom}(X,Y)$ with the image of it in $\mathrm{Hom}(G(X),G(Y))$, then we obtain the decomposition $\mathrm{Hom}(G(X),G(Y)) = \mathrm{Hom}(X,Y) \oplus \mathrm{Hom}(G(X),Y_0)$.*

(3) *It follows from the inclusion $X \subseteq \widehat{X}$ that $X_0 = \widehat{X}_0$ and $\mathrm{Hom}(G(\widehat{X}),Y_0) = \mathrm{Hom}(G(X),Y_0)$.*

Proof. By construction, $G(X)$ is a pure submodule in $X_0^\bullet$ containing X_0. However, the torsion submodule of the module $X_0^\bullet$ is equal to X_0 (Proposition 21.3). Therefore $t(G(X)) = X_0$. Since $G(X) \subseteq X_0^\bullet$, we have that $G(X)$ is a reduced module. Since $f(X) \subseteq (X_0^\bullet)^1$, the module $f(X)$ has elements of infinite order. Consequently, $G(X)$ is a mixed module. We also note that $X_0 \cap (X_0^\bullet)^1 = 0$, since $(X_0^\bullet)^1$ is a torsion-free module by Proposition 21.5. Let D be the divisible hull of the module X. Since $X_0 \cap f(X) = 0$, there exists a homomorphism $\alpha\colon G(X) \to D$ which annihilates X_0 and coincides on $f(X)$ with the inverse homomorphism for f. The module $G(X)/(X_0 + f(X))$ is primary. Therefore $\mathrm{Ker}(\alpha) = X_0$. We obtain that $\mathrm{Im}(\alpha)$ is a divisible module. By properties of divisible hulls, $\mathrm{Im}(\alpha) = D$. Thus, $G(X)/X_0 \cong D$.

(1) We assume that $X = f(X)$. Then

$$X \subseteq G(X) \cap (X_0^\bullet)^1 = G(X)^1.$$

Let $y \in G(X)^1$. Then $p^n y \in X$ for some n. Consequently, $p^n y = p^n x$, $x \in X$ by Lemma 22.1. Therefore $y - x \in G(X)^1 \cap X_0 = 0$ and $y \in X$.

(2) Under the natural mapping from $\mathrm{Hom}(X,Y)$ into $\mathrm{Hom}(G(X),G(Y))$, the homomorphism φ corresponds to the homomorphism $\varphi_0^\bullet$. By construction, $\varphi_0^\bullet \neq 0$ for $\varphi \neq 0$. According to (1), every homomorphism $\psi\colon G(X) \to G(Y)$ induces the homomorphism $\xi\colon X \to Y$. It follows from $\psi = \xi + (\psi - \xi)$ that $\mathrm{Hom}(G(X),G(Y)) = \mathrm{Hom}(X,Y) \oplus \Lambda$, where Λ consists of all mappings annihilating X. Let $\mu \in \mathrm{Hom}(G(X),Y_0)$. It follows from $X = G(X)^1$ that $\mu X \subseteq Y_0 \cap Y = 0$. Therefore $\mu \in \Lambda$. Let $\lambda \in \Lambda$. Since $G(X)/X$ is a primary module, $\lambda(G(X))$ is also a primary module. Consequently, $\lambda \in \mathrm{Hom}(G(X),Y_0)$. As a result, we obtain $\Lambda = \mathrm{Hom}(G(X),Y_0)$.

(3) A basic submodule B of the module X is a basic submodule of the comple-

tion $\widehat{X}$. By property (2) from Section 9, we have

$$X/p^iX \cong B/p^iB \cong \widehat{X}/p^i\widehat{X}$$

for every $i \geq 1$. Therefore

$$X_0 = \widehat{X}_0 \qquad \text{and} \qquad G(X) \subseteq G(\widehat{X}).$$

We take the mapping $h\colon \operatorname{Hom}(G(\widehat{X}), Y_0) \to \operatorname{Hom}(G(X), Y_0)$ which maps the homomorphism $G(\widehat{X})$ in Y_0 into the restriction of it to $G(X)$. We verify that h is an isomorphism. Since $G(\widehat{X})/G(X)$ is a divisible module, h is a monomorphism. Let $\mu \in \operatorname{Hom}(G(X), Y_0)$. Then $\mu(X) = 0$ (see (2)). The restriction of μ to X_0 can be extended to a homomorphism $\mu^\bullet\colon X_0^\bullet \to Y_0^\bullet$. In addition, the restriction of $\mu^\bullet$ to $G(X)$ coincides with μ. Consequently, $\mu^\bullet\widehat{X} = 0$. Let ν be the restriction $\mu^\bullet$ to $G(\widehat{X})$. Then

$$\nu \in \operatorname{Hom}(G(\widehat{X}), Y_0) \quad \text{and} \quad h(\nu) = \mu.$$

We can define ν in another manner. We have $G(X) \cap \widehat{X} = X$ and $G(\widehat{X}) \subseteq G(X) + \widehat{X}$. Let $y \in G(\widehat{X})$ and $y = x + c$, where $x \in G(X)$, $c \in \widehat{X}$. We assume that $\nu(y) = \mu(x)$. Then $h(\nu) = \mu$. Thus, h is an isomorphism and two considered homomorphism groups can be identified with the use of h. □

We specialize the theorem and obtain corollaries for endomorphism rings.

Corollary 22.3. *There exists an additive functor G from the category of reduced torsion-free modules into the category of reduced mixed modules such that the following properties hold.*

(1) *If X is a reduced torsion-free module, then $G(X)/t(G(X))$ is a divisible torsion-free module and*

$$\operatorname{End}(G(X)) = \operatorname{End}(X) \oplus \operatorname{Hom}(G(X), t(G(X))),$$

i.e., the ring $\operatorname{End}(G(X))$ is a split extension of the ring $\operatorname{End}(X)$.

(2) *If some reduced torsion-free modules X_i with $i \in I$ are pure dense submodules of the same complete module and $\operatorname{End}(X_i) = \operatorname{End}(X_j)$ for all $i, j \in I$, then $\operatorname{End}(G(X_i)) = \operatorname{End}(G(X_j))$ for all $i, j \in I$.*

Proof. Only the proof of (2) requires some explanation. We obtain that all modules X_i have the same completion. It follows from Theorem 22.2 (3) that $\operatorname{Hom}(G(X_i), t(G(X_i))) = \operatorname{Hom}(G(X_j), t(G(X_j)))$ for any two subscripts $i, j \in I$. We also note that all modules X_i are contained in the same cotorsion hull. All considered endomorphisms and homomorphisms are endomorphisms of this hull. Now it is clear, the sense in which we consider the relation and the inclusion for End and Hom. □

Using the functor G, we can extend various properties of torsion-free modules X stated in terms of their endomorphism rings to mixed modules $G(X)$ (for example, see Corollary 22.5). Thus, realization theorems for torsion-free modules provide split realization theorems for mixed modules.

A mixed module M is said to be *essentially indecomposable* if for every direct decomposition $M = A \oplus B$ of M, one of the summands A or B is a primary module (cf., the definition of an essentially indecomposable torsion-free module before Corollary 20.1).

Corollary 22.4. *There exists a reduced essentially indecomposable mixed module M such that $t(M) = \bigoplus_{i \geq 1} R(p^i)$ and $M/t(M) \cong K$.*

Proof. We take the module R as the module X from Corollary 22.3. We set $M = G(R)$. Since $\operatorname{End}(R) \cong R$, we obtain that $\operatorname{End}(M) = R \oplus \operatorname{Hom}(M, t(M))$. The remaining argument is similar to the proof of Corollary 20.1. □

Sometimes it is more convenient to use the so-called *Walker category* denoted by Walk instead of the ordinary category of modules. The objects of the category Walk are modules and the set of morphisms from a module M into a module N is the set

$$\operatorname{Hom}_{\overline{W}}(M, N) = \operatorname{Hom}(M, N) / \operatorname{Hom}(M, t(N)).$$

This category is considered more explicitly in Section 29. In particular, it is shown that Walk is an additive category. Let $\operatorname{End}_{\overline{W}}(M) = \operatorname{End}(M)/\operatorname{Hom}(M, t(M))$ be the endomorphism ring of the module M in this category.

For arbitrary reduced torsion-free modules X and Y, we set

$$\bar{G}(X) = G(X) \qquad \text{and} \qquad \bar{G}(\varphi) = G(\varphi) + \operatorname{Hom}(G(X), t(G(Y)))$$

for a homomorphism $\varphi \colon X \to Y$. We obtain the functor $\bar{G}$ from the category of reduced torsion-free modules into the Walker category of reduced mixed modules. Theorem 22.2 states that $\bar{G}$ is an embedding, i.e., the category of reduced torsion-free modules is equivalent to some full subcategory of the Walker category of reduced mixed modules. In particular, $\operatorname{End}_{\overline{W}}(G(X)) \cong \operatorname{End}(X)$.

Franzen and Goldsmith have constructed another functor H from the category of torsion-free modules into the category of mixed modules. The functors G and H are naturally equivalent. This follows from the property that each of the functors is equivalent to some functor of the tensor product (details are given in [105, 88] and Exercise 1).

For mixed modules, there exist analogues of the notions of a rigid system and an A-rigid system of modules considered in Section 20. The set of mixed modules $\{M_i\}_{i \in I}$ is usually called a rigid system if $\operatorname{Hom}(M_i, M_j) = \operatorname{Hom}(M_i, t(M_j))$ for any two distinct subscripts $i, j \in I$. This is equivalent to the property that

$\mathrm{Hom}_{\overline{W}}(M_i, M_j) = 0$ in the category Walk. It is useful to extend this notion with the use of the method from Section 20.

Let A be a torsion-free R-algebra. A set of mixed modules $\{M_i\}_{i\in I}$ is called an *A-rigid system* if $\mathrm{End}(M_i) \cong A \oplus \mathrm{Hom}(M_i, t(M_i))$ (a split extension) and $\mathrm{Hom}(M_i, M_j) = \mathrm{Hom}(M_i, t(M_j))$ for any two distinct $i, j \in I$. In addition, if there exists a ring H (without identity element) which is an A-A-bimodule, such that

$$\mathrm{End}(M_i) \cong A \oplus H \qquad \text{and} \qquad \mathrm{Hom}(M_i, M_j) \cong H$$

for any two distinct $i, j \in I$, then we call such system *A-H-rigid*.

Applying the functor G to rigid systems of torsion-free modules, we can obtain rigid systems of mixed modules with prescribed properties. For example, we take one of the rigid systems of torsion-free modules from Theorem 19.10. Precisely, assume that we take an R-algebra A, the numbers s and d from Theorem 19.10, and the corresponding A-rigid system $\{X_i\}_{i\in I}$ of torsion-free modules. We have a system $\{G(X_i)\}_{i\in I}$ of reduced mixed modules which satisfy assertions (1)–(3) of Theorem 22.2.

We recall that if Y is a module, then Y_0 denotes the sum $\bigoplus_{i\geq 1} Y/p^iY$. We also note that $\widehat{\bigoplus_s A} \cong \bigoplus_s \widehat{A}$. We set $H = \mathrm{Hom}(G(\bigoplus_s \widehat{A}), \bigoplus_s A_0)$. By Theorem 22.2, the torsion submodule of the module $G(\bigoplus_s \widehat{A})$ is equal to $\bigoplus_s A_0$. Consequently, H is a ring and an A-A-bimodule.

Corollary 22.5. *$\{G(X_i)\}_{i\in I}$ is an A-H-rigid system of reduced mixed modules. In addition, $t(G(X_i)) \cong \bigoplus_s A_0$ and the factor module $G(X_i)/t(G(X_i))$ is isomorphic to the divisible hull of the module $\bigoplus_d A$.*

Proof. It follows from Theorem 22.2 that

$$\mathrm{Hom}(G(X_i), (X_j)_0) = \mathrm{Hom}(G(\widehat{X}_i), (X_j)_0).$$

Since $B \subseteq X_j \subseteq \widehat{B}$ and X_j is a pure submodule in $\widehat{B}$, we have $(X_j)_0 = B_0 = \bigoplus_s A_0$. Furthermore, we have $G(\widehat{X}_i) = G(\widehat{B}) = G(\bigoplus_s A)$. Thus, $\mathrm{Hom}(G(X_i), (X_j)_0) = H$ for all subscripts $i, j \in I$. Now the property that $\{G(X_i)\}_{i\in I}$ is an A-H-rigid system follows from Theorem 22.2. By the same theorem, $t(G(X_i)) = (X_i)_0 = \bigoplus_s A_0$ and $G(X_i)/t(G(X_i))$ is isomorphic to the divisible hull of the module X_i. This divisible hull is isomorphic to the divisible hull of the module $\bigoplus_d A$. □

In particular, we obtain that any two modules $G(X_i)$ and $G(X_j)$ are not isomorphic to each other for $i \neq j$, but their endomorphism rings are isomorphic to each other.

As the ring R, we take the ring $\mathbb{Z}_p$ of all rational numbers whose denominators are coprime to a given prime integer p. A $\mathbb{Z}_p$-module is called a *p-local* (Abelian) group (see Section 4). We also set $s = 1$ and $d = 2$. We obtain the following result.

Corollary 22.6. *There exists a continual set of non-isomorphic reduced mixed p-local groups G_i with isomorphic endomorphism rings such that*

$$t(G_i) \cong \bigoplus_{i \geq 1} \mathbb{Z}(p^i) \quad \textit{and} \quad G_i/t(G_i) \cong \mathbb{Q} \oplus \mathbb{Q}$$

for all i, where $\mathbb{Z}(p^i)$ is the cyclic group of order p^i and $\mathbb{Q}$ is the field of rational numbers.

Exercise 1 (Göbel–May [105]). There exists another way of the construction of the functor G from Theorem 22.2. We take the mixed module $A = G(R)$. Prove that the functors G and $A \otimes (-)$ are naturally equivalent. Here we have $(A \otimes (-))X = A \otimes_R X$ for a reduced torsion-free module X and

$$(A \otimes (-))\varphi = 1 \otimes \varphi \colon A \otimes_R X \to A \otimes_R Y$$

for homomorphism $\varphi \colon X \to Y$ of reduced torsion-free modules X and Y.

Exercise 2. Let t be a positive integer. There exists a mixed module M such that for any two positive integers m and n, we have

$$M^m \cong M^n \qquad \Longleftrightarrow \qquad m \equiv n \pmod t$$

(cf., Exercise 1 in Section 20).

Remarks. Every ring R is the endomorphism ring of some module (e.g., $_RR$). However, the question about the representations of rings by endomorphism rings becomes very difficult if we either fix the ring, whose modules are considered, or consider only special classes of modules. By the familiar Wedderburn–Artin theorem, the ring S is isomorphic to the operator ring of a finite-dimensional vector space over the division ring if and only if S is a simple Artinian ring. Wolfson [318] described linear operator rings of vector spaces of any dimension. The works [249] of Pierce and [205] of Liebert contain characterizations of endomorphism rings of Abelian p-groups without elements of infinite height. It is interesting that Theorem 18.4 resembles these characterizations. This emphasizes similarity between primary modules and torsion-free modules over a complete domain.

In [212] Liebert have characterized endomorphism rings of simply presented p-groups. Franzsen and Schultz [89] and Liebert [211] considered the characterization problem for endomorphism rings of some free or locally free modules. It seems that there were not serious attempts to describe endomorphism rings of some mixed modules.

It is easy to indicate the moment, when realization theorems appeared. In 1963, Corner [44] has proved that every countable reduced torsion-free ring is isomorphic to the endomorphism ring of some countable reduced Abelian torsion-free group. In another paper [46] Corner has first proved a split realization theorem for endomorphism rings of Abelian p-groups. Before we formulate the theorem, we define the ideal of small endomorphisms. For primary modules, this ideal plays the role which coincides with the role played by the ideal $\mathrm{Ines}(M)$ for torsion-free modules. A homomorphism φ between Abelian p-groups G and G' is said to be *small* if for every positive integer e, there exists a positive integer n such that $\varphi((p^n G)[p^e]) = 0$ (Pierce [248]). All small homomorphisms from G into G' form the group $\mathrm{Small}(G, G')$. All small endomorphisms of the group G form the ideal $\mathrm{Small}(G)$ of the ring $\mathrm{End}(G)$, and the factor ring $\mathrm{End}(G)/\mathrm{Small}(G)$ is a complete $\widehat{\mathbb{Z}}_p$-algebra (Pierce [248]). Corner has proved the following result. Let A be a complete torsion-free $\widehat{\mathbb{Z}}_p$-algebra of p-rank $\leq\aleph_0$. Then there exists an Abelian p-group G without elements of infinite height such that $\mathrm{End}(G)$ is a split extension of the ring A with the use of the ideal $\mathrm{Small}(G)$; i.e., $\mathrm{End}(G) \cong A \oplus \mathrm{Small}(G)$.

Within about 20 years, Dugas and Göbel [57, 58] removed the countability assumption in the above-considered theorem of Corner [44] about torsion-free rings (see Theorem 19.8). We note that Dugas and Göbel considered cotorsion-free modules and rings. As for the cited theorem of Corner [46] about endomorphism rings of p-groups, we note that Dugas and Göbel have obtained a generalization of this theorem in another paper [59] (also without restrictions to cardinalities). In [59] and other works, "large" rigid systems of essentially indecomposable p-groups were constructed. An Abelian p-group is said to be *essentially indecomposable* if every nontrivial decomposition of the group into a direct sum of two summands has at least one bounded summand (cf., the definitions of essentially indecomposable torsion-free modules and mixed modules in Sections 20 and 22). For every group G in such a system, we have $\mathrm{End}(G) \cong A \oplus \mathrm{Small}(G)$, where A is a preassigned complete $\widehat{\mathbb{Z}}_p$-algebra. Every homomorphism between two distinct elements of the system is small (cf., Theorem 20.2 and Corollary 22.5). It follows from Section 19 that for torsion-free modules, there exist split realization theorems and "complete" realization theorems. The situation with mixed modules is considered in Section 22.

In 1980th, the study of realization theorems was the main part of the theory of endomorphism rings. Many results for modules over some commutative rings were obtained; in particular, many results for Abelian groups and modules over discrete valuation domains were obtained. As a result powerful remarkable theorems has been proved. The review paper of Göbel [98] and the paper of Corner and Göbel [49] contain the history of the question, methods, applications, and the bibliography. In the second work, a unified approach to realization theorems is developed; it unites the "primary case", the "torsion-free case", and the "mixed case". The use of inessential homomorphisms plays a crucial role in this method. The theory of "inessentiality" is developed. Inessential homomorphisms of primary modules coincide with small homomorphisms. For mixed modules, inessential homomorphisms coincide with homomorphisms $\varphi\colon M \to N$ such that $\varphi M \subseteq t(N)$ or φM is a bounded module.

The book of Göbel–Trlifaj [109] contains quite general realization theorems for cotorsion-free algebras and some algebras close to cotorsion-free algebras. In this book, necessary set-theoretical methods are also developed.

In fact, realizations for endomorphism rings obtained in several works are topological (e.g., see [45, 61, 49]). Precisely, a topological ring (or an algebra) is topologically isomorphic to the endomorphism ring of a certain module (as a topological ring in the finite topology) or it is topologically embeddable in such an endomorphism ring.

In other works, rings are realized as endomorphism rings of modules which are "almost free" in some sense. This requires the use of additional axioms of the set theory. Such realizations can be found in the works [102, 104] and other works (see also the books of Göbel–Trlifaj [109], Eklof [63] and Eklof–Mekler [65]).

There are many papers, where representations of rings by endomorphism rings of modules with distinguished submodules are studied. This field is considered in the review paper [99] and the book [109].

Realization theorems have important applications to many module-theoretical questions. Examples of such questions are test problems of Kaplansky and their generalizations (see Exercise 1 in Section 20 and Exercise 2 in Section 22), pathologies in direct decompositions, the existence of large (essentially) indecomposable modules, and the existence of "very" decomposable modules. The work [49, Section 9] of Corner and Göbel contains details about this topic and a guide to the literature. The book of Göbel–Trlifaj [109] contains results related to the topic.

Problem 7. Obtain characterizations of endomorphism rings of torsion-free modules over a complete discrete valuation domain which are more simple than the characterizations obtained in Theorem 18.4.

Problem 8. Prove characterization theorems for endomorphism rings of some mixed modules (of Warfield modules, of modules M such that $t(M)$ does not have elements of infinite height and $M/t(M)$ is a divisible module).

Problem 9. (a) Present quite large classes of torsion-free modules M such that the ring $\mathrm{End}(M)$ is a split extension of some torsion-free R-algebra with the use of the ideal $\mathrm{Ines}(M)$ or $\mathrm{Fin}(M)$.

(b) Present quite large classes of mixed modules M such that $\mathrm{End}(M)$ is a split extension of some torsion-free R-algebra by the ideal $\mathrm{Hom}(M, t(M))$ or $t(\mathrm{End}(M))$.

A module M over the ring S is called a module *with unique addition* or a *UA-module* if it is impossible to define on M another addition such that M is a module over S with the same multiplication by elements of S (see [234]). It is not very difficult to describe UA-modules M over a discrete valuation domain R. It is interesting to consider modules obtained from M.

Problem 10. (a) Characterize R-modules M such that there exists only a finite number of addition operations on M satisfying the property that M is an R-module with respect to the given module multiplication.

(b) Which R-modules are UA-modules over their endomorphism rings or over the centers of their endomorphism rings?

(c) Describe modules M such that the K-space KM is a UA-module over the quasiendomorphism ring $K\,\mathrm{End}(M)$ (these K-spaces are introduced in Chapter 5).

(d) What can we say about mixed modules M such that $M/t(M)$ is a UA-module over the endomorphism ring $\mathrm{End}_{\overline{W}}(M)$ of the module M in the category Walk?

Chapter 5

Torsion-free modules

In Chapter 5, we consider the following topics:

- elementary properties of torsion-free modules (Section 23);
- the category of quasihomomorphisms (Section 24);
- purely indecomposable and copurely indecomposable modules (Section 25);
- indecomposable modules over Nagata valuation domains (Section 26).

Chapter 5 is devoted to torsion-free modules that form one of the three main classes of modules. Two other classes are the class of primary modules and the class of mixed modules (see Section 4). Features of the studies of primary modules are mentioned in Introduction and mixed modules are studied in the next chapter. Modules of finite rank are mainly considered. Only the case of an incomplete discrete valuation domain R is interesting, since every torsion-free module of finite rank over a complete domain is free (Corollary 11.7).

There is an essential distinction between primary modules and torsion-free modules. As a rule, primary modules have a quite large (in a certain sense) number of endomorphisms (see Corollary 7.5 and Corollary 7.6 and Proposition 37.5). On the contrary, torsion-free modules very often have few endomorphisms and the endomorphism ring weakly affects the original module in the general case (for example, in view of the topics considered in Chapter 7). Any indecomposable primary module is a module of rank 1 (Corollary 11.8), whereas indecomposable torsion-free modules can have arbitrarily large finite ranks (see Example 11.9 and remarks at the end of Section 26). As for direct decompositions, the existence of large indecomposable torsion-free modules is the main difference between torsion-free modules and primary modules. However, direct decompositions of primary or torsion-free modules (even in the case of torsion-free modules of finite rank) can have various unexpected properties (see remarks to Chapter 4 and Chapter 5; see also Sections 75, 90, and 91 of the book of Fuchs [93]).

In Section 23 we present several standard properties of torsion-free modules. In Section 24, we consider the category of torsion-free modules; morphisms of this category are called quasihomomorphisms. This category differs from the ordinary category of modules considered in Sections 13 and 22, since we consider other morphisms. In the next chapter, we also consider two similar categories. In

Section 25, we study modules of "smallest" or "largest" (in some sense) p-rank. In Section 26, indecomposable torsion-free modules of finite rank are considered. In addition, examples of indecomposable modules are also obtained.

In this chapter and all remaining chapters, we assume that any discrete valuation domain R is commutative. The advantages of such an assumption were discussed in the beginning of Section 19. We preserve the notation $\widehat{R}$, K, and $\widehat{K}$. Precisely, $\widehat{R}$ is the p-adic completion of the ring R; K and $\widehat{K}$ are the fields of fractions of the rings R and $\widehat{R}$, respectively. These symbols are used without additional explanations.

23 Elementary properties of torsion-free modules

Here, we collect some results on torsion-free modules considered earlier in Sections 4, 9 and present several new results.

As earlier, denote by $\operatorname{Hom}(M, N)$ and $M \otimes N$ the homomorphism group and the tensor product for the R-modules M and N, respectively. We canonically consider both these groups as R-modules (see Sections 2 and 13). In particular, the ring $\operatorname{End}(M)$ is an R-algebra; this was already used in Chapter 4. The K-space $K \otimes M$ and the R_p-space $R_p \otimes M$ will be very useful (R_p is the residue field of the ring R). We can identify the K-spaces $K \otimes M$ and KM, where KM is defined in Section 4 (it is the set of fractions of certain form). If M is a torsion-free module, then we also identify M with the image of the embedding $M \to KM$, $x \to 1 \cdot x$, $x \in M$ (Lemma 4.3). Then KM is a minimal K-space containing M and KM is the divisible hull of the module M (Proposition 6.8). Therefore, an arbitrary module M is a torsion-free module of finite rank if and only if M is isomorphic to a submodule of some finite-dimensional vector space over K. By Proposition 4.6, the modules R and K are the only torsion-free modules of rank 1. Every proper submodule of the module K is isomorphic to R.

If M and N are torsion-free modules, then it is obvious that $\operatorname{Hom}(M, N)$ is also a torsion-free module. Consequently, we can take the divisible hull $K \operatorname{Hom}(M, N)$ of it. The divisible module KN is injective by Theorem 6.5. Therefore, every homomorphism from M into N can be extended to a unique K-homomorphism (a linear operator) from KM into KN. The meaning of the inclusion $\operatorname{Hom}(M, N) \subseteq \operatorname{Hom}_K(KM, KN)$ is clear. In this case,

$$\operatorname{Hom}(M, N) = \{f \in \operatorname{Hom}_K(KM, KN) \mid fM \subseteq N\}.$$

Since $\operatorname{Hom}_K(KM, KN)$ is a K-space, we have that $K \operatorname{Hom}(M, N)$ can be naturally embedded in $\operatorname{Hom}_K(KM, KN)$. Precisely, any element $(r/p^n)\varphi \in K \operatorname{Hom}(M, N)$ ($r/p^n \in K$, $\varphi \in \operatorname{Hom}(M, N)$) is corresponded to a homomorphism $KM \to KN$ such that

$$((r/p^n)\varphi)((s/p^m)x) = (rs/p^{n+m})\varphi(x)$$

for all $s/p^m \in K$ and $x \in M$. Under such a correspondence, we have

$$K\operatorname{Hom}(M,N) = \{f \in \operatorname{Hom}_K(KM,KN) \mid \exists n \geq 0 : p^n f \in \operatorname{Hom}(M,N)\}.$$

The rank $r(M)$ of the module M was defined in Section 4. For the torsion-free module M, we have $r(M) = \dim_K(KM)$ (Lemma 4.5). The p-rank $r_p(M)$ of the module M was also introduced in Section 4; it is equal to the dimension of the R_p-space M/pM (see also property (6) in Section 9 and the remark after it). If A is a pure submodule of the module M, then it is easy to see that

$$r(M) = r(A) + r(M/A) \quad \text{and} \quad r_p(M) = r_p(A) + r_p(M/A).$$

We find the relation between the rank and the p-rank. First, it follows from property (6) of Section 9 that $r_p(M) = r(B)$ for every basic submodule B of the module M. Therefore, the relation $r_p(M) = 0$ holds if and only if M is a divisible module, and $r_p(M) = r(M)$ if and only if M is a free module.

Proposition 23.1. *Let M be a torsion-free module of finite rank and let F be some free submodule in M of maximal rank. Then $M/F = C \oplus D$, where C is a direct sum of a finite number of cyclic primary modules, D is a divisible primary module, and $r(M) = r_p(M) + r(D)$.*

Proof. First, we note that every such a submodule F is generated by some maximal linearly independent system of elements of the module M. The converse assertion is also true (see Section 4). Consequently, $r(F) = r(M)$ and M/F is a primary module. We choose some basic submodule B of the module M. A free basis of the module B (it is called a *p-basis* of the module M) can be extended to a maximal linearly independent system of elements of the module M (see Sections 5 and 9). Therefore, it is clear that there exists a free module B_0 such that $B \oplus B_0$ is a free submodule of maximal rank in M. We denote it by F_0; we also consider the factor module M/F_0. It follows from the relations $r(M) = r(F)$ and $r_p(M) = r(B)$ that $r(M) - r_p(M) = r(B_0)$. The factor module M/B is a divisible torsion-free module (i.e., a K-space) of rank $r(M) - r_p(M)$. We have

$$M/F_0 \cong (M/B)/(F_0/B),$$

where F_0/B is a free module of rank $r(M) - r_p(M)$. Therefore M/F_0 is a divisible primary module of rank $r(M) - r_p(M)$.

We pass to the factor module M/F. Let $a_1, \dots, a_n$ and $b_1, \dots, b_n$ be free bases of the modules F_0 and F, respectively, where $n = r(M)$. There exists an integer

$k \geq 0$ such that $p^k a_i \in F$ and $p^k b_i \in F_0$ for all i. Then we have $p^k F_0 \subseteq F$ and $p^k F \subseteq F_0$. Consequently, $F_0/(F \cap F_0)$ and $F/(F \cap F_0)$ are direct sums of a finite number of cyclic primary modules. Considering the isomorphisms

$$M/(F \cap F_0) \cong (M/F_0)/(F_0/(F \cap F_0))$$

and

$$M/(F \cap F_0) \cong (M/F)/(F/(F \cap F_0)),$$

we obtain that the module $M/(F \cap F_0)$ is isomorphic to the divisible module M/F_0. The module M/F is equal to $C \oplus D$, where $D \cong M/F_0$ and C is the module C indicated in the proposition. □

Let M be a torsion-free module of finite rank and let A be a submodule of M such that M/A is a primary module. Then $r(M) = r(A)$ and $r_p(M) \leq r_p(A)$. The relation $r(M) = r(A)$ is obvious. We use Proposition 23.1 to verify the inequality.

Lemma 23.2. *For every module M, there exists an isomorphism $R_p \otimes M \cong M/pM$. Therefore*

$$r_p(M) = \dim_{F_p}(R_p \otimes M).$$

Proof. Let us consider the induced exact sequence of modules

$$R_p \otimes pM \to R_p \otimes M \to R_p \otimes M/pM \to 0.$$

Here, the first mapping is the zero mapping. Consequently, we obtain

$$R_p \otimes M \cong R_p \otimes M/pM \cong M/pM.$$ □

Proposition 23.3. *If M and N are two torsion-free modules of finite rank, then*

(1) $r(\mathrm{Hom}(M, N)) \leq r(M) \cdot r(N)$ *and* $r_p(\mathrm{Hom}(M, N)) \leq r_p(M) \cdot r_p(N)$;

(2) $M \otimes N$ *is a torsion-free module,*

$$r(M \otimes N) = r(M) \cdot r(N), \quad \textit{and} \quad r_p(M \otimes N) = r_p(M) \cdot r_p(N).$$

Proof. (1) We have

$$\begin{aligned} r(\mathrm{Hom}(M, N)) &= \dim_K(K\,\mathrm{Hom}(M, N)) \leq \dim_K(\mathrm{Hom}(KM, KN)) \\ &= \dim_K(KM) \cdot \dim_K(KN) = r(M) \cdot r(N). \end{aligned}$$

There exists an embedding of R_p-spaces $R_p \otimes \mathrm{Hom}(M, N) \to \mathrm{Hom}_{R_p}(R_p \otimes M, R_p \otimes N)$. An arbitrary element of $R_p \otimes \mathrm{Hom}(M, N)$ is equal to $1 \otimes \varphi$. With any element $1 \otimes \varphi$, we associate the induced homomorphism $R_p \otimes M \to R_p \otimes N$ defined by the relation $s \otimes m \to s \otimes \varphi(m)$, $s \in R_p$, $m \in M$. In such a manner, we obtain an embedding. Then we consider Lemma 23.2 and repeat the calculations used for Hom.

(2) Taking induced exact sequences, we can embed $M \otimes N$ in the K-space $KM \otimes KN$. We have

$$\begin{aligned} r(M \otimes N) &= \dim_K(K \otimes M \otimes N) = \dim_K(K \otimes M) \cdot \dim_K(K \otimes N) \\ &= r(M) \cdot r(N). \end{aligned}$$

Replacing K by R_p, we obtain similar relations for p-ranks. □

For the modules M and N from the proposition, we obtain that $\mathrm{Hom}(M, N)$ is a torsion-free module of finite rank and $K \,\mathrm{Hom}(M, N)$ is a finite-dimensional K-space. In particular, $K \,\mathrm{End}(M)$ is an finite-dimensional K-algebra.

Exercise 1. For every torsion-free R-module M of finite rank, there exists a pure submodule A of finite rank in $\widehat{R}$ with $r_p(M) = r_p(\mathrm{Hom}(M, A))$.

24 Category of quasihomomorphisms

The main result of the section is Theorem 24.9 on the existence of some duality for torsion-free modules of finite rank. In addition, category-theoretical notions related to this theorem are defined in terms of the modules. Additive categories and direct sums of objects were defined in Section 1. We also define projections for direct sums.

Lemma 24.1. *An object A of the additive category $\mathcal{E}$ and morphisms $e_i \in \mathrm{Hom}_{\mathcal{E}}(A_i, A)$ $(i = 1, \dots, n)$ form the direct sum of objects $A_1, \dots, A_n$ with embeddings $e_1, \dots, e_n$ if and only if for every i, there exists a morphism $q_i \in \mathrm{Hom}_{\mathcal{E}}(A, A_i)$ such that $e_j q_i = 0$ for $i \neq j$, $e_i q_i = 1_{A_i}$, and $1_A = q_1 e_1 + \cdots + q_n e_n$. In this case, $q_i e_i$ is an idempotent of the ring $\mathrm{End}_{\mathcal{E}}(A)$ for every $i = 1, \dots, n$.*

Proof. Necessity. For a fixed subscript i, we define $d_{ji} \in \mathrm{Hom}_{\mathcal{E}}(A_j, A_i)$ by setting $d_{ii} = 1_{A_i}$ and $d_{ji} = 0$ if $j \neq i$, $j = 1, \dots, n$. By the definition of the direct sum, there exists a morphism $q_i \in \mathrm{Hom}_{\mathcal{E}}(A, A_i)$ with $e_j q_i = d_{ji}$ for every j. Then $q_i e_i$ is an idempotent of the ring $\mathrm{End}_{\mathcal{E}}(A)$ for every $i = 1, \dots, n$. It follows from the relations

$$e_i(q_1 e_1 + \cdots + q_n e_n) = e_i = e_i 1_A$$

and the uniqueness property for direct sums that $1_A = q_1 e_1 + \cdots + q_n e_n$.

Sufficiency. Let $f_i \in \operatorname{Hom}_{\mathcal{E}}(A_i, B)$ be morphisms for every $i = 1, \dots, n$. We set

$$f = q_1 f_1 + \cdots + q_n f_n : A \to B.$$

Then $e_i f = f_i$ for all i. Furthermore, if the morphism $g \in \operatorname{Hom}_{\mathcal{E}}(A, B)$ has the property $e_i g = f_i$ for every i, then

$$1_A g = q_1(e_1 g) + \cdots + q_n(e_n g) = q_1 f_1 + \cdots + q_n f_n = f.$$

Consequently, A is the direct sum of objects $A_1, \dots, A_n$ with embeddings $e_1, \dots, e_n$. □

The morphisms $q_1, \dots, q_n$ are called the *projections* of the direct sum $A = A_1 \oplus \cdots \oplus A_n$ with embeddings $e_1, \dots, e_n$.

We say that idempotents are *split* in the category $\mathcal{E}$ if for every object A and every idempotent $e \in \operatorname{End}_{\mathcal{E}}(A)$, there exist an object B in $\mathcal{E}$ and morphisms $f \in \operatorname{Hom}_{\mathcal{E}}(B, A)$, $g \in \operatorname{Hom}_{\mathcal{E}}(A, B)$ such that $fg = 1_B$ and $gf = e$.

If idempotents are split in the category $\mathcal{E}$, then it is easy to verify that an object A of $\mathcal{E}$ is indecomposable if and only if 0 and 1_A are the only idempotents of the ring $\operatorname{End}_{\mathcal{E}}(A)$ (see Bass [32], Faith [70], Arnold [6]). We formulate some known theorem on uniqueness of decompositions for additive categories. The theorem deals with local rings which were defined before Proposition 3.1. In Section 27, two stronger results are proved. Remarks to this chapter and the end of Section 27 contain some information about the topics related to uniqueness and other properties of direct decompositions.

Theorem 24.2 (Bass [32], Faith [70], Arnold [6]). *Let $\mathcal{E}$ be an additive category and let idempotents be split in $\mathcal{E}$. We assume that the direct sum $A = A_1 \oplus \cdots \oplus A_n$ is given in $\mathcal{E}$, where every $\operatorname{End}_{\mathcal{E}}(A_i)$ is a local ring. Then*

(1) *if $A = B_1 \oplus \cdots \oplus B_m$, then every object B_j is a finite direct sum of indecomposable objects in $\mathcal{E}$;*

(2) *in (1) if every B_j is indecomposable, then $m = n$ and there exists a permutation θ of the set $\{1, 2, \dots, n\}$ such that A_i is isomorphic to $B_{\theta(i)}$ for every i.*

We return to torsion-free modules. Let M and N be two torsion-free modules. The elements of the K-space $K \operatorname{Hom}(M, N)$ are called *quasihomomorphisms* (see the previous section in connection to the notation $K \operatorname{Hom}(M, N)$) and the space $K \operatorname{Hom}(M, N)$ is called the *quasihomomorphism group*. We define the category $\mathcal{TF}$ as follows. Torsion-free modules of finite rank are objects of $\mathcal{TF}$ and quasihomomorphisms are morphisms of $\mathcal{TF}$.

Proposition 24.3. *The category $\mathcal{TF}$ is additive and idempotents are split in $\mathcal{TF}$.*

Proof. We define the composition

$$K\operatorname{Hom}(M,N) \times K\operatorname{Hom}(N,L) \to K\operatorname{Hom}(M,L),$$

by the relation $(k_1\varphi, k_2\psi) \to k_1k_2(\varphi\psi)$. This composition is associative and $1{\cdot}1_M$ is the identity element which is also denoted by 1_M. We know that $K\operatorname{Hom}(M,N)$ is an Abelian group, and it is easy to verify that the composition of quasihomomorphisms is bilinear.

There exist finite direct sums in $\mathcal{TF}$. Let $A_1,\dots,A_n$ be objects in $\mathcal{TF}$. We take the module direct sum $M = A_1 \oplus \dots \oplus A_n$ and natural embeddings $\varkappa_i : A_i \to M$, $i = 1,\dots,n$. We verify that M is the direct sum in $\mathcal{TF}$ of the objects $A_1,\dots,A_n$ with embeddings $\varkappa_1,\dots,\varkappa_n$. Let $f_i \in K\operatorname{Hom}(A_i,N)$ be quasihomomorphisms for some torsion-free module N, where $i = 1,\dots,n$. We choose a nonnegative integer n with $p^n f_i \in \operatorname{Hom}(A_i,N)$. There exists a unique homomorphism $\varphi : M \to N$ such that $\varkappa_i\varphi = p^n f_i$ for all i. Then $(1/p^n)\varphi \in K\operatorname{Hom}(M,N)$ and $\varkappa_i((1/p^n)\varphi) = f_i$ for all i. The uniqueness of $(1/p^n)\varphi$ follows from the uniqueness of φ. Thus, $\mathcal{TF}$ is an additive category.

Let $\varepsilon = (r/p^n)\alpha \in K\operatorname{End}(M)$ for some $r/p^n \in K$ and let $\alpha \in \operatorname{End}(M)$ be an idempotent of the ring $K\operatorname{End}(M)$. We need to prove the existence of a torsion-free module A of finite rank and quasihomomorphisms $\varkappa \in K\operatorname{Hom}(A,M)$, $\pi \in K\operatorname{Hom}(M,A)$ such that $\pi\varkappa = \varepsilon$ and $\varkappa\pi = 1_A$. Let $A = \alpha M$, $i : A \to M$ be the natural embedding, $\varkappa = (1/p^n)i$, and let $\pi = r\alpha$. Then $\pi\varkappa = (r/p^n)(\alpha i) = \varepsilon$. It follows from the relation $\varepsilon^2 = \varepsilon$ that $r^2\alpha^2 = rp^n\alpha$. Now let $a \in A$ and $a = \alpha x$, where $x \in M$. We have

$$(rp^n)(\varkappa\pi)a = r^2\alpha^2(x) = rp^n\alpha(x) = rp^n a.$$

Therefore $\varkappa\pi = 1_A$ and the idempotent ε is split. □

The category $\mathcal{TF}$ is called the *category of quasihomomorphisms* (of torsion-free modules of finite rank). The following terminology is used in the literature. The isomorphisms of the category $\mathcal{TF}$ are called *quasiisomorphisms*. If two modules M and N are isomorphic to each other in $\mathcal{TF}$, then we say that M and N are quasiisomorphic to each other and write $M \sim N$. If $M = A_1 \oplus \dots \oplus A_n$ in $\mathcal{TF}$, then M is a *quasidirect sum* of the modules A_i which are called *quasisummands*. Modules which are indecomposable in $\mathcal{TF}$ are called *strongly indecomposable* modules. The projections from Lemma 24.1 are called *quasiprojections*. The endomorphism ring $K\operatorname{End}(M)$ of a module M in $\mathcal{TF}$ is called the *quasiendomorphism ring* of M. For the subsequent presentation, it is important that the K-algebra $K\operatorname{End}(M)$ is finite-dimensional (see the end of Section 23).

We formulate the main notions and properties of the category $\mathcal{TF}$ in purely module-theoretical terms.

Corollary 24.4. *For any two torsion-free modules M and N of finite rank, the following conditions are equivalent*:

(1) M *is quasiisomorphic to* N;

(2) *there exist homomorphisms* $\varphi\colon M \to N$, $\psi\colon N \to M$ *and an integer* $n \geq 0$ *such that* $\varphi\psi = p^n 1_M$ *and* $\psi\varphi = p^n 1_N$;

(3) *there exist monomorphisms* $\varphi\colon M \to N$ *and* $\psi\colon N \to M$ *such that the modules* $N/\operatorname{Im}\varphi$ *and* $M/\operatorname{Im}\psi$ *are bounded*;

(4) *there exist a submodule* N' *and an integer* n *such that* $p^n N \subseteq N' \subseteq N$ *and* $M \cong N'$;

(5) *there exist submodules* M', N' *and integers* m, n *such that* $p^m M \subseteq M' \subseteq M$, $p^n N \subseteq N' \subseteq N$, *and* $M' \cong N'$.

Proof. (1) $\Longrightarrow$ (2). The modules M and N are isomorphic to each other in $\mathcal{TF}$. There exist $f \in K\operatorname{Hom}(M,N)$ and $g \in K\operatorname{Hom}(N,M)$ with $fg = 1_M$ and $gf = 1_N$ in $\mathcal{TF}$. We choose an integer $m \geq 0$ such that

$$p^m f = \varphi \in \operatorname{Hom}(M,N) \quad \text{and} \quad p^m g = \psi \in \operatorname{Hom}(N,M).$$

Then $\varphi\psi = p^{2m}1_M$ and $\psi\varphi = p^{2m}1_N$.

(2) $\Longrightarrow$ (3). It is clear that φ and ψ are monomorphisms and $p^n N = \operatorname{Im}(\psi\varphi) \subseteq \operatorname{Im}(\varphi)$. Similarly, $p^n M \subseteq \operatorname{Im}(\psi)$.

(3) $\Longrightarrow$ (4). We take an integer n with $p^n N \subseteq \operatorname{Im}(\varphi)$ and set $N' = \operatorname{Im}(\varphi)$.

The implication (4) $\Longrightarrow$ (5) is obvious.

(5) $\Longrightarrow$ (1). Let $\varphi\colon M' \to N'$ and $\psi\colon N' \to M'$ be mutually inverse isomorphisms. Then $(1/p^m)(p^m\varphi)$ and $(1/p^n)(p^n\psi)$ are morphisms from $K\operatorname{Hom}(M,N)$ and $K\operatorname{Hom}(N,M)$, respectively; the compositions of these morphisms in $\mathcal{TF}$ are equal to 1_M and 1_N, respectively. □

It is obvious that two isomorphic modules are quasiisomorphic to each other. If the both modules M and N are free or divisible and they are quasiisomorphic to each other, then M and N are isomorphic to each other. This is not always true for arbitrary modules (see Example 25.7). The rank and the p-rank of the module are preserved under quasiisomorphisms. For the rank, this follows from property (4); for the p-rank, this follows from Proposition 23.1.

Corollary 24.5. *For a torsion-free module M of finite rank, the following assertions* (a), (b), *and* (c) *hold.*

(a) *Let* $M = A_1 \oplus \cdots \oplus A_n$ *be a direct sum in* $\mathcal{TF}$, e_i *and* q_i *be the morphisms from Lemma* 24.1, *and let* $q_i e_i = (r_i/p^{k_i})\varepsilon_i$, *where* $r_i/p^{k_i} \in K$, $\varepsilon_i \in \operatorname{End}(M)$, $i = 1, \dots, n$. *There exists an integer* $m \geq 0$ *such that*

$$p^m M \subseteq \varepsilon_1(M) \oplus \cdots \oplus \varepsilon_n(M).$$

Furthermore, if $\varkappa_i : \varepsilon_i(M) \to M$ *is the natural embedding, then* M *is the direct sum in* $\mathcal{TF}$ *of the modules* $\varepsilon_1(M), \dots, \varepsilon_n(M)$ *with embeddings* $\varkappa_1, \dots, \varkappa_n$.

(b) *The following conditions are equivalent*:

(1) M *is a strongly indecomposable module*;

(2) $A = 0$ *or* $B = 0$ *in every case, where* $p^m M \subseteq A \oplus B \subseteq M$;

(3) $K\,\mathrm{End}(M)$ *is a local ring.*

(c) *There exist an integer* $m \geq 0$ *and strongly indecomposable modules* $A_1, \dots, A_n$ *such that*

$$p^m M \subseteq A_1 \oplus \cdots \oplus A_n \subseteq M.$$

Proof. (a) By Lemma 24.1, we have $1_M = q_1 e_1 + \cdots + q_n e_n$, where $q_i e_i$ is a pairwise orthogonal idempotents of the ring $K\,\mathrm{End}(M)$. It is easy to prove that the sum $\varepsilon_1(M) + \cdots + \varepsilon_n(M)$ is a direct sum. Let m be the maximum of the integers $k_1, \dots, k_n$. Then for every $x \in M$, we have

$$p^m x = p^m(q_1 e_1 + \cdots + q_n e_n)(x) \in \varepsilon_1(M) \oplus \cdots \oplus \varepsilon_n(M).$$

The second assertion is directly verified. Therefore, the original direct sum $M = A_1 \oplus \cdots \oplus A_n$ can be replaced in $\mathcal{TF}$ by the direct sum $M = \varepsilon_1(M) \oplus \cdots \oplus \varepsilon_n(M)$.

(b) The equivalence of (1) and (2) follows from (a). If we have (1), then M is indecomposable in $\mathcal{TF}$. Consequently, 0 and 1_M are the only idempotents of the ring $K\,\mathrm{End}(M)$. This implies that the finite-dimensional K-algebra $K\,\mathrm{End}(M)$ is local. Conversely, the local ring $K\,\mathrm{End}(M)$ does not have nontrivial idempotents and M is indecomposable in $\mathcal{TF}$.

The proof of assertion (c) uses (b) and the induction on the rank of the module M. □

It follows from assertions 24.2–24.5 that the category $\mathcal{TF}$ satisfies the isomorphism theorem for direct decompositions with indecomposable summands. We formulate this theorem without mention the category $\mathcal{TF}$.

Corollary 24.6. *Let* M *be a torsion-free module of finite rank* $k \geq 0$ *and let* $A_1, \dots, A_n$ *be strongly indecomposable modules such that*

$$p^k M \subseteq A_1 \oplus \cdots \oplus A_n \subseteq M$$

(*see Corollary* 24.5 (c)).

(1) *If* $l \geq 0$ *and*

$$p^l M \subseteq B_1 \oplus \cdots \oplus B_m \subseteq M,$$

then every module B_j *is quasiisomorphic to the direct sum of some modules* A_i.

(2) *If each of the modules* B_j *is strongly indecomposable, then* $m = n$ *and there exists a permutation* θ *of the set* $\{1, \ldots, n\}$ *such that* A_i *is quasiisomorphic to* $B_{\theta(i)}$ *for every* i.

Any strongly indecomposable module is indecomposable. In general, the converse is not true.

Before we consider the duality mentioned in the beginning of the section, we make a little preliminary work. We will use the p-adic completion $\widehat{R}$ of the domain R and the field of fractions $\widehat{K}$ of the domain $\widehat{R}$. We keep in mind the known ring inclusions $R \subseteq \widehat{R} \subset \widehat{K}$ and $R \subset K \subseteq \widehat{K}$, where $\widehat{R} \cap K = R$. Furthermore $\widehat{R}/R$ is a divisible torsion-free R-module (see the paragraph before Corollary 11.15). Every torsion-free $\widehat{R}$-module of finite rank is free. If X is a subset of a module A over some ring S, then SX denotes the S-submodule in A generated by X. Assuming a homomorphism of S-modules, we say an "S-homomorphism". All subsequent studies are considered in some finite-dimensional vector $\widehat{K}$-spaces.

Let $\widehat{V}$ be a finite-dimensional vector $\widehat{K}$-space, X be some basis of $\widehat{V}$, $V = KX$, and let $X = X_1 \cup X_2$, where $X_1 \cap X_2 = \varnothing$. Then $\widehat{V} = \widehat{K}X_1 \oplus \widehat{K}X_2$. We assume that $\alpha \colon \widehat{R}X_2 \to \widehat{R}X_1$ is an $\widehat{R}$-homomorphism. Since $\widehat{K}X_i$ is the divisible hull of $\widehat{R}X_i$, we have that α can be extended to a unique $\widehat{K}$-homomorphism (a linear operator) $\widehat{K}X_2 \to \widehat{K}X_1$ which is also denoted by α. With respect to the above decomposition, the matrix

$$\begin{pmatrix} 1 & 0 \\ \alpha & 1 \end{pmatrix}$$

defines the automorphism $\widehat{\alpha}$ of the $\widehat{K}$-space $\widehat{V}$. By the symbol $(X_1, X_2, \widehat{V}, \widehat{\alpha})$, we denote the R-module $\widehat{\alpha}(V) \cap (\widehat{R}X_1 \oplus \widehat{K}X_2)$.

Lemma 24.7. *Let* $M = (X_1, X_2, \widehat{V}, \widehat{\alpha})$. *Then* M *is a torsion-free module of finite rank,* $\widehat{\alpha}(RX)$ *is a free submodule in* M *of maximal rank,* $KM = \widehat{\alpha}(V)$, RX_1 *is a basic submodule of the module* M, *and* $\widehat{R}M = \widehat{R}X_1 \oplus \widehat{K}X_2$. *If* M *is a reduced module, then* $\alpha(y) \neq 0$ *for all nonzero* $y \in KX_2$.

Proof. We have that V and $\widehat{\alpha}(V)$ are K-spaces of dimension $|X|$. Consequently, M is an torsion-free R-module of finite rank. Since RX is a free module, we have that $\widehat{\alpha}(RX)$ is also a free module of rank $|X|$ and $\widehat{\alpha}(RX) \subseteq M$. It follows from the inclusion $M \subseteq \widehat{\alpha}(V)$ that $KM \subseteq \widehat{\alpha}(V)$. Consequently, these K-spaces coincide, since they have equal dimensions. Since $\widehat{\alpha}(RX_1) = RX_1$, we have that $RX_1 \subseteq M$ and RX_1 is a free module of rank $|X_1|$. Furthermore, M is a pure submodule in $\widehat{R}X_1 \oplus \widehat{K}X_2$ and $(\widehat{R}X_1 \oplus \widehat{K}X_2)/RX_1$ is a divisible torsion-free module which is isomorphic to $\widehat{R}X_1/RX_1 \oplus \widehat{K}X_2$ (see Exercise 5 (b)). Therefore M/RX_1 is a divisible torsion-free module and RX_1 is a basic submodule of the module M. The purity of M in $\widehat{R}X_1 \oplus \widehat{K}X_2$ implies the purity of $\widehat{R}M$ in $\widehat{K}X_1 \oplus \widehat{K}X_2$ (Exercise 6). Consequently, $\widehat{R}M$ coincides with $\widehat{R}X_1 \oplus \widehat{K}X_2$, since they have equal ranks as $\widehat{R}$-modules.

Let M be a reduced module, $y \in KX_2$, and let $\alpha(y) = 0$. Then

$$\widehat{\alpha}(y) = (1+\alpha)(y) = y.$$

Consequently, $Ky \subset M$ and $y = 0$, since $Ky \cong K$ for $y \neq 0$. □

Every torsion-free module of finite rank can be constructed by the construction considered before Lemma 24.7.

Lemma 24.8. *Every torsion-free module M of finite rank is equal to $(X_1, X_2, \widehat{V}, \widehat{\alpha})$ for some X_1, X_2, V, and α.*

Proof. We choose some p-basis X_1 of the module M. Then RX_1 is a basic submodule of the module M. We enlarge X_1 to a maximal linearly independent set $X_1 \cup Y$ of the module M. We define the vector $\widehat{K}$-space $\widehat{V} = \widehat{K}X_1 \oplus \widehat{K}Y$ with basis $X_1 \cup Y$. Then $KM = KX_1 \oplus KY$. Since RX_1 is a basic submodule of the module M, we have that $\widehat{R}X_1$ is a basic submodule of the $\widehat{R}$-module $\widehat{R}M$ (Exercise 5 (c)). We have a decomposition $\widehat{R}M = \widehat{R}X_1 \oplus D$, where D is the maximal divisible submodule of the $\widehat{R}$-module $\widehat{R}M$ (consider that $\widehat{R}X_1$ is a complete module).

We find a linearly independent subset X_2 in $\widehat{V}$ and an $\widehat{R}$-homomorphism $\alpha\colon \widehat{R}X_2 \to \widehat{R}X_1$ such that

$$\widehat{R}M = \widehat{R}X_1 \oplus \widehat{K}X_2 \quad \text{and} \quad (1+\alpha)(X_2) = Y.$$

We denote by π_1 and π_2 the projections from $\widehat{R}M$ onto $\widehat{R}X_1$ and D, respectively. The restriction of π_2 to $\widehat{R}Y$ is injective. We set $X_2 = \pi_2(Y)$. Then X_2 is a basis of the $\widehat{K}$-space D, since Y is a linearly independent set and $|Y| = \dim D$. Consequently,

$$\widehat{K}X_2 = D \quad \text{and} \quad \widehat{R}M = \widehat{R}X_1 \oplus \widehat{K}X_2.$$

Let α be the composition of the mappings $\pi_2^{-1} : \widehat{R}X_2 \to \widehat{R}Y$ and $\pi_1\colon \widehat{R}Y \to \widehat{R}X_1$. If $y \in Y$ and $y = z + x$, where $z \in \widehat{R}X_1$, $x \in D$, then $\pi_2(y) = x \in X_2$. Then $(1+\alpha)x = y$ and it is easy to verify that $(1+\alpha)(X_2) = Y$.

Finally, let $V = KX_1 \oplus KX_2$. We consider the module $(X_1, X_2, \widehat{V}, \widehat{\alpha})$ which is denoted by N. We have

$$N = \widehat{\alpha}(V) \cap (\widehat{R}X_1 \oplus \widehat{K}X_2)$$

where $\widehat{\alpha} = \begin{pmatrix} 1 & 0 \\ \alpha & 1 \end{pmatrix}$. It follows from $(1+\alpha)(X_2) = Y$ that

$$\widehat{\alpha}(V) = KX_1 \oplus KY = KM \quad \text{and} \quad M \subseteq \widehat{\alpha}(V).$$

Thus, $M \subseteq N$. However, N/M is a primary module, since $(KX_1 \oplus KY)/M$ is primary. On the other hand, N/M is a torsion-free module, since the module $\widehat{R}M/M$ is torsion-free (Exercise 5 (b)). Consequently, $N = M$. □

We take two modules

$$M = (X_1, X_2, \widehat{V}, \widehat{\alpha}) \quad \text{and} \quad N = (Y_1, Y_2, \widehat{U}, \widehat{\beta}),$$

and we determine the interrelations between quasihomomorphisms $M \to N$ and $\widehat{K}$-homomorphisms $\widehat{V} \to \widehat{U}$. Let $f\colon M \to N$ be some quasihomomorphism. Since $KM = \widehat{\alpha}(V)$ and $KN = \widehat{\beta}(U)$, there exists a unique homomorphism of K-spaces $f' : \widehat{\alpha}(V) \to \widehat{\beta}(U)$ which extends f. We have that $\widehat{\alpha}(X_1 \cup X_2)$ and $\widehat{\beta}(Y_1 \cup Y_2)$ are bases of $\widehat{K}$-spaces $\widehat{V}$ and $\widehat{U}$, respectively. Therefore, there exists a unique $\widehat{K}$-homomorphism $\widehat{f}\colon \widehat{V} \to \widehat{U}$ which extends f'. Consequently, $\widehat{f}$ extends f. We choose a nonzero element $r \in R$ with $(rf)M \subseteq N$. Then $(rf)(\widehat{R}M) \subseteq \widehat{R}N$. Therefore $\widehat{f}(\widehat{K}X_2) \subseteq \widehat{K}Y_2$ ($\widehat{K}X_2$ and $\widehat{K}Y_2$ are maximal divisible submodules of modules $\widehat{R}M$ and $\widehat{R}N$, respectively).

Conversely, let $g\colon \widehat{V} \to \widehat{U}$ be a $\widehat{K}$-homomorphism such that $g(\widehat{\alpha}(V)) \subseteq \widehat{\beta}(U)$ and $g(\widehat{K}X_2) \subseteq \widehat{K}Y_2$. For $\widehat{R}$-modules, we have the relations

$$\widehat{R}M = \widehat{R}X_1 \oplus \widehat{K}X_2, \qquad \widehat{R}N = \widehat{R}Y_1 \oplus \widehat{K}Y_2.$$

We recall that $\widehat{\alpha}(V) = KM$ and $\widehat{\beta}(U) = KN$. It is clear that there exists a nonzero element $r \in R$ such that $(rg)(X_1) \subseteq \widehat{R}N$ and $(rg)(\widehat{R}M) \subseteq \widehat{R}N$. Therefore $(rg)M \subseteq N$ (see Exercise 7). We set $f = g_{|M}$. Then $f\colon M \to N$ is a quasihomomorphism.

We note that a quasihomomorphism $f\colon M \to N$ is a quasiisomorphism if and only if the both mappings $\widehat{f}\colon \widehat{\alpha}(V) \to \widehat{\beta}(U)$ and $\widehat{f}\colon \widehat{K}X_2 \to \widehat{K}Y_2$ are isomorphisms. Finally, f is an isomorphism if and only if $\widehat{f}\colon \widehat{\alpha}(V) \to \widehat{\beta}(U)$ and $\widehat{f}\colon \widehat{R}M \to \widehat{R}N$ are isomorphisms.

We have completed the preliminary work and return to the category $\mathcal{TF}$. Below, we use the notions of a contravariant functor, a natural equivalence, and other notions; they are presented in Section 1. A contravariant functor $F : \mathcal{E} \to \mathcal{E}$ is called a *duality* if F^2 is naturally equivalent to the identity functor of the category $\mathcal{E}$.

We recall the known duality for finite-dimensional vector spaces. Let $\widehat{K}$-mod denote the category of finite-dimensional vector $\widehat{K}$-spaces. For two finite-dimensional $\widehat{K}$-spaces V and U and a $\widehat{K}$-homomorphism $f\colon V \to U$, we set $G(V) = V^* = \operatorname{Hom}_{\widehat{K}}(V, \widehat{K})$ and $G(f) = f^*$, where f^* is the induced homomorphism $U^* \to V^*$, $f^*(\varphi) = f\varphi$ ($\varphi \in U^*$; see Section 2). We obtain the duality G of the category $\widehat{K}$-mod. The corresponding natural equivalence $h\colon 1_{\widehat{K}\text{-mod}} \to G^2$ is defined as follows. For the $\widehat{K}$-space V, the natural isomorphism $h_V : V \to V^{**}$ is defined by the relation $(h_V(a))(\psi) = \psi(a)$ for all $a \in V$ and $\psi \in V^*$. The $\widehat{K}$-space V^* is said to be *dual* to V and the homomorphism f^* is said to be dual to f.

Let $X = (x_1, x_2, \dots, x_n)$ be some basis of the $\widehat{K}$-space V. For every $i = 1, \dots, n$, we denote by x_i^* the element of the $\widehat{K}$-space V^* such that $x_i^*(x_i) = 1$ and $x_i^*(x_j) = 0$ for $i \neq j$ (x_i^* is a homomorphism $V \to \widehat{K}$). We denote $(x_n^*, \dots, x_2^*, x_1^*)$ by X^*. Here X^* is a basis of the space V^*.

Theorem 24.9 (Arnold [3]). *There exists a duality F of the category $\mathcal{TF}$ such that for every torsion-free module M of finite rank, the following properties hold*:

(1) $r(F(M)) = r(M)$;

(2) $r_p(F(M)) = r(M) - r_p(M)$;

(3) *M is a free module if and only if $F(M)$ is a divisible module.*

Proof. We define the required contravariant functor F. Let M be a torsion-free module of finite rank and let $M = (X_1, X_2, \widehat{V}, \widehat{\alpha})$ (cf., Lemma 24.8). We set $F(M) = (X_2^*, X_1^*, (\widehat{V})^*, \bar{\alpha})$, where $X_2^* \cup X_1^*$ is the basis of the dual $\widehat{K}$-space $(\widehat{V})^*$ defined in the manner indicated before the theorem. The mapping $\bar{\alpha}$ coincides with $((\widehat{\alpha})^*)^{-1} : (\widehat{V})^* \to (\widehat{V})^*$. The matrix form of $\bar{\alpha}$ is

$$\begin{pmatrix} 1 & -\alpha^* \\ 0 & 1 \end{pmatrix},$$

where $\alpha^* : \widehat{R}X_1^* \to \widehat{R}X_2^*$ is the mapping which is dual to $\alpha\colon \widehat{R}X_2 \to \widehat{R}X_1$. We note that $KX^* = KX_2^* \oplus KX_1^*$; KX^* can be identified with $V^* = \operatorname{Hom}_K(V, K)$, since $X = X_1 \cup X_2$ is a basis of the K-space V. By Lemma 24.7, $F(M)$ is a torsion-free module of finite rank.

Now assume that we have two torsion-free modules $M = (X_1, X_2, \widehat{V}, \widehat{\alpha})$ and $N = (Y_1, Y_2, \widehat{U}, \widehat{\beta})$ of finite rank and a quasihomomorphism $f\colon M \to N$. We can assume that f is a K-homomorphism $V \to U$ and a $\widehat{K}$-homomorphism $\widehat{V} \to \widehat{U}$. Let $f^* : (\widehat{U})^* \to (\widehat{V})^*$ be the dual homomorphism of $\widehat{K}$-spaces. We show that the restriction of the homomorphism f^* to $F(N)$ is a quasihomomorphism $F(N) \to F(M)$. For this purpose, we use remarks after the proof of Lemma 24.8. Since $f(V) \subseteq U$, we have $f^*(U^*) \subseteq V^*$. We need to verify that $f^*(\widehat{K}Y_1^*) \subseteq \widehat{K}X_1^*$. It is sufficient to prove that $f^*(Y_1^*) \subseteq \widehat{K}X_1^*$. Let $X_1 = (x_1, \dots, x_s)$, $X_2 = (x_{s+1}, \dots, x_m)$, $Y_1 = (y_1, \dots, y_t)$, and let $Y_2 = (y_{t+1}, \dots, y_n)$. We take some $y_j^* \in Y_1^*$, $j = 1, \dots, t$. We have $f^*(y_j^*) = q_1x_1^* + \dots + q_mx_m^*$, where $q_1, \dots, q_m \in \widehat{K}$. Let $x_i \in X_2$ and $s+1 \le i \le m$. Then we have

$$(f^*(y_j^*))(x_i) = (fy_j^*)(x_i) = y_j^*(f(x_i)) = 0,$$

since $f^*(X_2) \subseteq \widehat{K}Y_2$ and y_j^* annihilates $\widehat{K}Y_2$. On the other hand,

$$(f^*(y_j^*))(x_i) = q_1x_1^*(x_i) + \dots + q_mx_m^*(x_i) = q_i.$$

Consequently, $q_i = 0$ for all $i = s+1, \dots, m$ and $f^*(y_j^*) \in \widehat{K}X_1^*$.

Now we set $F(f) = f^* : F(N) \to F(M)$. The property that F is a contravariant functor on the category $\mathcal{TF}$ follows from the property that F is induced by the contravariant functor G on the category $\widehat{K}\text{-mod}$.

We prove that the functor F^2 is naturally equivalent to the identity functor of the category $\mathcal{TF}$. For the module $M = (X_1, X_2, \widehat{V}, \widehat{\alpha})$, the module $F^2(M)$ has the form $(X_1^{**}, X_2^{**}, (\widehat{V})^{**}, \bar{\bar{\alpha}})$. We note that the definition of $\bar{\bar{\alpha}}$ is similar to the definition of $\bar{\alpha}$ and $\bar{\bar{\alpha}} = (\widehat{\alpha})^{**}$. We take the natural $\widehat{K}$-homomorphism $h_{\widehat{V}} : \widehat{V} \to (\widehat{V})^{**}$. Then $h_{\widehat{V}}(\widehat{K}X_2) = \widehat{K}X_2^{**}$. Furthermore, we have

$$\begin{aligned} h_{\widehat{V}}(\widehat{\alpha}(V)) &= h_{\widehat{V}}(\widehat{\alpha}(KX_1 \oplus KX_2)) = (\widehat{\alpha})^{**}(h_{\widehat{V}}(KX_1 \oplus KX_2)) \\ &= (\widehat{\alpha})^{**}(KX_2^{**} \oplus KX_1^{**}) = \bar{\bar{\alpha}}(KX_2^{**} \oplus KX_1^{**}). \end{aligned}$$

Let g_M be the restriction of $h_{\widehat{V}}$ to M. By remarks after Lemma 24.8, we have that $g_M : M \to F^2(M)$ is a quasiisomorphism, i.e., it is an isomorphism in $\mathcal{TF}$.

Let $f \colon M \to N = (Y_1, Y_2, \widehat{U}, \widehat{\beta})$ be some quasihomomorphism. As above, we consider it as a $\widehat{K}$-homomorphism $\widehat{V} \to \widehat{U}$. The functor G^2 and the identity functor of the category $\widehat{K}\text{-mod}$ are naturally equivalent. Therefore $fh_{\widehat{U}} = h_{\widehat{V}}G^2(f)$. Therefore $fg_N = g_MF^2(f)$. Consequently, F^2 is naturally equivalent to the identity functor of the category $\mathcal{TF}$.

Properties (1) and (2) follow from Lemma 24.7 and the definition of the functor F. This implies Property (3). □

The duality F from Theorem 24.9 is called the *Arnold duality*. A more brief construction of the Arnold duality is given in the paper of Lady [194].

Exercise 1. Construct an indecomposable torsion-free module of finite rank which has a nontrivial quasidecomposition.

Exercise 2. If M is a reduced module, then $\widehat{R}M$ is the completion of M in p-adic topology.

Exercise 3 (Arnold [3]). Construct an example of isomorphic torsion-free modules M and N of rank 3 and p-rank 1 such that the modules $F(M)$ and $F(N)$ are not isomorphic to each other. (Since F is a duality, the modules $F(M)$ and $F(N)$ are quasiisomorphic to each other.)

Exercises 4–7 are related to some details of the proofs of Lemma 24.7 and Lemma 24.8. In these exercises, M and N are torsion-free modules of finite rank.

Exercise 4. The $\widehat{R}$-modules $\widehat{R} \otimes M$ and $\widehat{R}M$ are canonically isomorphic to each other.

Exercise 5. Prove that

(a) the rank of the $\widehat{R}$-module $\widehat{R}M$ is equal to the rank of the R-module M;

(b) $\widehat{R}M/M$ is a divisible torsion-free module;

(c) every p-basis of the module M is a p-basis for the module $\widehat{R}M$;

(d) $\widehat{R}M \cap KM = M$.

Exercise 6. If A is a pure submodule in the module M, then $\widehat{R}A$ is a pure submodule of the module $\widehat{R}M$.

Exercise 7. Let $g\colon \widehat{V} \to \widehat{U}$ be a homomorphism of $\widehat{K}$-spaces considered in remarks after Lemma 24.8. If $g(KM) \subseteq KN$ and $(rg)(\widehat{R}M) \subseteq \widehat{R}N$ for some nonzero element $r \in R$, then $(rg)M \subseteq N$.

Exercise 8. Prove that two torsion-free modules M and N of finite rank are quasiisomorphic to each other if and only if $r(M) = r(N)$, $r_p(M) = r_p(N)$, and M is isomorphic to some submodule in N.

25 Purely indecomposable and copurely indecomposable modules

Under the Arnold duality, divisible modules correspond to free modules; this follows from Theorem 24.9. In this section, we consider and study two larger classes of modules which are dual to each other. For a torsion-free module M of finite rank, we always have the inequality $r_p(M) \leq r(M)$. For $r_p(M) = 0$, we obtain divisible modules. For $r_p(M) = r(M)$, we obtain free modules. The object of our attention are modules M such that $r_p(M) = 1$ or $r_p(M) = r(M) - 1$. They have various remarkable properties.

As earlier, $\widehat{M}$ denotes the completion of a reduced torsion-free module M in the p-adic topology. If M and N are two such modules, then every homomorphism from M into N can be extended to a unique homomorphism from $\widehat{M}$ in $\widehat{N}$ (Proposition 11.12). In addition, M and N are isomorphic to each other if and only if there exists an $\widehat{R}$-isomorphism $\varphi\colon \widehat{M} \to \widehat{N}$ with $\varphi(KM) \subseteq KN$ (we assume that $\widehat{M} \cap KM = M$; see also Exercises 5 and 7 in Section 24).

A reduced torsion-free module M of finite rank is said to be *purely indecomposable* if every pure submodule of the module M is indecomposable. There are many characterizations of purely indecomposable modules.

Proposition 25.1. *For a reduced torsion-free module M of finite rank, the following conditions are equivalent*:

(1) *M is a purely indecomposable module*;

(2) *M is isomorphic to some pure submodule of the module $\widehat{R}$*;

(3) $r_p(M) = 1$;

(4) *if A is a proper pure submodule of the module M, then M/A is a divisible module*;

(5) *for any reduced torsion-free module N, every nonzero homomorphism $M \to N$ is a monomorphism.*

Proof. (1) $\Longrightarrow$ (2). Every basic submodule B of the module M is isomorphic to R. Consequently, $\widehat{M} \cong \widehat{B} \cong \widehat{R}$ which implies (2) (see Section 11).

(2) $\Longrightarrow$ (3). Since $0 < r_p(M) \leq r_p(\widehat{R}) = 1$, we have $r_p(M) = 1$.

(3) $\Longrightarrow$ (4). Since $1 = r_p(M) = r_p(A) + r_p(M/A)$ and $r_p(A) \neq 0$, we have that $r_p(M/A) = 0$ and M/A is a divisible module.

(4) $\Longrightarrow$ (5). We assume that there exists a nonzero homomorphism $\varphi\colon M \to N$ with $\operatorname{Ker}(\varphi) \neq 0$. Then $M/\operatorname{Ker}(\varphi)$ and $\operatorname{Im}(\varphi)$ are divisible modules, which is impossible.

(5) $\Longrightarrow$ (1). We assume that M is not a purely indecomposable module. Then there exists a pure submodule A of the module M which has a nontrivial direct decomposition $A = X \oplus Y$. The projection from the module A onto X can be extended to a homomorphism $M \to \widehat{M}$ (Theorem 11.4) with nonzero kernel, which contradicts to (5). Consequently, M is a purely indecomposable module. □

If M is a purely indecomposable module, then it follows from $r_p(M) = 1$ that $M/p^kM \cong R(p^k)$ for all $k \geq 1$. The nonreduced torsion-free module M of finite rank with $r_p(M) = 1$ is equal to $N \oplus D$, where N is a purely indecomposable module and D is a divisible module.

Proposition 25.2. *If M and N are purely indecomposable modules, then the following conditions are equivalent*:

(1) $M \cong N$;

(2) $M \sim N$;

(3) $\operatorname{Hom}(M,N) \neq 0$ *and* $\operatorname{Hom}(N,M) \neq 0$;

(4) $r(M) = r(N)$ *and* $\operatorname{Hom}(M,N) \neq 0$.

Proof. The implication (1) $\Longrightarrow$ (2) is obvious. The implication (2) $\Longrightarrow$ (3) follows from Corollary 24.4.

(3) $\Longrightarrow$ (4). It follows from Proposition 25.1 that there exist monomorphisms $M \to N$ and $N \to M$. Consequently,

$$r(M) \leq r(N) \leq r(M) \quad \text{and} \quad r(M) = r(N).$$

(4) $\Longrightarrow$ (1). As earlier, there exists a monomorphism from M into N. We assume that M is a submodule in N. Let G be a free submodule in M of maximal rank. It follows from Proposition 23.1 that $M/G = C_1 \oplus D_1$, where C_1 is the direct sum of a finite number of cyclic primary modules and D_1 is a divisible primary module. Since $r(M) = r(N)$, we have that $N/G = C_2 \oplus D_2$, where the structures of the summands C_2 and D_2 are similar to the structures of C_1 and D_1. Furthermore, we have

$$r(M) = r_p(M) + r(D_1) \quad \text{and} \quad r(N) = r_p(N) + r(D_2).$$

Since $r_p(M) = r_p(N)$, we have $r(D_1) = r(D_2)$, i.e., $D_1 \cong D_2$. It follows from

$$N/M \cong (N/G)/(M/G)$$

that N/M is the direct sum of a finite number of cyclic primary modules. Therefore, there exists an integer $k \geq 0$ with $p^k N \subseteq M$. Since $N/p^k N$ is a cyclic module $R(p^k)$, we have that $M/p^k N = p^m N/p^k N$, where $m \leq k$ and $M = p^m N$. However, $N \cong p^m N$ under the correspondence $x \to p^m x$. Consequently, $M \cong N$. □

We can obtain a sufficiently complete information about the endomorphism ring of a purely indecomposable module. As earlier (see Section 19), we assume that the domain R is contained in the endomorphism ring of this torsion-free module (the element $r \in R$ is identified with the endomorphism $x \to rx$ ($x \in M$) of the module M). If we use module terms for some R-algebra, we consider the algebra as an R-module. The symbol $[P : K]$ denotes the degree of the extension of the field P over the field K.

Proposition 25.3. *Let M be a purely indecomposable module and let $S = \operatorname{End}(M)$. Then the R-algebra S is isomorphic to some pure subalgebra in $\widehat{R}$ and S is a discrete valuation domain with prime element p. The field of fractions P of the domain S satisfies $K \subseteq P \subseteq \widehat{K}$, $[P : K]$ is finite, $P = KS$, and* $\operatorname{End}_R(S) = \operatorname{End}_S(S)$.

Proof. By Proposition 25.1, we can assume that M is a pure submodule in $\widehat{R}$. It is clear that M is a dense submodule. Now we have the following property. An arbitrary endomorphism $\alpha \in S$ can be extended to an endomorphism $\widehat{\alpha}$ of the module $\widehat{R}$ and the extension is unique (Proposition 11.12). In addition, $\widehat{\alpha}$ is an $\widehat{R}$-module endomorphism (Example 12.4). Consequently, $\widehat{\alpha}$ coincides with the multiplication on the ring $\widehat{R}$ by some element $r \in \widehat{R}$. Conversely, every element $r \in \widehat{R}$ with $rM \subseteq M$ gives an endomorphism of the module M. The algebra S can be identified with the set $\{r \in \widehat{R} \mid rM \subseteq M\}$. In addition, S is a pure subalgebra in $\widehat{R}$, since M is pure in $\widehat{R}$. Let r be a nonzero element of S. We have $r = p^n v$, where $n \geq 0$ and v is an invertible element in $\widehat{R}$. In fact, $v \in S$, since S is a pure subalgebra. Furthermore, vM is a pure submodule in M which is isomorphic to M. Consequently, $vM = M$, since the rank M is finite. Thus, $r = p^n v$, where v is an invertible element in S. Now it is clear that the nonzero ideals of the ring S have the form $p^n S$, $n \geq 0$. Consequently, S is a discrete valuation domain with prime element p. Since $R \subseteq S \subseteq \widehat{R}$, we have $K \subseteq P \subseteq \widehat{K}$. It follows from the relation $P = \{r(1/p^n) \mid r \in S, n \geq 0\}$ that $P = KS$. The dimension of P over K is equal to the rank of the R-module S; this rank is finite (Proposition 23.3). Let α be some endomorphism of the R-module S. As above, α coincides with the multiplication on S by some element $r \in \widehat{R}$. It follows from $\alpha(1) = r \cdot 1 = r$ that $r \in S$. Consequently, α is an S-endomorphism of the module S. □

Purely indecomposable modules are easily described up to isomorphism. We fix an integer $n \geq 1$. Let $1, a_1, \dots, a_n$ be some linearly independent elements of the R-module $\widehat{R}$. We denote by $A(a_1, \dots, a_n)$ the pure submodule in $\widehat{R}$ generated by these elements. It is easy to verify that

$$A(a_1, \dots, a_n) = A(r_1 a_1, \dots, r_n a_n)$$

for all nonzero elements $r_i \in R$ and

$$A(a_1, \dots, a_n) = KA(a_1, \dots, a_n) \cap \widehat{R},$$

where

$$KA(a_1, \dots, a_n) = K \oplus Ka_1 \oplus \dots \oplus Ka_n$$

is the K-subspace generated by $A(a_1, \dots, a_n)$ in $\widehat{K}$ (see Exercise 5 in Section 24).

Let M be a purely indecomposable module of rank $n+1$. Then M is isomorphic to the module $A(a_1,\dots,a_n)$ for some linearly independent elements $1, a_1,\dots,a_n$ in $\widehat{R}$. Indeed, it follows from Proposition 25.1 that we can assume that M is a pure submodule in $\widehat{R}$. We choose some nonzero element in M. The element is equal to $p^k v$, where $k \geq 0$ and v is some invertible element in $\widehat{R}$. Since M is pure, we also have $v \in M$. We have $M \cong v^{-1}M$, where $v^{-1}M$ is a pure submodule in $\widehat{R}$. In addition, $1 = v^{-1}v \in v^{-1}M$. It is clear that $v^{-1}M$ coincides with $A(a_1,\dots,a_n)$ for some elements $a_1,\dots,a_n$.

It follows from the proof of Proposition 25.3 that

$$\operatorname{End}(A(a_1,\dots,a_n)) = \{r \in \widehat{R} \mid rA(a_1,\dots,a_n) \subseteq A(a_1,\dots,a_n)\}.$$

If $r \in \operatorname{End}(A(a_1,\dots,a_n))$, then $r \in A(a_1,\dots,a_n)$, since $A(a_1,\dots,a_n)$ contains the identity element. Therefore, we also have

$$\begin{aligned} &K\operatorname{End}(A(a_1,\dots,a_n)) \\ &= \{r \in K \oplus Ka_1 \oplus \cdots \oplus Ka_n \mid ra_i \in K \oplus Ka_1 \oplus \cdots \oplus Ka_n,\ 1 \leq i \leq n\}. \end{aligned}$$

Theorem 25.4 (Arnold [3], Arnold–Dugas [12]). *Two purely indecomposable modules $A(a_1,\dots,a_n)$ and $A(b_1,\dots,b_n)$ are isomorphic to each other if and only if K has elements s_i and t_{ji} $(0 \leq i,j \leq n)$ such that $u = s_0 + s_1a_1 + \cdots + s_na_n$ is an invertible element in $\widehat{R}$ and $ub_j = t_{j0} + t_{j1}a_1 + \cdots + t_{jn}a_n$.*

Proof. Let $\varphi\colon A(b_1,\dots,b_n) \to A(a_1,\dots,a_n)$ be some isomorphism and let $u = \varphi(1)$. Similar to the proof of Proposition 25.3, we can obtain that φ coincides with the multiplication by $\varphi(1)$, i.e., $\varphi(y) = u \cdot y$ for all $y \in A(b_1,\dots,b_n)$. It is clear that u is an invertible element and $u, ub_j \in A(a_1,\dots,a_n)$; this implies the existence of the elements s_i and t_{ji}.

Conversely, we assume that there exist indicated elements s_i and t_{ji}. We consider the multiplication of the ring $\widehat{R}$ by the element u. We choose p^k such that $p^k u, p^k(ub_j) \in A(a_1,\dots,a_n)$ for all j. Then $u, ub_j \in A(a_1,\dots,a_n)$, since the submodule $A(a_1,\dots,a_n)$ is pure. Considering the purity, we obtain

$$uA(b_1,\dots,b_n) \subseteq A(a_1,\dots,a_n).$$

Since $uA(b_1,\dots,b_n)$ is a pure submodule in $A(a_1,\dots,a_n)$, the above inclusion is an equality (see also Proposition 25.2). Consequently, the multiplication by u is an isomorphism between $A(b_1,\dots,b_n)$ and $A(a_1,\dots,a_n)$. □

A torsion-free module M is said to be *coreduced* if M does not have direct summands which are free modules. The module M is coreduced if and only if $\operatorname{Hom}(M,R) = 0$. Under the Arnold duality, reduced modules correspond to coreduced modules. If M is an arbitrary torsion-free module of finite rank, then $M = N \oplus G$, where N is a coreduced module and either G is a free module or $G = 0$.

A coreduced module M of finite rank is said to be *copurely indecomposable* if the factor module M/A is indecomposable for every pure submodule A.

Characterizations of copurely indecomposable modules are dual to characterizations of purely indecomposable modules from Proposition 25.1.

Proposition 25.5. *For a coreduced torsion-free module M of finite rank, the following properties are equivalent*:

(1) *M is a copurely indecomposable module*;

(2) $r(M) = r_p(M) + 1$;

(3) *every submodule A of the module M with $r(A) < r(M)$ is free*;

(4) *$F(M)$ is a purely indecomposable module, where F is the Arnold duality from Theorem* 24.9;

(5) *for each coreduced torsion-free module N and every nonzero homomorphism $\varphi\colon N \to M$, the factor module $M/\operatorname{Im}(\varphi)$ is a bounded module.*

Proof. (1) $\Longrightarrow$ (2). We assume that $r(M) = r_p(M) + k$ and $k \geq 2$. We take some basic submodule B of the module M. Then we have

$$r(M/B) = r(M) - r(B) = r_p(M) + k - r_p(M) = k.$$

Therefore, the divisible module M/B of rank k is a decomposable module, which contradicts to (1). Consequently, $k = 1$.

(2) $\Longrightarrow$ (3). Let $A \subseteq M$ and $r(A) < r(M)$. We assume that A is pure in M; otherwise, we can take the pure submodule generated by A in M, see property (7) from Section 7. We have the relations

$$r(M) = r_p(M) + 1 = r(A) + r(M/A) \quad \text{and} \quad r_p(M) = r_p(A) + r_p(M/A).$$

We also have $r(M/A) \neq r_p(M/A)$. Otherwise, M/A is a free module and $M = A \oplus C$ with $C \cong M/A$ (Theorem 5.4), which contradicts to the property that M is a coreduced module. Using the above relations, it is easy to verify that $r(A) = r_p(A)$. Therefore A is free.

(3) $\Longrightarrow$ (1). We assume that A is a pure submodule in M with $A \neq M$ and there exists a nontrivial decomposition $M/A = X/A \oplus Y/A$. We have

$$M/X \cong (M/A)/(X/A) \cong Y/A,$$

where Y/A is a free module, since Y is a finitely generated free module. This leads to the decomposition $M = X \oplus Z$, where Z is a free module. This contradicts to the property that M is a coreduced module. Consequently, M is a copurely indecomposable module.

The equivalence of (2) and (4) can be proved with the use of the duality F and the equivalence of properties (1) and (3) from Proposition 25.1.

(3) $\Longrightarrow$ (5). Let $\varphi\colon N \to M$ be a nonzero homomorphism, where N is a coreduced torsion-free module. If $r(\mathrm{Im}(\varphi)) \neq r(M)$, then $\mathrm{Im}(\varphi)$ and $N/\mathrm{Ker}(\varphi)$ are free modules. Consequently, $N = \mathrm{Ker}(\varphi) \oplus G$, where G is a free module. This contradicts to the property that N is a coreduced module. Therefore $r(\mathrm{Im}(\varphi)) = r(M)$.

We also have that $\mathrm{Im}(\varphi)$ is a coreduced module (the proof of (3) $\Longrightarrow$ (1) has a similar point). Thus, we have the following situation. There is a submodule A of the module M such that $r(A) = r(M)$ and A is a coreduced module. We need to prove that M/A is a bounded module. Let X be a free submodule of the module A of maximal rank. It follows from Proposition 23.1 and (2) that $M/X = C \oplus D$, where C is a direct sum of a finite number of cyclic primary modules and D is the cocyclic module $R(p^\infty)$. Similarly, $A/X = C_1 \oplus D_1$. Here $D_1 \neq 0$; otherwise, $p^k A \subseteq X$ for some k and A is a free module. Consequently, $D_1 = D$. We can also assume that $C_1 \subseteq C$. In this case, $M/A \cong C/C_1$ and M/A is a bounded module.

(5) $\Longrightarrow$ (3). We assume that A is a nonzero submodule of the module M such that $r(A) < r(M)$ and A is not free. Then $A = N \oplus X$, where N is a coreduced module and either X is a free module or $X = 0$. The existence of the embedding $N \to M$ contradicts to (5). Consequently, (3) holds. □

Thus, purely indecomposable modules and copurely indecomposable modules correspond to each other under the Arnold duality. Consequently, if we know properties of purely (resp., copurely) indecomposable modules, then we can obtain properties of copurely (resp., purely) indecomposable modules using the duality.

Using Proposition 25.2, Proposition 25.5, and the functor F, we obtain the following result.

Corollary 25.6. *For any two copurely indecomposable modules M and N, the following conditions are equivalent*:

(1) $M \sim N$;

(2) $\mathrm{Hom}(M, N) \neq 0$ *and* $\mathrm{Hom}(N, M) \neq 0$;

(3) $r(M) = r(N)$ *and* $\mathrm{Hom}(M, N) \neq 0$.

The information about the endomorphism ring of a copurely indecomposable module M is considerably less than the information about the endomorphism ring of a purely indecomposable module. Since F is a duality, the rings $K\,\mathrm{End}(M)$ and $K\,\mathrm{End}(F(M))$ are anti-isomorphic to each other (the anti-isomorphism is defined in remarks to Chapter 7). However, $F(M)$ is a purely indecomposable module and $K\,\mathrm{End}(F(M))$ is a field (Proposition 25.3 and Proposition 25.5). Consequently, in fact, we have anti-isomorphism. Therefore $\mathrm{End}(M)$ is a commutative domain and $K\,\mathrm{End}(M)$ is a field which is isomorphic to some field P, where $K \subseteq P \subseteq \widehat{K}$ and the extension degree $[P : K]$ is finite (see Proposition 25.3).

A torsion-free module, which is quasiisomorphic to a copurely indecomposable module, is copurely indecomposable. For example, this follows from Proposition 25.5. Consequently, copurely indecomposable modules can be classified up to quasiisomorphism by the use of the Arnold duality and Theorem 25.4. Two quasiisomorphic copurely indecomposable modules are not necessarily isomorphic to each other (Example 25.7). The classification of copurely indecomposable modules up to isomorphism is a considerably more difficult problem. In the remaining part of the section, we partially study this problem.

The elements of a copurely indecomposable module of rank $n+1$, $n \geq 1$, can be represented by vectors in $(\widehat{R})^n$ (the vectors are elements of the free $\widehat{R}$-module $(\widehat{R})^n$ of rank n). We call attention to the property that all subsequent events occur in the $\widehat{K}$-space $(\widehat{K})^n$ of dimension n. In particular, all considered modules and spaces (for example, $(\widehat{R})^n$ and K^n) are contained in this space.

If $a_1, \dots, a_n \in \widehat{R}$, then the symbol $A[\Gamma]$ denotes the pure R-submodule in $(\widehat{R})^n$ generated by R^n and the vector $\Gamma = (a_1, \dots, a_n)$ from $(\widehat{R})^n$. It is easy to prove that $A[\Gamma]$ is a coreduced module if and only if the elements $1, a_1, \dots, a_n$ are linearly independent over R. In this case $A[\Gamma]$ is a copurely indecomposable module of p-rank n and the rank of $A[\Gamma]$ is equal to $n+1$ (see Exercise 8).

Now let M be a copurely indecomposable module of rank $n+1$ with basic submodule B of rank n. We note that $n = r_p(M)$. The module M is a pure submodule of the completion $\widehat{B}$, where $\widehat{B}$ is a free $\widehat{R}$-module of rank n. Clearly, we can assume that M is isomorphic to some pure submodule in $(\widehat{R})^n$ and B is isomorphic to R^n. Since $r(M) = n+1$, there exist elements $a_1, \dots, a_n$ in $\widehat{R}$ such that M is isomorphic to the module $A[\Gamma]$, where $\Gamma = (a_1, \dots, a_n)$. Therefore, the formulated classification problem is reduced to the description of copurely indecomposable modules of the form $A[\Gamma]$. An essentially equivalent method of constructing copurely indecomposable modules is presented in Exercise 7.

We present several simple properties of the copurely indecomposable modules $A[\Gamma]$ and their endomorphism rings. It is clear that R^n is a basic submodule of the module $A[\Gamma]$. Furthermore, we have the relations $KA[\Gamma] = K^n \oplus K\Gamma$ (where $K\Gamma$ is the K-subspace generated by the vector Γ) and

$$A[\Gamma] = (K^n \oplus K\Gamma) \cap (\widehat{R})^n.$$

(Here we use our agreement that all considered objects are contained in the same

$\widehat{K}$-space.) It follows from the method of the construction of the functor F that

$$F(A(a_1,\dots,a_n)) \sim A[\Gamma] \quad \text{and} \quad F(A[\Gamma]) \cong A(a_1,\dots,a_n).$$

The rank of the purely indecomposable module $A(a_1,\dots,a_n)$ is equal to $n+1$. The quasiendomorphism algebra $K\operatorname{End}(A[\Gamma])$ is isomorphic to the K-algebra $K\operatorname{End}(A(a_1,\dots,a_n))$ contained in $K\oplus Ka_1\oplus\dots\oplus Ka_n$ (see the remark before Theorem 25.4). Consequently, $\operatorname{End}(A[\Gamma])$ is a domain which is isomorphic to some subring of the subfield $K\operatorname{End}(A(a_1,\dots,a_n))$ in $\widehat{K}$ (some information about this subfield is contained in Proposition 25.3).

Example 25.7 (Arnold–Dugas [12]). *Let $\{a_1,a_2\}$ be a subset in $\widehat{R}$ which is algebraically independent over K* (an algebraic independence is defined in Lemma 19.9), *$\Gamma = (a_1,a_2)$, and let $\Delta = (pa_1,a_2)$. Then $A[\Gamma]$ and $A[\Delta]$ are quasiisomorphic copurely indecomposable modules which are not isomorphic to each other.*

Proof. We have

$$A[\Gamma] = (K^2 \oplus K(a_1,a_2)) \cap (\widehat{R})^2,$$
$$A[\Delta] = (K^2 \oplus K(pa_1,a_2)) \cap (\widehat{R})^2,$$

and

$$F(A[\Gamma]) \cong A(a_1,a_2) = A(pa_1,a_2) \cong F(A[\Delta]).$$

Consequently, $A[\Gamma]$ and $A[\Delta]$ are quasiisomorphic to each other. We denote by φ the endomorphism $(p,1)$ of the space $\widehat{K}\oplus\widehat{K}$ such that φ coincides with the multiplication by p on the first summand and φ coincides with the identity mapping on the second summand. In a similar manner, we define the endomorphism $\psi = (1,p)$. Then $\varphi(\widehat{R})^2 \subseteq (\widehat{R})^2$ and $\varphi(KA[\Gamma]) \subseteq KA[\Delta]$. Therefore $\varphi \in \operatorname{Hom}(A[\Gamma],A[\Delta])$. Similarly, we have that $\psi \in \operatorname{Hom}(A[\Delta],A[\Gamma])$ and $\varphi\psi = p(1,1) = \psi\varphi$.

Since F is a duality and $A(a_1,a_2) = A(pa_1,a_2)$, we have

$$K\operatorname{Hom}(A[\Gamma],A[\Delta]) \cong K\operatorname{End}(A(a_1,a_2)).$$

However,

$$K\operatorname{End}(A(a_1,a_2)) = \{r \in K\oplus Ka_1\oplus Ka_2 \mid ra_1, ra_2 \in K\oplus Ka_1\oplus Ka_2\}$$

(see the paragraph before Theorem 25.4). Considering this property and the algebraic independence of the elements a_1 and a_2, we have $K\operatorname{End}(A(a_1,a_2)) = K$. Consequently, $\operatorname{Hom}(A[\Gamma],A[\Delta]) = R\varphi$, since φ is not divisible by p; similarly, we have $\operatorname{Hom}(A[\Delta],A[\Gamma]) = R\psi$. If we assume that $A[\Gamma]\cong A[\Delta]$, then φ and ψ are isomorphisms, since they are not divisible by p. Consequently, modules $A[\Gamma]$ and $A[\Delta]$ are not isomorphic to each other. □

We specialize Proposition 2.4 on the representation of endomorphisms by matrices. Let $M = \bigoplus_{i=1}^{n} A_i$ be a direct sum of modules. If $a \in M$ and $a = a_1 + \cdots + a_n$ ($a_i \in A_i$), then we represent the element a (if it is necessary) as the vector $(a_1, \ldots, a_n)$. Let $\alpha \in \mathrm{End}(M)$ and let (α_{ij}) be the matrix corresponding to α, where $\alpha_{ij} \in \mathrm{Hom}(A_i, A_j)$. Using the proof of Proposition 2.4, it is easy to verify the matrix relation $\alpha a = a(\alpha_{ij})$, where the vector $a = (a_1, \ldots, a_n)$ and the matrix (α_{ij}) are multiplied by the ordinary rule of matrix multiplication and αa is the corresponding vector.

In what follows the considered matrices are $n \times n$ matrices over the ring R. Such a matrix is invertible if and only if the determinant of the matrix is an invertible element of R.

The following lemma contains a simple isomorphism criterion of modules of the form $A[\Gamma]$. It was already mentioned that matrices are multiplied according to the standard rule.

Lemma 25.8. *We assume that $M = A[\Gamma]$ and $N = A[\Delta]$ are copurely indecomposable modules of rank $n+1$. If $\Delta = s\Gamma V + \Lambda$ for some nonzero element $s \in K$, an invertible matrix V, and the vector Λ from K^n, then M and N are isomorphic to each other.*

Proof. We take the automorphism φ of the space $(\widehat{K})^n$ such that $\varphi(x) = xV^{-1}$ for every vector x in $(\widehat{K})^n$. It is clear that φ is also an automorphism of the module $(\widehat{R})^n$. Furthermore, we have $KA[\Delta] = K^n \oplus K\Delta$, $(K^n)V^{-1} = K^n$,

$$\Delta V^{-1} = s\Gamma + \Lambda V^{-1} \in K\Gamma \oplus K^n = KA[\Gamma], \quad \text{and} \quad \varphi(KA[\Delta]) \subseteq KA[\Gamma].$$

Since $A[\Delta] = KA[\Delta] \cap (\widehat{R})^n$ and $A[\Gamma] = KA[\Gamma] \cap (\widehat{R})^n$, we have $\varphi(A[\Delta]) \subseteq A[\Gamma]$. We can assert that $\varphi\colon A[\Delta] \to A[\Gamma]$ is an isomorphism, since $\varphi(A[\Delta])$ is a pure submodule in $A[\Gamma]$. □

The meaning of the next lemma is the following property: if the modules $A[\Gamma]$ and $A[\Delta]$ are quasiisomorphic to each other, then the vectors Γ and Δ are related to each other in a certain manner.

Lemma 25.9 (Arnold–Dugas [12]). *Let $M = A[\Gamma]$, $r(M) = n+1$, and let N be a torsion-free module which is quasiisomorphic to M. Then*

(1) *N is isomorphic to the module $A[\Gamma W]$ for some upper triangular matrix $W = (w_{ij})$ with nonzero determinant and $h(w_{ii}) > h(w_{ij})$ in every case, where $1 \leq i < j \leq n$ and $w_{ij} \neq 0$;*

(2) *N is isomorphic to the module $A[\Gamma V D]$ for some invertible matrix V and a diagonal matrix D with diagonal elements $p^{e(1)}, \ldots, p^{e(n)}$ and $e(1) \geq \cdots \geq e(n)$;*

(3) *there exists a vector* $\Delta \in (\widehat{R})^n$ *such that* M *is isomorphic to* $A[\Delta]$ *and* N *is isomorphic to* $A[\Delta D]$ *for some diagonal matrix* D *with diagonal elements* $p^{e(1)}, \dots, p^{e(n)}$ *and* $e(1) \geq \dots \geq e(n)$.

Proof. (1) Since M and N are quasiisomorphic to each other, there exists a submodule L of the module M such that N is isomorphic to L and $KL = KM = K^n \oplus K\Gamma$ (Corollary 24.4). First, we prove that L is isomorphic to $A[\Gamma V]$ for some matrix V with nonzero determinant. We note that R^n is a basic submodule of the module M and $B = L \cap R^n$ is a basic submodule of the module L with $KB = K^n$ (Exercise 3). We take a K-automorphism of the space K^n, which extends some isomorphism from B onto R^n; we also take the corresponding invertible $n \times n$ matrix V' over K. We choose p^k ($k \geq 0$) such that $p^k V' = V$ is a matrix over R. It is clear that $\{bV \mid b \in B\} = p^k R^n$. This relation allows us to define $\varphi\colon \widehat{B} \to (p^k \widehat{R})^n$ by the relation $\varphi(x) = xV$. Here φ is an $\widehat{R}$-isomorphism and

$$\varphi\colon KL = K^n \oplus K\Gamma \to KA[\Gamma V] = K^n \oplus K\Gamma V.$$

Since

$$L = \widehat{B} \cap KL \quad \text{and} \quad A[\Gamma V] = (\widehat{R})^n \cap (K^n \oplus K\Gamma V),$$

we have that φ is an isomorphism from L onto $p^k A[\Gamma V]$. Then $(1/p^k)\varphi$ is an isomorphism from L onto $A[\Gamma V]$, which is required.

Now we can apply the standard procedure to the matrix V. Precisely, we multiply the matrix V from the right by some invertible matrices and obtain the required upper triangular matrix W (Exercise 1). By Lemma 25.8, such multiplications do not change the isomorphism class of the module $A[\Gamma V]$. Thus, $A[\Gamma V]$ is isomorphic to $A[\Gamma W]$. Since N is isomorphic to $A[\Gamma V]$, the proof of (1) is completed.

(2) By (1), N is isomorphic to $A[\Gamma W]$ for some matrix W with nonzero determinant. Since R is a discrete valuation domain, there exist invertible matrices V and U with $V^{-1}WU = D$, where D is a diagonal matrix with diagonal elements $p^{e(1)}, \dots, p^{e(n)}$ and $e(1) \geq \dots \geq e(n)$ (Exercise 2). Then $WU = VD$ and it follows from Lemma 25.8 that $A[\Gamma W]$ is isomorphic to $A[\Gamma WU] = A[\Gamma VD]$.

(3) We set $\Delta = \Gamma V$, where V is the matrix from (2). Then the module $M = A[\Gamma]$ is isomorphic to $A[\Delta]$ by Lemma 25.8. Now we use (2) and obtain that N is isomorphic to $A[\Delta D]$. □

Lemma 25.9 reduces the isomorphism problem for two quasiisomorphic copurely indecomposable modules M and N to the case, where $M = A[\Gamma]$ and $N = A[\Gamma D]$ for some diagonal matrix D with diagonal elements $p^{e(1)}, \dots, p^{e(n)}$ and $e(1) \geq \dots \geq e(n)$.

We present an isomorphism criterion for copurely indecomposable modules M with $r(\mathrm{End}(M)) = 1$. In this case, $\mathrm{End}(M) = R$. There are many such modules provided the extension degree $[\widehat{K} : K]$ is sufficiently large (see Exercise 10).

Theorem 25.10 (Arnold–Dugas [12]). *We assume that $M = A[\Gamma]$ is a copurely indecomposable module with $r(\mathrm{End}(M)) = 1$ and $N = A[\Gamma D]$ for some diagonal matrix D with diagonal elements $p^{e(1)}, \dots, p^{e(n)}$ and $e(1) \geq \dots \geq e(n)$. Then M and N are isomorphic to each other if and only if $e(1) = \dots = e(n)$.*

Proof. We define a mapping $\varphi\colon (\widehat{R})^n \to (\widehat{R})^n$ by setting $\varphi(x) = xD$. Then φ induces a homomorphism from $M = A[\Gamma]$ into $N = A[\Gamma D]$, since

$$\varphi\colon KM = K^n \oplus K\Gamma \to KN = K^n \oplus K\Gamma D.$$

The height of the homomorphism φ (as an element of the R-module $\mathrm{Hom}(M, N)$) is equal to $e(n)$. Furthermore,

$$K\,\mathrm{Hom}(M, N) \cong K\,\mathrm{End}(M) = K.$$

Consequently, $\mathrm{Hom}(M, N) = R(1/p^{e(n)})\varphi$, since the homomorphism $(1/p^{e(n)})\varphi$ is not divisible by p. Consequently, the homomorphism is a generator of the R-module $\mathrm{Hom}(M, N)$. Therefore, it is clear that M and N are isomorphic to each other if and only if $(1/p^{e(n)})\varphi$ is an isomorphism. Since $(1/p^{e(n)})\varphi$ coincides with the multiplication by the matrix $(1/p^{e(n)})D$, we have that the modules M and N are isomorphic to each other if and only if this matrix is invertible. This is equivalent to the existence of the relations $e(1) = \dots = e(n)$. □

In the paper [12], Arnold and Dugas also study the classification problem for copurely indecomposable modules M such that $r(\mathrm{End}(M)) = r(M)$ or $1 < r(\mathrm{End}(M)) < r(M)$. We note that $r(\mathrm{End}(M)) \leq r(M)$, since $\mathrm{End}(M)$ is a domain. Indeed, for a fixed nonzero element $a \in M$, the mapping $\alpha \to \alpha(a)$ ($\alpha \in \mathrm{End}(M)$) is an R-module embedding $\mathrm{End}(M) \to M$. The paper contains interesting examples of copurely indecomposable modules of rank 3. They demonstrate obstructions to the search for an exact description of isomorphism classes of copurely indecomposable modules which are quasiisomorphic to the given copurely indecomposable module. Arnold, Dugas, and Rangaswamy [14] consider finite direct sums of purely indecomposable modules. In [14], such sums are called *pi-decomposable* modules. In this paper, the following topics are studied: the search for number invariants with respect to isomorphism of *pi*-decomposable modules, torsion-free homomorphic images of *pi*-decomposable modules, and modules which are quasiisomorphic to *pi*-decomposable modules (in addition, see [17]).

Exercise 1. Let V be an $n \times n$ matrix over a ring R with nonzero determinant. Prove that we can multiply the matrix V from the right by some invertible matrices and obtain an upper triangular matrix $W = (w_{ij})$ with nonzero determinant, and

$h(w_{ii}) > h(w_{ij})$ in every case, where $1 \le i < j \le n$ and $w_{ij} \ne 0$. (The multiplication of a column by an invertible element and interchanging columns can be obtained by indicated multiplications.)

Exercise 2. If W is a matrix with nonzero determinant, then there exist invertible matrices V and U such that $V^{-1}WU$ is a diagonal matrix with diagonal elements $p^{e(1)}, \dots, p^{e(n)}$ and $e(1) \ge \dots \ge e(n)$.

Exercise 3. Let B be a basic submodule of a torsion-free module M of finite rank. If $M' \subset M$ and M/M' is a bounded module, then $B' = B \cap M'$ is a basic submodule of the module M'. Furthermore if $B' \subset B$ and B/B' is bounded, then there exists a $M' \subset M$ such that M/M' is a bounded module and $B' = B \cap M'$.

Exercise 4. If M is a copurely indecomposable module and A is a pure submodule in M, then the factor module M/A is a strongly indecomposable module.

Exercise 5. (a) Let M be a reduced torsion-free module such that $r_p(M) = n$ and $r(M) = n + k$. Then there exists an exact sequence of modules

$$0 \to M \to A_1 \oplus \dots \oplus A_n \to D \to 0,$$

where D is a divisible torsion-free module and A_i is a purely indecomposable module such that $r(A_i) \le k + 1$ for $i = 1, \dots, n$.

(b) Formulate and prove the dual assertion for a coreduced module M.

Exercise 6. Verify the relations

$$A(a_1, \dots, a_n) = KA(a_1, \dots, a_n) \cap \widehat{R},$$
$$KA(a_1, \dots, a_n) = K \oplus Ka_1 \oplus \dots \oplus Ka_n,$$

and

$$K\,\mathrm{End}(A(a_1, \dots, a_n))$$
$$= \{r \in K \oplus Ka_1 \oplus \dots \oplus Ka_n \mid ra_i \in K \oplus Ka_1 \oplus \dots \oplus Ka_n,\ 1 \le i \le n\}.$$

Exercise 7 (Kaplansky [166], Arnold–Dugas [12]). Let V be a vector K-space of dimension $n+1$ with basis $x, y_1, \dots, y_n$. We choose some elements $a_1, \dots, a_n \in \widehat{R}$. Let $a_j = \lim a_{ij}$, where $a_{ij} \in R$, $j = 1, \dots, n$, $i = 0, 1, 2, \dots$. We set

$$w_i = x + \sum_{j=1}^{n} a_{ij} y_j, \quad i \ge 0.$$

Let M be the R-submodule in V generated by the elements $y_1, \dots, y_n, w_0, w_1/p, \dots, w_i/p^i, \dots$. Then

(1) $r_p(M) = n$, $r(M) = n + 1$, and the elements $y_1, \dots, y_n$ generate a basic submodule of the module M;

(2) M is a coreduced module if and only if the system $\{1, a_1, \dots, a_n\}$ is linearly independent; in this case, M is a copurely indecomposable module;

(3) every module M such that $r_p(M) = n$ and $r(M) = n+1$ can be constructed (up to isomorphism) by the given method.

Exercise 8. Prove that the module $A[\Gamma]$ is copurely indecomposable if and only if $\{1, a_1, \dots, a_n\}$ is a linearly independent system of elements of the R-module $\widehat{R}$.

Exercise 9. Prove that $F(A(a_1, \dots, a_n)) \sim A[\Gamma]$ and $F(A[\Gamma]) \cong A(a_1, \dots, a_n)$.

Exercise 10. Let Γ be a vector $(a_1, \dots, a_n)$ such that

$$\{1, a_1, \dots, a_n, a_1 a_i, \dots, a_n a_i\}$$

is a linearly independent system for every i. Then $r(\mathrm{End}(A[\Gamma])) = 1$.

Exercise 11. Let $M \subseteq N$, where M is a purely indecomposable module and N is a reduced torsion-free module of finite rank. Then $M \cong M_*$, where M_* is the pure submodule in N generated by M (see property (7) in Section 7).

26 Indecomposable modules over Nagata valuation domains

In this section, we touch on the classification problem for indecomposable torsion-free modules of finite rank. In complete generality, the problem is very difficult (some information about this topic is contained in remarks at the end of the section and in remarks at the end of the chapter). We call attention to the property that purely indecomposable and copurely indecomposable modules are always indecomposable. We consider this problem for modules over Nagata valuation domains. For such discrete valuation domains, the field extension $K < \widehat{K}$ has several special properties. We will use some of the properties. Before we consider the properties, we present several general results on field extensions. All they are well known and are contained in books of Bourbaki [35] and Zariski–Samuel [326, 327]. The reader certainly knows basic notions of the theory of field extensions. We only mention separable and purely inseparable extensions and properties of valuations in fields.

Let F be a field. A polynomial $f(x)$ over the field F is said to be *separable* if $f(x)$ does not have multiple roots in any extension of the field F. We assume that the field E is an algebraic (in particular, finite) extension of the field F. An element a of the field E is said to be *separable* over F if a is a root of a polynomial which is separable over F. This is equivalent to the property that a minimal polynomial of the element a over the field F is separable. If every element of E is separable over F, then we say that E is a *separable extension* of the field F. An element a is said to be *purely inseparable* over F if there exists an integer $n \geq 0$ such that a^{q^n} is contained in F for some prime integer q. If every element of E is purely inseparable over F, then we say that E is a *purely inseparable extension* of the field F. In this case, the characteristic of the field F is necessarily finite, say q, and a minimal polynomial of every element of E over F has the form $x^{q^n} - c$ for some $n \geq 0$ and $c \in F$. If E is an arbitrary algebraic extension of the field F, then all F-separable elements of the field E form a field which is a maximal separable extension of the field F contained in E.

We define several new notions related to valuations in fields and the corresponding valuation rings (valuations and valuation rings were defined in Section 3). Let F be a field, $(\Gamma, \leq)$ be a linearly ordered group, and let ν be a valuation of the field F with values in Γ. The image of ν under the a mapping $F^* \to \Gamma$ is an ordered subgroup in Γ which is called the *value group* of the valuation ν. Let ν_1, ν_2 be two valuations of the field F with values in Γ. They are said to be *equivalent* if there exists an order-preserving isomorphism φ from the value group $\nu_1 F^*$ onto the value group $\nu_2 F^*$ (i.e., φ is an isomorphism of ordered groups) such that $\nu_2 = \nu_1 \varphi$. The valuations ν_1 and ν_2 are equivalent if and only if the corresponding valuation rings, considered as subrings of the field F, coincide. Therefore, there exist a bijective correspondence between valuation rings in the field F and equivalence classes of valuations.

We briefly recall several results on the behavior of valuations under field extensions. Let E be an extension of a field F and let R' be the valuation ring for ν' in the field E (with values in Γ). The restriction ν of the mapping ν' to F is a valuation of the field F with valuation ring R which is equal to $F \cap R'$. The value group of the valuation ν is a subgroup in the value group of the valuation ν'. Conversely, every valuation of the field F has an extension to a valuation of the field E. The index $[\nu' E^* : \nu F^*]$ of the value group of the valuation ν in the value group of the valuation ν' is called the *ramification index* of the valuation ν' with respect to the valuation ν; the index is denoted by $e(\nu'/\nu)$.

In regards to Section 3, the rings R and R' are local. The fields $R/J(R)$ and $R'/J(R')$ are called *residue fields* of the rings R and R', respectively. Since $J(R) = J(R') \cap F$, we have the field embedding

$$R/J(R) \to R'/J(R')$$

$(r+J(R) \to r+J(R'),\ r \in R)$. Using this embedding, the residue field $R/J(R)$ is identified with some subfield of the residue field $R'/J(R')$. The extension degree $[R'/J(R') : R/J(R)]$ of the field $R'/J(R')$ with respect to the field $R/J(R)$ is called the *residue degree* of the valuation ν' with respect to the valuation ν; it is denoted by $f(\nu'/\nu)$.

If $(\nu'_i)_{i\in I}$ is a family of valuations of the field E extending the valuation ν and every valuation of the field E extending ν is equivalent to exactly one of the valuations ν'_i, then $(\nu'_i)_{i\in I}$ is called a *complete extension system* of the valuation ν on the field E. Important properties of extensions of valuations are gathered in the following theorem (see Bourbaki [35, Chapter VI, Section 8]).

Theorem. *Let F be a field, ν be the valuation of the field F, and let E be a finite extension of degree n of the field F. Then*

(a) *every complete system $(\nu'_i)_{i\in I}$ of extensions of the valuation ν on E is finite;*

(b) *the inequality $\sum_{i\in I} e(\nu'_i/\nu) f(\nu'_i/\nu) \le n$ holds;*

(c) *the valuation rings for ν'_i are pairwise incomparable with respect to inclusion.*

In some cases, the inequality in (b) can be replaced by an equality.

Corollary. *If the valuation ν is discrete and the extension E is separable over F, then $\sum_{i=1}^{s} e(\nu'_i/\nu_i) f(\nu'_i/\nu) = n$, where $(\nu'_i)_{1\le i\le s}$ is a complete system of extensions of the valuation ν to the field E.*

With regard to discrete valuations, we also note that if ν' is some extension of the valuation ν of the field F to the field E, then the valuation ν' is discrete if and only if the valuation ν is discrete.

Let $V_\gamma = \{x \in F \mid \nu(x) > \gamma\}$, where $\gamma \in \Gamma$. The set of all subgroups V_γ $(\gamma \in \Gamma)$ of the additive group of the field F forms a basis of neighborhoods of zero of the topology $\mathcal{T}_\nu$ in the field F; this topology is called the *topology determined by the valuation ν*. The topology $\mathcal{T}_\nu$ is a Hausdorff topology and the mapping $\nu\colon F^* \to \Gamma$ is continuous if we assume that the group Γ is equipped by the discrete topology. If ν is discrete, then the topology on the valuation ring for ν induced by the topology $\mathcal{T}_\nu$ coincides with the p-adic topology. (However, this is not true in the general case (Exercise 2).) The completion $\widehat{F}$ of the field F with respect to the topology $\mathcal{T}_\nu$ is a topological field; the valuation ν is extended by continuity to a unique valuation $\widehat{\nu}\colon \widehat{F}^* \to \Gamma$. The topology of the field $\widehat{F}$ is the topology defined by the valuation $\widehat{\nu}$. The valuation ring for $\widehat{\nu}$ is the completion of the valuation ring for ν. For a discrete valuation ν, the completion of $\widehat{F}$ coincides with the field of fractions of the valuation ring $\widehat{\nu}$.

Let R be a discrete valuation domain. Then K and $\widehat{K}$ are topological fields with respect to the corresponding discrete valuation ν, $\widehat{K}$ is the completion of the field K, where K still is the field of fractions of the domain R and $\widehat{K}$ is the field of fractions of the p-adic completion $\widehat{R}$ of the domain R. The p-adic topology on R or $\widehat{R}$ coincides with the induced topology. The valuation ν of the field K has a unique extension to a valuation of the field $\widehat{K}$.

Now we define Nagata valuation domains. Since $R = \widehat{R} \cap K$, it is easy to verify that the dimension of the K-space $\widehat{K}$ is equal to the rank of the R-module $\widehat{R}$. The domain R is called a *Nagata valuation domain* if the extension degree $[\widehat{K} : K]$ of the field $\widehat{K}$ over K is finite and exceeds 1 (therefore R is not a complete domain). The existence of such valuation domains was proved by Nagata [244, Example 3.3].

Let R be a Nagata valuation domain. We prove that $\widehat{K}$ is a purely inseparable extension of the field K (see Ribenboim [254]). Let $\langle p \rangle$ be an infinite cyclic group, where p is the prime element of the ring R, and let $\nu \colon K^* \to \langle p \rangle$ be the canonical valuation mentioned in Section 3. The valuation can be extended to the canonical valuation $\widehat{\nu} \colon \widehat{K}^* \to \langle p \rangle$. It was mentioned earlier that $\widehat{\nu}$ is a unique extension ν to a valuation of the field $\widehat{K}$.

Now we assume that F is a separable extension of the field K contained in $\widehat{K}$. Every valuation of the field F can be extended to a valuation of the field $\widehat{K}$. Consequently, ν has a unique extension to a valuation of the field F and $s = 1$ in the formula from the above corollary. Let ν' be an extension of ν to a valuation of the field F. Then

$$e(\nu'/\nu)f(\nu'/\nu) = [F : K].$$

It follows from the choice of ν that $e(\nu'/\nu) = 1$. Furthermore, since

$$R/pR \subseteq R'/pR' \subseteq \widehat{R}/p\widehat{R} = R/pR,$$

where R' is the valuation ring ν' (see the point before the definition of the residue degree of the valuation), we have $R/pR = R'/pR'$, whence $f(\nu'/\nu) = 1$. We obtain $[F : K] = 1$ and $F = K$; this implies that $\widehat{K}$ is purely inseparable over K. We can make an important conclusion that the fields K and $\widehat{K}$ have finite characteristic q and $[\widehat{K} : K]$ is a power of q. For every $a \in \widehat{K}$, there exists an integer $n \geq 0$ with $a^{q^n} \in K$.

Before we pass to modules over Nagata valuation domains, we make some remarks. First, if the degree $[\widehat{K} : K]$ is infinite, then the rank of the R-module $\widehat{R}$ is infinite. In such a situation, there exist indecomposable torsion-free modules of any finite rank (Example 11.9). If $[\widehat{K} : K] = 1$, then $R = \widehat{R}$. In this case, every indecomposable torsion-free module is equal to R or K (Corollary 11.8). In what follows the results of the previous section are essentially used. We especially note that all reduced indecomposable modules appeared below are purely indecomposable or copurely indecomposable.

Till the end of the section, R is a Nagata valuation domain. We begin with the minimally possible case, where $[\widehat{K} : K] = 2$. For such domains R, we have $r(\widehat{R}) = 2$ and $r_p(\widehat{R}) = 1$. Theorem 26.1 and Theorem 26.2 are obtained by Zanardo [325] with the use of matrix Kurosh invariants for torsion-free modules of finite rank (see also remarks to the chapter in connection to these invariants).

Theorem 26.1. *Let M be an indecomposable module of finite rank. Then M is isomorphic to one of the modules R, K, $\widehat{R}$.*

Proof. If $r(M) = 1$, then M is isomorphic to R or K. Let $r(M) = 2$. For $r_p(M) = 2$, M is a free module (see the beginning of Section 23); for $r_p(M) = 0$, M is divisible. In any case, M is a decomposable module, which is impossible. Therefore $r_p(M) = 1$. Consequently, M is a purely indecomposable module (purely indecomposable and copurely indecomposable modules were studied in the previous section). It is isomorphic to some pure submodule N of the module $\widehat{R}$. Since $r(N) = r(\widehat{R})$, we have $N = \widehat{R}$, whence $M \cong \widehat{R}$.

We assume that $r(M) = 3$. Similar to the previous paragraph, we have that either $r_p(M) = 1$ or $r_p(M) = 2$. In the first case, M need to be isomorphic to some pure submodule in $\widehat{R}$ which contradicts to the conditions for ranks. In the case $r_p(M) = 2$, the module M is copurely indecomposable. Then $F(M)$ is a purely indecomposable module of rank 3 (Proposition 25.5). As above, this is impossible. Thus, indecomposable modules of rank 3 do not exist. Now we assume that every reduced torsion-free module of rank $<n$ (where $n \geq 3$) is the direct sum of modules of the form R or $\widehat{R}$. We take a reduced torsion-free module M of rank n. It is clear, we can assume that M is a coreduced module. The module M cannot be copurely indecomposable; otherwise, $F(M)$ is a purely indecomposable module of rank n. Consequently, there exists a pure nonfree submodule N in M of rank $n-1$ (Proposition 25.5). By the induction hypothesis,

$$N = R \oplus \cdots \oplus R \oplus \widehat{R} \oplus \cdots \oplus \widehat{R}$$

where the summands of the form $\widehat{R}$ are necessarily present in the decomposition. The pure complete submodule $\widehat{R} \oplus \cdots \oplus \widehat{R}$ is a direct summand:

$$M = \widehat{R} \oplus \cdots \oplus \widehat{R} \oplus X.$$

We can apply the induction hypothesis to the module X. □

We have that for $[\widehat{K} : K] = 2$, every torsion-free module of finite rank is the direct sum of a divisible module, a free module, and a complete module.

We pass to a considerably more complicated situation, where $[\widehat{K} : K] = 3$. Then we have that the characteristic of the field $\widehat{K}$ is equal to 3 and $a^3 \in K$ for every $a \in \widehat{K}$. We fix some invertible in $\widehat{R}$ element u which is not contained in R. Then $u^3 = w \in R$; therefore w is an invertible element in R. Furthermore, we have $\widehat{K} = K(u)$ and the elements 1, u, u^2 form a basis of the K-space $\widehat{K}$ and a maximal linearly independent system of elements of the module $\widehat{R}$. Here we denote by $K(u)$ the simple extension of the field K obtained by adjoining the element u.

We consider indecomposable modules of rank ≤ 2. Let a be some element of the complement $\widehat{R} \setminus R$. We recall that $A(a)$ denotes the pure submodule in $\widehat{R}$ generated by the elements 1 and a. (This submodule can be also denoted by $A[a]$; in fact, $A(a) = A[a]$, see Section 25.) We have that $A(a)$ is a purely indecomposable module of rank 2. The above properties hold for the module $A(u)$, where u is the chosen element. The role of this module is clarified by the following result.

Theorem 26.2 (Zanardo [325]). *If M is an indecomposable module of rank ≤ 2, then M is isomorphic to one of the modules R, K, $A(u)$.*

Proof. Similar to the previous theorem, we can assume that $r(M) = 2$. We similarly obtain that $r_p(M) = 1$ and M is a purely indecomposable module. Then $M \cong A(a)$ for some element $a \in \widehat{R} \setminus R$ (see Section 25). We verify that $A(a) \cong A(u)$ for every such element a. Considering that 1, u, u^2 form a basis of the K-space $\widehat{K}$, we have $ra = r_0 + r_1 u + r_2 u^2$ for some $r, r_i \in R$, where $r \neq 0$. Since $A(a) = A(ra)$, we can assume that $r = 1$. Furthermore, it is obvious that $A(a) = A(r_1 u + r_2 u^2)$. To simplify calculations, we denote the module $A(r_1 u + r_2 u^2)$ by $A(su + tu^2)$, where $s, t \in R$. The multiplication of the ring $\widehat{R}$ by the element $s - tu$ is an endomorphism of the R-module $\widehat{R}$. Since the multiplication by $s - tu$ is a nonzero homomorphism from $A(su + tu^2)$ into $A(u)$, these modules are isomorphic to each other by Proposition 25.2. Consequently, $A(a) \cong A(u)$. □

Let $a, b \in \widehat{R} \setminus R$ and let $\Gamma = (a, b)$ be a vector. All vectors appeared later are vectors of length 2. We recall that the symbol $A[\Gamma]$ denotes the pure submodule in $\widehat{R} \oplus \widehat{R}$ generated by the module $R \oplus R$ and the vector Γ (see Section 25). Here we denote this submodule by $A[a, b]$. The module $A[a, b]$ is indecomposable if and only if 1, a, and b are linearly independent elements in $\widehat{R}$ (see Exercise 8 in Section 25). In this case $A[a, b]$ is a copurely indecomposable module of p-rank 2 and the rank of $A[a, b]$ is equal to 3. We have relations

$$KA[a, b] = K(1, 0) \oplus K(0, 1) \oplus K(a, b)$$

and

$$A[a, b] = KA[a, b] \cap (\widehat{R} \oplus \widehat{R})$$

(Exercise 6 in Section 25). We also recall the notation $A(a,b)$ from Section 25. We denote by $A(a,b)$ the pure submodule in $\widehat{R}$ generated by the elements 1, a, b. If these elements are linearly independent, then $A(a,b) = \widehat{R}$. Consequently,

$$A[a,b] \sim F(A(a,b)) = F(\widehat{R})$$

(here F is the Arnold duality; see Sections 24, 25 and Exercise 9 from Section 25). Therefore, the modules of the form $A[a,b]$, where 1, a, and b are linearly independent elements, pairwise are quasiisomorphic to each other. In particular, $A[a,b] \sim A[u,u^2]$. In fact, Arnold and Dugas have proved the following result in the paper [10]. If M is an indecomposable torsion-free module of rank 3, then M is isomorphic to one of the modules $\widehat{R}$ or $A[u,p^j u^2]$ for some $j \geq 0$. An arbitrary indecomposable torsion-free module of finite rank is isomorphic to some module from the following list: R, K, $\widehat{R}$, $A(u)$, and $A[u,p^j u^2]$ for all $j \geq 0$. In addition, Arnold and Dugas essentially use the result of Lady [194] that the modules from this list describe up to quasiisomorphism all strongly indecomposable torsion-free modules of finite rank (we recall that $[\widehat{K} : K] = 3$). At the end of their paper, Arnold and Dugas prove that if $[\widehat{K} : K] \geq 4$, then for a given $m \geq 2$, there exists a strongly indecomposable (in particular, indecomposable) module of p-rank m whose rank is equal to $2m$.

Exercise 1. Let E be a field, a be an element of E which is algebraic over a subfield F of the field E, and let f be a minimal polynomial of the element a of degree n. Then we have:

(1) the simple extension $F(a)$ is isomorphic to the factor ring $F[x]/(f)$, where (f) is the ideal generated by f;

(2) $[F(a) : F] = n$ and $\{1, a, \dots, a^{n-1}\}$ is a basis of the vector space $F(a)$ over F.

Exercise 2. Let F be a field, ν be a valuation of F, R be the valuation ring for ν, and let J be the maximal ideal of R. The topology defined on the ring R by the valuation ν, coincides with the J-adic topology if and only if either R is a field or R is a discrete valuation domain. (The J-adic topology is defined in Section 14.)

The remaining exercises are based on the Arnold–Dugas papers [10] and [12]. In these papers, modules are considered over a domain R with $[\widehat{K} : K] = 3$. We preserve the notation from Theorem 26.2 and the text after the proof of the theorem.

Exercise 3. (a) $A[u, p^i u^2] \cong A[p^i u, u^2]$ for every $i \geq 1$.

(b) If $A[u, p^i u^2] \cong A[u, p^j u^2]$, then $i = j$.

(c) There exist embeddings

$$A[u, p^i u^2] \to A[u, p^{i-1} u^2] \quad \text{and} \quad A[u, p^{i-1} u^2] \to A[u, p^i u^2].$$

In every case the image has index p.

Exercise 4. If M and N are indecomposable torsion-free modules of p-rank 2 and of rank 3, then $\operatorname{tr}_M(N) = N$, where $\operatorname{tr}_M(N) = \sum \operatorname{Im}(\varphi)$ and φ runs over all homomorphisms from M into N; this is the so-called trace of M in N.

Exercise 5. We assume that N is a torsion-free module of finite rank and X is a submodule in N such that either $A = N/X \cong A(u)$ or $A = N/X \cong A[u, p^i u^2]$ for some $i \geq 0$. If $\operatorname{tr}_A(N) = N$, then X is a direct summand of the module N.

Exercise 6. The R-modules $A[u, u^2]$ and $\operatorname{End}(A[u, u^2])$ are isomorphic to each other.

Remarks. Ribenboim [254] studied Nagata valuation domains. The paper of Facchini–Zanardo [68] contains other results in a more general case.

Arnold [4] transfered his duality presented in Section 24 to the category of quasihomomorphisms of factor-divisible torsion-free Abelian groups of finite rank. Earlier, Warfield [302] discovered a duality for locally free torsion-free groups. The duality is induced by the functor $\operatorname{Hom}(-, A)$, where A is a group of rank 1. Fomin [81, 82] and Vinsonhaler and Wickless [296] continued these studies. They have proved dualities which unite dualities of Warfield and Arnold. Other dualities and equivalences for Abelian groups were discovered in papers of Fomin and Wickless [83, 84, 85, 86].

In remarks to Chapter 3, it was already noted that equivalences and dualities (in category-theoretical sense) are very important. Theorem 13.1 and Theorem 22.2 provide examples of equivalences. In the paper [194], Lady has proved equivalences between some categories of quasihomomorphisms of torsion-free modules of finite rank and some categories of finitely generated modules over some hereditary Artinian K-algebra. Furthermore, Lady used known properties of this algebra and modules over it. Morita-equivalences and Morita-dualities were samples for such theorems (see Faith [69]).

We recommend to the reader the deep Lady's study [194, 195, 196]. In the papers, the notion of a "splitting field" of a torsion-free module is introduced and successfully used. Below, M denotes a torsion-free module of finite rank over a discrete valuation domain R. Let F be a field with $K \subseteq F \subseteq \widehat{K}$. We set $S = \widehat{R} \cap F$. The field F is called a *splitting field* for the module M or, equivalently, the module

M is said to be F*-decomposable* if $S \otimes M$ is a direct sum of a free module and a divisible S-module. In this case, S is a discrete valuation domain with prime element p. The module M always has a splitting field F that is a finite extension of K. Every purely indecomposable or copurely indecomposable module M has a unique least splitting field. The main Lady's studies are related to the situation, where the degree of a splitting field F over K is finite. Under such an assumption, for the category of quasi-homomorphisms of K-decomposable modules, the equivalence indicated above was proved. This lead Lady to the classification (up to a quasi-isomorphism) of strongly indecomposable F-decomposable modules for $[F : K] = 2, 3$. There are finitely many of such modules. If $[F : K] \geq 4$, then there exist strongly indecomposable F-decomposable modules of arbitrarily large rank. It is clear that for a Nagata valuation domain, the field $\widehat{K}$ is a splitting field for every module M. Consequently, the results of Lady are applicable. In such way, they were already mentioned in Section 26.

The known Kurosh matrix invariants (see [191] and [93, Sec. 93]) define an Abelian torsion-free group of finite rank up to an isomorphism. Arnold presented a modified version of these classical invariants for torsion-free modules of finite rank; he used them in [3]. This version was used by Zanardo [325] lateron.

The study from Section 26 is related to the existence and uniqueness problem of direct decompositions of modules (see also Sections 24, 27, and 29). For torsion-free modules over a commutative domain, a natural problem related to this problem is the search for the largest possible rank of indecomposable torsion-free modules of finite rank. Vámos made a major contribution to the solution of this problem for modules over (not necessarily discrete) valuation domains. The reader certainly needs to take a look to his incentive paper [292]. This paper contains solutions of several problems; in addition, related results are presented. If R is a commutative domain, then we follow Vámos and set $\mathrm{fr}(R) = n$ provided there exists an indecomposable torsion-free R-module of rank n and every torsion-free R-module of finite rank $> n$ is decomposable; $\mathrm{fr}(R) = \infty$ means that for any positive integer m, there exists an indecomposable torsion-free R-module whose rank is finite and exceeds m. Matlis studied domains with $\mathrm{fr}(R) = 1$. If R is a valuation domain, then by the result of Kaplansky [168] and Matlis [221], we have that $\mathrm{fr}(R) = 1$ if and only if R is a maximal domain. Now we can represent the results, which are proved or mentioned in Section 26, as follows. If the extension degree $[\widehat{K} : K]$ is infinite, then $\mathrm{fr}(R) = \infty$ and $\mathrm{fr}(R) = n$ for $[\widehat{K} : K] = n$, where $n = 1, 2, 3$. If $[\widehat{K} : K] \geq 4$, then $\mathrm{fr}(R) = \infty$.

Let R be the ring of all rational numbers whose denominators are coprime to integers from some fixed finite set of prime integers (R is the intersection of a finite number of domains $\mathbb{Z}_p$). Yakovlev [324] described categories of direct summands of direct sums of several copies of a torsion-free R-module A of finite rank (as a monoid with the operation induced by the direct addition); see also [323] and the book [67] of Facchini.

The uniqueness of a direct decomposition of a given module usually means that the Krull–Schmidt theorem holds for the decomposition. The end of Section 27 contains some information about this theorem and related studies. Vámos [292] and Lady [194] have proved that the Krull–Schmidt theorem holds for every torsion-free module of finite rank over a discrete valuation domain if and only if the domain is a Henselian ring. A general theory of Henselian rings was developed in the book of Nagata [244].

At the end of Section 25, the results of Arnold, Dugas, and Rangaswamy on *pi*-decomposable modules were mentioned. A module is said to be *pi-decomposable* if it is a finite direct sum of purely indecomposable modules. The class of *pi*-decomposable modules can be enlarged in the following direction. We denote by $\mathcal{A}$ the class of all finite direct sums of torsion-free modules M of finite rank such that $\mathrm{End}(M)$ is a discrete valuation domain (see, for example, Exercise 7 in Section 37), $\mathcal{A}'$ is the class of all modules which are quasiisomorphic to modules in $\mathcal{A}$.

Problem 11. Study properties of modules in the classes $\mathcal{A}$ and $\mathcal{A}'$.

Problem 12. (a) Describe the Jacobson radical of the endomorphism ring of a module in the class $\mathcal{A}'$.

(b) Describe automorphism groups of modules in $\mathcal{A}'$.

In the case of Abelian groups, the modules in the class $\mathcal{A}$ correspond to finite direct sums of torsion-free groups A of finite rank such that $\mathrm{End}(A)$ is a strongly homogeneous ring. (The material related to similar groups is presented in [183, Chapter 7] and in exercises from [183, Section 41].) The *pi*-decomposable modules correspond to direct sums of torsion-free groups of finite rank with cyclic p-basis subgroups (such groups are considered in [183, Section 44]). All almost completely decomposable groups (i.e., subgroups of finite index in completely decomposable torsion-free groups of finite rank) are contained in the analogue of the class $\mathcal{A}'$. An explicit theory of these groups is presented in the book of Mader [218].

Chapter 6

Mixed modules

In Chapter 6, we consider the following topics:

- the uniqueness and refinements of decompositions in additive categories (Section 27);
- isotype, nice, and balanced submodules (Section 28);
- the categories Walk and Warf (Section 29);
- simply presented modules (Section 30);
- decomposition bases and extending of homomorphisms (Section 31);
- Warfield modules (Section 32).

In Sections 11, 21 and 22 we already dealt with mixed modules. This chapter is completely devoted to such modules. For some time, the theory of mixed modules was in stagnation as compared with the theories of primary modules and torsion-free modules. At the present time, the theory is intensively studied. This situation is already obvious in many studied subclasses of direct sums of mixed modules of torsion-free rank 1.

Mixed modules "are assembled" from primary modules and torsion-free modules. Consequently, they inherit the complexity of the structure of these objects. The "assembling" process also increase difficulties.

According to the definition presented in Section 4, a mixed module M necessarily contains nonzero elements of finite order and elements of infinite order. We recall, that $t(M)$ is the torsion submodule (or the *primary submodule*) of the module M, i.e., the set of all elements in M of finite order. The rank of the torsion-free factor module $M/t(M)$ was called earlier the torsion-free rank of the module M. In the present chapter, the torsion-free rank of the mixed module M is called, for brevity, the *rank* of the module M. This does not lead to confusion.

A mixed module M is said to be *split* if $M = t(M) \oplus F$ for some torsion-free module F. The module M is always split provided $t(M)$ is a bounded module or $M/t(M)$ is a free module.

The chapter is very rich in content. In Section 27, we work with an arbitrary additive category. In Section 29, we work with two categories of module origin. We consider several types of submodules playing an important role for mixed modules (Section 28). In the remaining Sections 30–32, the theory of one class of mixed modules is developed. Theorem 32.6 is the culmination of the theory.

Without loss of generality, we will assume that all modules are reduced (unless otherwise stated). This considerably reduces proofs.

The symbol ω always denotes the first infinite ordinal number.

27 Uniqueness and refinements of decompositions in additive categories

We consider two quite general isomorphism theorems for direct decompositions and refinements of direct sums in an additive category. (These notions were introduced before Theorem 8.6.) The section is based on the paper of Walker–Warfield [298]. We work in some additive category $\mathcal{E}$. In Section 1 and Section 24, some information about additive categories is contained. We also recall several simple category-theoretical notions. Other more special notion will be defined as needed.

A morphism $f\colon A \to B$ of the category $\mathcal{E}$ is called a *monomorphism* if for every object X and any morphisms $g, h\colon X \to A$, the relation $gf = hf$ implies the relation $g = h$. The composition of two monomorphisms is a monomorphism; if fg is a monomorphism, then f is a monomorphism.

If $f\colon A \to M$ is a monomorphism, then A is called a *subobject* of the object M; we write $A \subset M$ in this case. Strictly speaking, subobjects coincide with pairs (A, f). To simplify the presentation, we will everywhere write A instead of (A, f) or $f\colon A \to B$ if this does not lead to confusion. In this situation, we are based on the notion of an equivalence for monomorphisms. Two monomorphisms $f\colon A \to M$ and $g\colon B \to M$ are said to be *equivalent* if there exist morphisms $e : A \to B$ and $h\colon B \to A$ such that $f = eg$ and $g = hf$. It is easy to verify that such morphisms e and h are uniquely defined isomorphisms. Now we can choose a representative in every class of equivalent monomorphisms; we assume that the representative is a subobject.

The relation $\subset$ between subobjects is a partial order. The greatest lower bound of two subobjects A and B of the object M (if it exists) is called the *intersection* of A and B; it is denoted by $A \cap B$.

We pass to direct sums in the category $\mathcal{E}$. Many properties of the sums are similar to the corresponding properties of direct sums of modules. We also say that a subobject $A \subset M$ is a direct summand of the object M if there exists a subobject $B \subset M$ such that $M = A \oplus B$. We already have Lemma 24.1. We consider two additional properties.

Lemma 27.1. *Let M be an object of the category $\mathcal{E}$, $M = A \oplus B$, e_A and e_B be the corresponding embeddings, and let p_A and p_B be the corresponding projections. Then*

(1) *if $e_C : C \to M$ is a monomorphism and $e_C p_A : C \to A$ is an isomorphism,*

then $M = C \oplus B$ *with embeddings* e_C, e_B *and projections* $q_C = p_A(e_C p_A)^{-1}$, $q_B = (1_M - q_C e_C)p_B$;

(2) *if* $M = C \oplus D$ *with embeddings* e_C, e_D *and projections* p_C, p_D *and* $e_A p_C : A \to C$ *is an isomorphism, then* $B \cong D$.

Proof. (1) It follows from calculations that $e_C q_B = 0$, $e_B q_C = 0$, $e_C q_C = 1_C$, $e_B q_B = 1_B$, and $1_M = q_C e_C + q_B e_B$. It remains to refer to Lemma 24.1.

(2) By (1), we obtain a decomposition $M = A \oplus D$ with embeddings e_A, e_D and projections q_A, q_D, where $q_A = p_C(e_A p_C)^{-1}$ and $q_D = (1_M - q_A e_A)p_D$. We state that $e_D p_B : D \to B$ is an isomorphism with the inverse isomorphism $e_B q_D : B \to D$. Indeed, we have

$$e_B q_D e_D p_B = e_B(q_A e_A + q_D e_D)p_B = e_B 1_M p_B = e_B p_B = 1_B$$

and

$$e_D p_B e_B q_D = e_D(p_A e_A + p_B e_B)q_D = e_D q_D = 1_D. \qquad \square$$

The definitions of isomorphic direct decompositions of a module and refinements of a direct decomposition, which presented before Theorem 8.6, can be transfered to direct decompositions of objects of an additive category without changes.

First, we consider finite direct sums.

Theorem 27.2 (Walker–Warfield [298]). *Let* M *be an object of the additive category* $\mathcal{E}$ *such that* $M = \bigoplus_{i=1}^{n} A_i$, *where* $\operatorname{End}(A_i)$ *is a local ring for all* i. *If* $M = \bigoplus_{j=1}^{m} B_j$, *where the ring* $\operatorname{End}(B_j)$ *does not have nontrivial idempotents for* $j = 1, \dots, m$, *then* $m = n$ *and there exists a permutation* θ *of the set* $\{1, 2, \dots, n\}$ *such that* $A_i \cong B_{\theta(i)}$ *for* $i = 1, \dots, n$. *An arbitrary direct decomposition of the object* M *can be refined to a decomposition consisting of at most* n *indecomposable summands.*

Proof. Let $e_i : A_i \to M$ and $f_j : B_j \to M$ be the embeddings and let $p_i : M \to A_i$ and $q_j : M \to B_j$ be the projections for the given two direct sums. Then $\sum_{j=1}^{m} e_1 q_j f_j p_1$ is the identity morphism of the object A_1. Since $\operatorname{End}(A_1)$ is a local ring, one of the summands is an invertible element of this ring (Proposition 3.1). We can assume (perhaps after renumbering) that the summand is the element $\gamma = e_1 q_1 f_1 p_1$. Now we consider the endomorphism $f_1 p_1 \gamma^{-1} e_1 q_1$ of the object B_1. After calculation, we obtain that the endomorphism is an idempotent. By assumption, the endomorphism is equal to 0 or 1. It is easy to verify that $(e_1 q_1 f_1 p_1 \gamma^{-1})^2$ is the identity morphism of the object A_1. Consequently, the factor $f_1 p_1 \gamma^{-1} e_1 q_1$ of it is equal to the identity morphism of the object B_1. We obtain that the morphism $f_1 p_1$ is an isomorphism from B_1 onto A_1 with the inverse isomorphism $\gamma^{-1} e_1 q_1$. It follows from Lemma 27.1 (1) that

$$M = B_1 \oplus A_2 \oplus \cdots \oplus A_n.$$

By Part (2) of this lemma, we have

$$A_2 \oplus \cdots \oplus A_n \cong B_2 \oplus \cdots \oplus B_m.$$

The assertion is verified by the use of the induction on the number of the summands A_i.

We pass to the second assertion. Let S be the endomorphism ring of the object M. The elements $p_1 e_1, \ldots, p_n e_n$ form a complete orthogonal system of idempotents of the ring S (see Lemma 24.1). Consequently, we have a decomposition of the ring S into the direct sum of right ideals $S = \bigoplus_{i=1}^n (p_i e_i)S$. Equivalently, we have the direct decomposition of the right S-module S. Now we have

$$\mathrm{End}_S((p_i e_i)S) \cong (p_i e_i)S(p_i e_i) \cong \mathrm{End}(A_i)$$

(see Exercise 2 with regard to the second isomorphism). This implies that for every i, the endomorphism ring $\mathrm{End}_S((p_i e_i)S)$ is local. Now we take any another decomposition of the ring S into a direct sum of right ideals. We have two decompositions of the ring S and we can apply one of variants of the Krull–Schmidt theorem (e.g., Theorem 24.2) to these decompositions. As a result, we obtain that any another decomposition does not contain more than n summands. Therefore, the required assertion on refinements follows from Lemma 24.1. □

We cannot state that endomorphism rings of indecomposable direct summands from the theorem do not have nontrivial idempotents, since idempotents are not necessarily split. In particular, decompositions of the ring S do not necessarily induce decompositions of the object M. In general, the proved theorem is weaker than Theorem 24.2.

In the category $\mathcal{E}$, the direct sum of every infinite set of objects can be defined similarly to the definition of finite direct sums. For infinite direct sums, Lemma 24.1 holds in the necessity part (excluding the relation $1_A = q_1 e_1 + \cdots + q_n e_n$). If the category $\mathcal{E}$ has the direct sum of every infinite set of objects, then we say that $\mathcal{E}$ is a *category with infinite sums*. In such a category, many properties of direct sums of modules hold. For example, we can "amalgamate" or "disintegrate" direct summands. For example, let $M = \bigoplus_{i \in I} A_i$ and $J \subset I$. Then there exists a direct sum $M = M_1 \oplus M_2$, where $M_1 = \bigoplus_{i \in J} A_i$ and $M_2 = \bigoplus_{i \in I \setminus J} A_i$; the embeddings and the projections of this direct sum are naturally related to the embeddings and the projections of the original sum.

We also formulate the known notion of the kernel of a morphism. The pair (L, l), where $l : L \to A$ is a monomorphism, is called the *kernel* of the morphism $f\colon A \to B$ if $lf = 0$ and every morphism $g\colon X \to A$ with $gf = 0$ can be uniquely represented in the form $g = hl$ for some $h\colon X \to L$. If there exist different kernels of the morphism f, then they define the same subobject of the object A in the sense indicated in the beginning of the section. We denote it by $\operatorname{Ker}(f)$. In the category of modules, it is clear that we obtain the ordinary notion of the kernel of a homomorphism. We say that $\mathcal{E}$ is a *category with kernels* (or $\mathcal{E}$ *has kernels*) if every morphism of the category $\mathcal{E}$ has the kernel.

Lemma 27.3. *Let $\mathcal{E}$ be an additive category with kernels. The following assertions hold*:

(1) *if $\pi \in \operatorname{End}(M)$ is an idempotent, then $M = \operatorname{Ker}(\pi) \oplus \operatorname{Ker}(1 - \pi)$*;

(2) *if $A, C \subset M$ and A is the kernel, then the intersection $A \cap C$ exists*;

(3) *if $M = A \oplus B$ and $A \subset C \subset M$, then $C = A \oplus (C \cap B)$.*

Proof. (1) We use the definition of the kernel several times. Let $i : \operatorname{Ker}(\pi) \to M$ and $j : \operatorname{Ker}(1 - \pi) \to M$ be monomorphisms. Then $i\pi = 0$ and $j(1 - \pi) = 0$. Since $(1 - \pi)\pi = 0$, there exists a unique morphism $p : M \to \operatorname{Ker}(\pi)$ with $1 - \pi = pi$. Similarly, there exists a morphism $q : M \to \operatorname{Ker}(1 - \pi)$ with $\pi = qj$. Then i and j, p and q satisfy the conditions of Lemma 24.1, i.e., i, j are the embeddings and p, q are the projections for the sum presented in (1). It is obvious that $pi + qj = 1_M$. Furthermore, it follows from the relations $i = 1 \cdot i$ and $i = (ip)i$ that $ip = 1$; similarly, $jq = 1$. Finally, it follows from the relations $0 \cdot j = 0$ and $(iq)j = 0$ that $iq = 0$. Similarly, we have $jp = 0$.

(2) Let $i : A \to M$ be the kernel of the morphism $f\colon M \to N$ and let $j : C \to M$ be a monomorphism. We take the kernel $l : L \to C$ of the morphism $jf\colon C \to N$. We verify that the monomorphism $lj : L \to M$ is the intersection of the subobjects A and C. We have $L \subset C$. Since $(lj)f = 0$, there exists a morphism $h\colon L \to A$ with $lj = hi$. Therefore h is a monomorphism and $L \subset A$. We assume that $X \subset M$, $X \subset A$, $X \subset C$, g, g_A, g_C are the corresponding monomorphisms. We can assume that $g_A i = g = g_C j$, since we have equivalent monomorphisms. It follows from $g_A(if) = 0$ that $g_C(jf) = 0$. Consequently, there exists a unique morphism $k : X \to L$ with $g_C = kl$. Since the morphism k is a monomorphism, we have $X \subset L$. (We note that $g_A = kh$.) Thus, L is the intersection $A \cap C$.

(3) Let $e_A : A \to M$ and $p_A : M \to A$ be the embedding and the projection and let $k : A \to C$ and $e_C : C \to M$ be monomorphisms; we can assume that $ke_C = e_A$. The morphism $e_C p_A k$ is an idempotent of the ring $\operatorname{End}(C)$. Consequently, it follows from (1) that we have the decomposition $C = \operatorname{Ker}(1 -$

$\pi) \oplus \mathrm{Ker}(\pi)$. Let $l : C \cap B \to C$ be a monomorphism, where l is the kernel of the morphism $e_C p_A$ (see (2)). Then $\mathrm{Ker}(1 - \pi) = A$, and $\mathrm{Ker}(\pi) = C \cap B$. □

In particular, Part (2) of the proved lemma is applicable in the case, where A is an arbitrary direct summand of the object M, since any direct summand is the kernel of the corresponding projection.

To present the main results, we need some notion related to properties of direct sums; this notion itself is of great interest.

We say that an object M of an additive category has the *exchange property* if M satisfies the following condition: if the object M is a direct summand of some object A which is the direct sum of the subobjects A_i, $i \in I$, i.e.,

$$A = M \oplus N = \bigoplus_{i \in I} A_i,$$

then there exist decompositions $A_i = A_i' \oplus B_i$ such that $A = M \oplus \bigoplus_{i \in I} A_i'$. If this condition holds in any case, where I is a finite set, then M has the *finite exchange property*.

If $M = A \oplus B$ with embeddings e_1, e_2 and projections p_1, p_2 and $f\colon M \to N$ is an isomorphism, then it is clear that we also have $N = A \oplus B$ with embeddings $e_1 f, e_2 f$ and projections $f^{-1} p_1, f^{-1} p_2$.

Lemma 27.4. *If $A \oplus B \oplus C = A \oplus \bigoplus_{i \in I} D_i$ and B has the exchange property, then*

$$A \oplus B \oplus C = A \oplus B \oplus \bigoplus_{i \in I} D_i',$$

where $D_i' \subset D_i$, $i \in I$.

Proof. Considering Lemma 27.1 and the above remark, we have $B \oplus C = \bigoplus_{i \in I} D_i$. The object B has the exchange property. Therefore

$$\bigoplus_{i \in I} D_i = B \oplus \bigoplus_{i \in I} D_i',$$

where $D_i' \subset D_i$. Therefore

$$A \oplus B \oplus C = A \oplus B \oplus \bigoplus_{i \in I} D_i',$$

which is required. □

Now it is easy to obtain the following corollary.

Corollary 27.5. *Any direct sum of a finite number of objects with exchange property has the exchange property.*

In terms of the endomorphism ring, we can consider one situation, where the finite exchange property holds.

Proposition 27.6. *Let $\mathcal{E}$ be an additive category with kernels and let M be an object in $\mathcal{E}$. If the endomorphism ring of M is local, then M has the finite exchange property.*

Proof. We assume that $A = M \oplus N = \bigoplus_{i=1}^{n} A_i$. Let $e : M \to A$ and $e_i : A_i \to A$ be the corresponding embeddings and let $p : A \to M$ and $p_i : A \to A_i$ be the corresponding projections. Then $1_M = \sum_{i=1}^{n} e p_i e_i p$. Since the endomorphism ring of M is local, we have that one of the summands is an invertible element of the ring $\operatorname{End}(M)$; let $\gamma = e p_1 e_1 p$ for definiteness. We set $\pi = p\gamma^{-1} e p_1 e_1 \colon A \to A$. Then $\pi^2 = \pi$ and by Lemma 27.3,

$$A = \operatorname{Ker}(\pi) \oplus \operatorname{Ker}(1 - \pi). \tag{$*$}$$

Let $k : \operatorname{Ker}(1 - \pi) \to A$ be an embedding. Since $0 = k(1 - \pi)$, we have $k = k\pi$. Furthermore, it follows from the relation $\pi p_j = 0$ that $kp_j = 0$ for $j > 1$. Indeed, the embedding $e_1 \colon A_1 \to A$ is the kernel for $p_2 + \cdots + p_n$. Consequently, the relations $kp_2 = \cdots = kp_n = 0$ imply the relation $k(p_2 + \cdots + p_n) = 0$. Therefore, there exists a morphism $l : \operatorname{Ker}(1 - \pi) \to A_1$ with $le_1 = k$. Thus, l is a monomorphism and $\operatorname{Ker}(1 - \pi) \subset A_1$. Therefore

$$A_1 = \operatorname{Ker}(1 - \pi) \oplus (A_1 \cap \operatorname{Ker}(\pi)) \tag{$**$}$$

(Lemma 27.3 (3)) and

$$A = \operatorname{Ker}(1 - \pi) \oplus (A_1 \cap \operatorname{Ker}(\pi)) \oplus \bigoplus_{i=2}^{n} A_i.$$

The projection $A \to \operatorname{Ker}(1-\pi)$ of this direct sum is equal to $p_1 e_1 q$, where $qk = \pi$ and q is of the projection $A \to \operatorname{Ker}(1 - \pi)$ for the sum $(*)$ (Lemma 27.3 (1)). The projection $A_1 \to \operatorname{Ker}(1 - \pi)$ with respect to $(**)$ is $e_1 q$ by Lemma 27.3 (3). By verifying, we obtain that $ep_1e_1q : M \to \operatorname{Ker}(1-\pi)$ and $kp\gamma^{-1} : \operatorname{Ker}(1-\pi) \to M$ are mutually inverse isomorphisms. However, ep_1e_1q is the composition of the embedding e from M in A with projection p_1e_1q from A onto $\operatorname{Ker}(1 - \pi)$. By Lemma 27.1 (1), we obtain

$$A = M \oplus (A_1 \cap \operatorname{Ker}(\pi)) \oplus \bigoplus_{i=2}^{n} A_i.$$

Therefore M has the finite exchange property. □

Exercise 5 contains an assertion which converses Proposition 27.6.

We say that an additive category $\mathcal{E}$ with infinite sums satisfies the *weak Grothendieck condition* if for every subscript set I, each nonzero object A, and for every monomorphism $A \to \bigoplus_{i\in I} B_i$, there exist a finite subset F of the set I and the commutative diagram

$$\begin{array}{ccc} C & \longrightarrow & \bigoplus_{i\in F} B_i \\ \downarrow & & \downarrow \\ A & \longrightarrow & \bigoplus_{i\in I} B_i \end{array}$$

with nonzero morphism $C \to A$.

If $\mathcal{E}$ is a category with kernels, then by Lemma 27.3, the intersection $A \cap (\bigoplus_{i\in F} B_i)$ always exists. Consequently, the existence of the indicated diagram is equivalent to the relation

$$A \cap \left(\bigoplus_{i\in F} B_i\right) \neq 0.$$

Proposition 27.7. *Let $\mathcal{E}$ be an additive category with kernels which satisfies the weak Grothendieck condition. If an indecomposable object M has the finite exchange property, then M has the exchange property.*

Proof. Let $A = M \oplus N = \bigoplus_{i\in I} A_i$. For every subset $J \subseteq I$, we set $A(J) = \bigoplus_{i\in J} A_i$. There exists a finite subset J in I such that $M \cap (\bigoplus_{i\in J} A_i) \neq 0$. By the finite exchange property, we have

$$A = M \oplus N = A(J) \oplus A(I \setminus J) = M \oplus \left(\bigoplus_{i\in J} A_i'\right) \oplus C,$$

where $A_i = A_i' \oplus B_i$, $i \in J$, $A(I \setminus J) = C \oplus C'$ and

$$M \oplus \left(\bigoplus_{i\in J} A_i'\right) \oplus C = \left(\bigoplus_{i\in J} B_i \oplus C'\right) \oplus \left(\bigoplus_{i\in J} A_i' \oplus C\right).$$

Now we have

$$M \cong \bigoplus_{i\in J} B_i \oplus C'$$

(Lemma 27.1). Since M is an indecomposable object, all objects B_i, $i \in J$, and C', except one, are equal to zero. We assume that $C' \neq 0$ and $B_i = 0$ for all $i \in J$. Since $A_i' = A_i$ for all $i \in J$, we have $M \cap A(J) = 0$; this contradicts to the choice of J. Consequently, $C' = 0$, $C = A(I \setminus J)$, and $A = M \oplus \left(\bigoplus_{i \in J} A_i'\right) \oplus A(I \setminus J)$. If we assume that $A_i' = A_i$ for the subscripts i in $I \setminus J$, then we obtain

$$A = M \oplus \left(\bigoplus_{i \in I} A_i'\right),$$

where $A_i' \subset A_i$, $i \in I$. Therefore M has the exchange property. □

Now we can present the main result on the uniqueness decompositions.

Theorem 27.8 (Walker–Warfield [298]). *Let $\mathcal{E}$ be an additive category with kernels and infinite sums, and let $\mathcal{E}$ satisfy the weak Grothendieck condition. If $M = \bigoplus_{i \in I} A_i$, where the endomorphism ring of every object A_i is local, then the following assertions hold.*

(a) *Every indecomposable direct summand of the object M is isomorphic to one of the summands A_i.*

(b) *If $M = \bigoplus_{j \in J} B_j$ and every summand B_j is indecomposable, then there exists a bijection $\theta : I \to J$ such that $A_i \cong B_{\theta(i)}$ for all $i \in I$.*

Proof. (a) Let $M = A \oplus B = \bigoplus_{i \in I} A_i$, where A is a nonzero indecomposable object. For a subset $J \subseteq I$, we denote by $M(J)$ the object $\bigoplus_{i \in J} A_i$. There exists a finite set $J \subset I$ such that $A \cap M(J) \neq 0$. It follows from Proposition 27.6, Proposition 27.7, and Corollary 27.5 that the object $M(J)$ has the exchange property. Consequently, $M = M(J) \oplus A' \oplus B'$, where $A' \subset A$ and $B' \subset B$. Therefore

$$A = A' \oplus (A \cap (M(J) \oplus B')) \quad \text{and} \quad A \cap (M(J) \oplus B') \neq 0.$$

Since A is indecomposable, we have $A' = 0$. Therefore $M = M(J) \oplus B'$, $B = B' \oplus (B \cap M(J))$, and $M = A \oplus B' \oplus (B \cap M(J))$. Therefore

$$M(J) \cong A \oplus (B \cap M(J)).$$

By Theorem 27.2, the object $B \cap M(J)$ is equal to $\bigoplus_{i=1}^{m} B_i$, where B_i is an indecomposable object for every i. It follows from Lemma 27.3 that $\operatorname{End}(A)$ and $\operatorname{End}(B_i)$ do not have nontrivial idempotents. It follows from Theorem 27.2 that $A \cong A_i$ for some i.

(b) By (a), every object B_j is isomorphic to some A_i. In particular, all A_i and B_j have the exchange property. For subsets $X \subseteq I$ and $Y \subseteq J$, we set $M(X) = \bigoplus_{i \in X} A_i$ and $N(Y) = \bigoplus_{j \in Y} B_j$. Furthermore, we set $I_k = \{i \in I \mid A_i \cong A_k\}$ for $k \in I$. Then the sets I_k $(k \in I)$ form a partition of the set I. Let $J_k = \{j \in J \mid B_j \cong A_k\}$ for $k \in I$. Then $\{J_k \mid k \in I\}$ is a partition of I. We take an arbitrary subscript k and assume that the set I_k is finite. In this case, the object $M(I_k)$ has the exchange property. Consequently,

$$M = M(I_k) \oplus M(I \setminus I_k) = M(I_k) \oplus N(J')$$

for some $J' \subset J$ (we consider the indecomposability of B_j). Now we have $M(I \setminus I_k) \cong N(J')$ (Lemma 27.1). This implies that for $j \in J'$, we have $B_j \not\cong A_k$; therefore $j \notin J_k$. Considering the existence of the isomorphism $M(I_k) \cong N(J \setminus J')$, we obtain $J_k = J \setminus J'$. Therefore $M(I_k) \cong N(J_k)$ and it follows from Theorem 27.2 that $|I_k| = |J_k|$.

Now we assume that I_k is an infinite set. We denote by Δ_k the set $\{j \in J \mid M = A_k \oplus N(J \setminus \{j\})\}$. There exists a finite set $F \subset J$ with $A_k \cap N(F) \neq 0$ (this follows from the use of the weak Grothendieck condition). If $j \notin F$, then the inclusion $F \subset J \setminus \{j\}$ implies the relation $A_k \cap N(J \setminus \{j\}) \neq 0$. Consequently, $\Delta_k \subset F$; therefore, every Δ_k is finite. Now let $t \in J_k$. Then $B_t \cap M(G) \neq 0$ for some finite subset G in I. In addition, it follows from the exchange property for $M(G)$ that $M = M(G) \oplus N(J \setminus H)$, where H is some subset in J. Since $M(G) \cong N(H)$, it follows from Theorem 27.2 that H is finite. It is clear that t is contained in H. Since $N(H \setminus \{t\})$ has the exchange property, we have

$$\begin{aligned} M &= N(H \setminus \{t\}) \oplus B_t \oplus N(J \setminus H) = M(G) \oplus N(J \setminus H) \\ &= N(H \setminus \{t\}) \oplus A_g \oplus N(J \setminus H) \end{aligned}$$

for some $g \in G$ (Lemma 27.4). Therefore $t \in \Delta_g$. In addition, $A_g \cong B_t \cong A_k$; therefore $g \in I_k$. We have proved the relation $J_k = \bigcup_{g \in I_k} \Delta_g$. Since I_k is infinite and every Δ_g is finite and nonempty, we have $|J_k| \leq |I_k|$. By using the symmetrical argument, we obtain $|J_k| = |I_k|$. □

Before we introduce several not very known notions, we agree in the following terminology. Let $f\colon A \to B$ and $g\colon X \to B$ be two morphisms. If there exists a morphism $h\colon X \to A$ with $g = hf$, then we say that *g factors through f* or A.

A *small* object of the category $\mathcal{E}$ is an object S such that every morphism $S \to \bigoplus_{i \in I} A_i$ factors through the embedding $\bigoplus_{i \in F} A_i \to \bigoplus_{i \in I} A_i$ for some finite subset F of the set I.

The following definition transfers the module property from Exercise 1 to categories.

An object M is said to be *finitely approximable* if every monomorphism $f\colon L \to M$ is an isomorphism if and only if for every small object S, every morphism $S \to M$ factors through f.

It is clear that any small object is a finitely approximable object.

Lemma 27.9. *Let $\mathcal{E}$ be an additive category with kernels. If M is a nonzero finitely approximable object and $M \subset \bigoplus_{i\in I} A_i$, then there exists a finite subset F in I such that $M \cap (\bigoplus_{i\in F} A_i) \neq 0$.*

Proof. Let e be a monomorphism from M into $\bigoplus_{i\in I} A_i$. Since the monomorphism $0 \to M$ is not an isomorphism, there exist a small object S and a nonzero morphism $f\colon S \to M$. The morphism fe factors through some finite sum $\bigoplus_{i\in F} A_i$, and f induces the nonzero morphism $S \to M \cap (\bigoplus_{i\in F} A_i)$. Consequently, the intersection is not equal to zero. □

Now we can assert that Proposition 27.7 and Theorem 27.8 remain be true if we replace the weak Grothendieck condition by the property that the object M is finitely approximable or all objects A_i ($i \in I$) are finitely approximable (see details in the paper of Walker–Warfield [298]).

To formulate another main result of the section, we define one more notion.

An object M is said to be *countably approximable* if there exist a countable family of small objects S_i and morphisms $g_i : S_i \to M$ ($i \geq 1$) such that every monomorphism $f\colon L \to M$ is an isomorphism if and only if every g_i factors through f.

The proof of the following theorem of Walker and Warfield is quite difficult. We recommend to the reader to see it in the original paper of Walker and Warfield [298].

Theorem 27.10. *Let $\mathcal{E}$ be an additive category with kernels and infinite sums, $\mathcal{E}$ satisfy the weak Grothendieck condition, and let $M = \bigoplus_{i\in I} A_i$, where for every i, the object A_i is countably approximable and the ring $\mathrm{End}(A_i)$ is local. Then every direct summand of the object M is isomorphic to the direct sum $\bigoplus_{i\in J} A_i$ for some subset $J \subseteq I$. Consequently, any two direct decompositions of the object M have isomorphic refinements.*

The Krull–Schmidt theorem usually means the assertion on an isomorphism of two direct decompositions of a module or an object of the category with indecomposable summands. In this case, we say that the module or the object satisfies the Krull–Schmidt theorem. The paper of Walker and Warfield [298] contains a detailed review of the studies related to this theorem. We have presented generalizations of two known theorems of Krull–Schmidt type. The first theorem is the following result of Azumaya [24]. Assume that a module M has a decomposition $\bigoplus_{i\in I} A_i$ (I is a finite or infinite set) such that the endomorphism ring of every A_i is local. Then any another indecomposable summand of the module M is isomorphic to A_i for some $i \in I$; in addition, if M is the direct sum of indecomposable submodules B_j ($j \in J$), then there exists a bijective mapping $f\colon I \to J$ such that $A_i \cong B_{f(i)}$ for all $i \in I$.

The second theorem is related to names like Crawley, Jónsson, and Warfield; the theorem can be formulated as follows (see the book of Anderson and Fuller [1, Theorem 26.5]). If the conditions of the previous theorem hold and each of the modules A_i is countably generated, then any two direct decompositions of the module M have isomorphic refinements. In particular, if N is a direct summand of the module M, then there exists a subset $J \subseteq I$ such that N is isomorphic to the direct sum of modules A_i, $i \in J$.

Crawley and Jónsson [53] have published some theorems on uniqueness of decompositions and isomorphisms of refinements for arbitrary algebraic systems. The main condition of their paper is the exchange property. This property and so-called cancellation property and substitution property are studied in [72, 303, 305, 306, 312] and other papers (see the bibliography in these papers).

Different variants of the Krull–Schmidt theorem for categories were obtained by Atiyah [22], Gabriel [97], Bass [32], Warfield [304], Arnold–Hunter–Richman [15], and Arnold [7].

The open question on the uniqueness of direct decompositions has two sides:

(1) Are two given decompositions isomorphic to each other?

(2) Is a direct summand of a given direct decomposition the direct sum of the corresponding summands?

Exercise 1. (a) Any finitely generated module over any ring is a small module. In general, the converse is not true. Every module is the least upper bound of submodules which are small modules.

(b) A small module over a discrete valuation domain is a finitely generated module.

Exercise 2. Transfer property (b) from Section 2 and Proposition 2.4 to additive categories.

Exercise 3. Any direct summand of an object with exchange property has the exchange property.

Exercise 4. If M is a finitely approximable object and $M = \bigoplus_{i \in I} A_i$, then every A_i is finitely approximable. If the category $\mathcal{E}$ has kernels, then the converse assertion also holds.

Exercise 5 (Warfield [303], Walker–Warfield [298]). An indecomposable finitely approximable object of an additive category with kernels has the exchange property if and only if the endomorphism ring of the object is local.

28 Isotype, nice, and balanced submodules

We define and study submodules of some special types. They are very useful for the theory which will be developed later. First, we specialize the notion of the height of an element and consider some accompanying material.

We call a sequence $\{\sigma_i\}_{i\geq 0}$ a height sequence provided every σ_i is either an ordinal number or the symbol ∞ and, in addition,

(1) if $\sigma_i = \infty$, then $\sigma_{i+1} = \infty$;

(2) if $\sigma_{i+1} \neq \infty$, then $\sigma_i < \sigma_{i+1}$.

The elements σ_i are called the *coordinates* of the given height sequence. We can define a natural partial order on the set of all height sequences. Let $u = \{\sigma_i\}$ and $v = \{\tau_i\}$ be two height sequences. We assume that $u \leq v$ if $\sigma_i \leq \tau_i$ for all $i \geq 0$. We also assume that $\sigma < \infty$ for every ordinal number σ. In fact, we obtain the lattice, where the greatest lower bound $\inf(u, v)$ is equal to the sequence $(\rho_0, \rho_1, \ldots)$ with $\rho_i = \min(\sigma_i, \tau_i)$. Similarly, the least upper bound $\sup(u, v)$ corresponds to the pointwise maximum. If $\sigma_i + 1 < \sigma_{i+1}$, then we say that the sequence u has a *jump* between σ_i and σ_{i+1}. Two height sequences u and v are said to be *equivalent* (the notation: $u \sim v$) if there exist two positive integers n and m such that $\sigma_{n+i} = \tau_{m+i}$ for $i = 0, 1, 2, \ldots$. This gives an equivalence relation on the set of all height sequences. We denote by $p^n u$ ($n \geq 0$) the height sequence $(\sigma_n, \sigma_{n+1}, \ldots)$. It is clear that $u \sim p^n u$.

Every element of the module M is related to some height sequence. First, we define the submodule $p^\sigma M$ for every ordinal number σ. For any nonnegative integer n, the submodule $p^n M$ was introduced in Section 4 as the set $\{p^n x \mid x \in M\}$. Now we set $p^\sigma M = p(p^{\sigma-1} M)$ provided $\sigma - 1$ exists, and we set $p^\sigma M = \bigcap_{\tau<\sigma} p^\tau M$ for any limit ordinal number σ. By considering cardinalities, we obtain that there exists the least ordinal number λ with $p^\lambda M = p^{\lambda+1} M$. This ordinal number λ is called the *length* of the module M; it is denoted by $l(M)$. Clearly, $p^\lambda M$ coincides with the maximal divisible submodule of the module M. Therefore M is a reduced module if and only if $p^\lambda M = 0$. The submodules $p^\sigma M$ are fully invariant in M (this can be easily proved by the transfinite induction) and we have a properly descending sequence

$$M \supset pM \supset \ldots \supset p^\sigma M \supset \ldots \supset p^\lambda M = p^{\lambda+1} M.$$

Using the submodules $p^\sigma M$, we "disintegrate" the notion of the height of an element in such a manner that we can make difference between elements of infinite height. We assumed earlier that $h(a) = \infty$ for an element $a \in p^\omega M$ (see Section 7). Now we proceed as follows. If $a \notin p^\lambda M$, where λ is the length of the module M, then there exists a unique ordinal number σ with $a \in p^\sigma M \setminus p^{\sigma+1} M$.

This ordinal number σ is sometimes called the *generalized height* of the element a; it is denoted by $h^*(a)$. By definition, the generalized height of an element $a \in p^\lambda M$ is equal to ∞. When we deal with mixed modules, it is convenient to assume that the height stands for the generalized height and the former notation $h(a)$ stands for the generalized height $h^*(a)$. This does not lead to confusion, since $h^*(a) = h(a)$ if the height is finite. In what follows $h(a)$ (or $h_M(a)$ if it is necessary to indicate the module in which the height is considered) denotes the generalized height. If M is a reduced module, then we have $h(a) = \infty$ only for $a = 0$.

For the "new" height, the main properties of the "ordinary" height hold (for example, see Exercise 1 in Section 7). First, "the triangle inequality" $h(a+b) \geq \min(h(a), h(b))$ holds; if $h(a) \neq h(b)$, then we have an equality. If $M = A \oplus B$ and $a \in A$, $b \in B$, then also an equality holds. The height is not decreased under a homomorphism. Precisely, $h_M(a) \leq h_N(\varphi a)$ for every homomorphism $\varphi\colon M \to N$ and an element $a \in M$.

The sequence $(h(a), h(p^2 a), \ldots)$ contains more information about an element a of the module M. It is clear that this sequence is a height sequence in the sense defined above. We denote the sequence by $U(a)$ or $U_M(a)$; it is called the *height sequence* (or the *Ulm sequence*) of the element a in the module M.

The following two properties of height sequences follow from the corresponding properties of heights:

$$U(a+b) \geq \inf(U(a), U(b)) \quad \text{and} \quad U(a) \leq U(\varphi a),$$

where φ is some homomorphism. We also make the following remarks. We can obtain the height sequence of the element pa from the height sequence of the element a by removing the first coordinate. The height sequence of an element of finite order consists of the symbols ∞ excluding of a finite number of positions. If M is a reduced module and a is an element of infinite order, then $U(a)$ does not contain the symbol ∞. Let $U(a) = (\sigma_0, \sigma_1, \ldots, \sigma_k, \ldots)$. If there is a jump between σ_k and σ_{k+1}, then M necessarily contains an element which has the order p and the height σ_k. If $o(a) = p^n$, then there always is a jump between $h(p^{n-1}a)$ and $h(p^n a)$. The maximal divisible submodule of the torsion-free module M is equal to $p^\omega M$. If $a \in M \setminus p^\omega M$, then $h(p^k a) = h(a) + k$ for all $k \geq 0$. Consequently, the height $h(a)$ contains complete information required for constructing the height sequence $U(a)$.

If a and b are linearly dependent elements (i.e., $ra = sb \neq 0$ with $r, s \in R$), then $U(a) \sim U(b)$. Let M be a mixed module of rank 1. Any two elements x and y of infinite order in M are linearly dependent; therefore $U(x) \sim U(y)$. Therefore, the equivalence class of the height sequence of every element of infinite order in M is an invariant of the module M; it is denoted by $U(M)$.

Every height sequence $u = \{\sigma_i\}_{i\geq 0}$ gives the fully invariant submodule

$$M(u) = \{x \in M \mid U(x) \geq u\}$$

(this follows from the triangle inequality).

The submodules $p^\sigma M$ with limit ordinal number σ play a special role. In Section 11, the first Ulm submodule M^1 of the module M was defined as the intersection $\bigcap_{n\geq 1} p^n M$ (therefore $M^1 = p^\omega M$). The submodule M^1 was intensively used in Section 22. We extend this notion to arbitrary ordinal numbers σ. We set $M^0 = M$,

$$M^\sigma = (M^{\sigma-1})^1 = \bigcap_{n\geq 1} p^n M^{\sigma-1}$$

if $\sigma - 1$ exists, and

$$M^\sigma = \bigcap_{\tau<\sigma} M^\tau$$

in the case of the limit ordinal number σ. There exists the least ordinal number ν with $M^{\nu+1} = M^\nu$. Such an ordinal number ν is called the *Ulm length* of the module M. Here M^ν is the maximal divisible submodule of the module M. If we consider only reduced modules M, then we obtain a properly descending sequence of fully invariant submodules

$$M = M^0 \supset M^1 \supset \cdots \supset M^\sigma \supset \cdots \supset M^\nu = 0.$$

The submodule M^σ is called the σth *Ulm submodule* of the module M, the factor module $M_\sigma = M^\sigma/M^{\sigma+1}$ is called the σth *Ulm factor* of the module M, and the sequence $M_0, M_1, \dots, M_\sigma, \dots$ ($\sigma < \nu$) is called the *Ulm sequence* of the module M. It follows from the definition of M^σ that $M^\sigma = p^{\omega\sigma} M$.

We consider several types of submodules and exact sequences related to the modules. We begin with isotype submodules which first appeared in the paper of Kulikov [189]. Such submodules are specially studied in the papers of Lane [200] and Hill–Ullery [152]. Let A be some submodule of the module M. If $p^\sigma A = A \cap p^\sigma M$ for every ordinal number σ, then we say that A is an *isotype* submodule in M. We have several characterizations of isotype submodules.

Lemma 28.1. *The following conditions are equivalent*:

(1) *A is an isotype submodule in the module M*;

(2) *the heights of elements of A coincide in the modules A and M*;

(3) *$A(u) = A \cap M(u)$ for all height sequences u*;

(4) *the induced sequence*

$$0 \to p^\sigma A \to p^\sigma M \to p^\sigma(M/A)$$

is exact for every ordinal number σ.

Proof. The equivalence of assertions (1)–(3) is directly verified. We clarify some details related to the sequence in (4). The homomorphism, which is second from the left, is an embedding. Furthermore, we have the restriction of the canonical homomorphism $M \to M/A$ to $p^\sigma M$; this is correct, since homomorphisms do not decrease heights. We need only to verify that the sequence is exact at the place $p^\sigma M$. It is equivalent to the relation $p^\sigma A = A \cap p^\sigma M$ (i.e., (1) holds). □

The notion of isotypeness is certainly stronger than the notion of purity. Every isotype submodule is pure. If $A^1 = 0$ or the submodule A is closed in the p-adic topology, then A is an isotype submodule if and only if A is a pure submodule (consider that $M^1 \subseteq A$ in the second case). We note that A is closed if and only if $(M/A)^1 = 0$ (Corollary 11.3).

We formulate three properties of isotypeness which easily follow from Lemma 28.1.

(a) Direct summands are isotype submodules.

(b) If $A \subseteq B \subseteq M$, where A is an isotype submodule in B and B is an isotype submodule in M, then A is an isotype submodule in M.

(c) The union of any ascending sequence of isotype submodules is also an isotype submodule.

Hill introduced another important class of submodules: the class of nice submodules. A submodule A of the module M is said to be *nice* if the relation

$$p^\sigma(M/A) = (p^\sigma M + A)/A$$

holds for all ordinal numbers σ. The submodule from the right is the image of the submodule $p^\sigma M$ under the canonical homomorphism $M \to M/A$. Since the height is not decreased under a homomorphism, the right part is always contained in the left part. We can also say that A is a nice submodule if and only if the induced mapping $p^\sigma M \to p^\sigma(M/A)$ from Lemma 28.1 is an epimorphism for every σ. For every integer $n \geq 0$ and the submodule A, the relation $p^n(M/A) = (p^n M + A)/A$ holds. Therefore, if the relation $p^\sigma(M/A) = (p^\sigma M + A)/A$ holds for all limit ordinal numbers σ, then the relation holds for all σ, i.e., A is a nice submodule in M.

Nice submodules are sometimes defined in another manner. It is clear that $h(x + a) \leq h(x + A)$ for all $x \in M \setminus A$ and $a \in A$ (the height of the residue class is considered in M/A). An element c of the residue class $x + A$ is said to be *proper* (with respect to A) if $h(c) = h(x + A)$.

Lemma 28.2. *A submodule A is nice if and only if every residue class with respect to the submodule A contains a proper with respect to A element.*

Proof. Let A be a nice submodule and let $x + A$ be an arbitrary residue class of height σ. Then
$$x + A \in p^\sigma(M/A) = (p^\sigma M + A)/A$$
and $x = c + a$, where $c \in p^\sigma M$, $a \in A$. The element c is the required proper element. Conversely, let every residue class contain a proper element and let $x + A \in p^\sigma(M/A)$. In particular, $h(x+A) \geq \sigma$. In the residue class $x+A$, we choose a proper element c. We have $h(c) \geq \sigma$; therefore $c \in p^\sigma M$. Therefore
$$x + A = c + A \in (p^\sigma M + A)/A. \qquad \square$$

We have three properties related to nice submodules. Considering the above, we can easy prove the following properties.

(d) Direct summands are nice submodules.

(e) Any submodule closed in the p-adic topology is a nice submodule.

Indeed, if A is closed in M, then it was mentioned above that $(M/A)^1 = 0$; therefore, the required relation holds for all limit ordinal numbers σ. Consequently, the relation holds for all σ.

(f) If $M = \bigoplus_{i\in I} M_i$ and $A = \bigoplus_{i\in I} A_i$, where $A_i \subseteq M_i$ $(i \in I)$, then A is a nice submodule in M if and only if A_i is a nice submodule in M_i for every i.

Now we formulate the most-used results on nice submodules.

Lemma 28.3. *Let $A \subseteq B \subseteq M$.*

(1) *If B is a nice submodule in M, then B/A is a nice submodule in M/A.*

(2) *If A is a nice submodule in M and B/A is a nice submodule in M/A, then B is a nice submodule in M.*

Proof. (1) Let $(x+A)+B/A$ be an arbitrary residue class of the module M/A with respect to the submodule B/A $(x \in M)$. Since B is a nice submodule in M, there exists an element $y \in x+B$ with $h(y) = h(x+B)$. Since the correspondence $m+B \to (m+A)+B/A$ $(m \in M)$ defines an isomorphism $M/B \cong (M/A)/(B/A)$, we have
$$h(x + B) = h((x + A) + B/A)$$
(it is clear to determine the factor modules in which the heights of the corresponding residue classes are considered). We have the relations
$$h(y + A) \leq h((y + A) + B/A) = h((x + A) + B/A)$$
and
$$h(y + A) \geq h(y) = h((x + A) + B/A).$$

Therefore

$$h(y+A) = h((x+A)+B/A)$$

and $y+A$ is a proper element with respect to B/A which is contained in the residue class $(x+A)+B/A$. By Lemma 28.2, B/A is a nice submodule in M/A.

(2) We take some residue class $x+B$. Since the submodule B/A is nice in the module M/A, we have $h(y+A) = h((x+A)+B/A)$, where $y+A \in (x+A)+B/A$. Since A is a nice submodule in M, there exists an element $z \in y+A$ with $h(z) = h(y+A)$. Similar to (1), we also have $h(z+B) = h((z+A)+B/A)$. In addition, $x+B = z+B$ and $(x+A)+B/A = (z+A)+B/A$. It follows from the relations that $h(z) = h(x+B)$ and $z \in x+B$. This means that z is a proper element. □

There are several sufficient conditions under which we can immediately assert that the submodule is nice.

Proposition 28.4. *A finitely generated submodule A of the module M is a nice submodule in each of the following cases*:

(1) *A is a primary module*;

(2) *R is a complete domain*;

(3) *the rank of the module M is equal to* 1.

Proof. (1) The module A is equal to the direct sum of a finite number of cyclic modules of finite order. It follows from Lemma 28.3 (2) and the induction on the number of cyclic summands and the order of elements of the module A that we can assume that A is a cyclic module of order p. All nonzero elements of such a module have the same height. It is easy to see that the given residue class with respect to the submodule A necessarily contains proper element.

(2) The module A is the direct sum of a finite number of cyclic modules. Considering (1) and Lemma 28.3 (2), it is sufficient to consider the case, where A is a cyclic module of infinite order. Thus, let $A = Ra$, where the element a has infinite order and $x \in M \setminus A$. The height of every nonzero element of the submodule Ra coincides with one of the following heights: $h(a)$, $h(pa)$, $h(p^2a), \dots$. If the residue class $x+A$ contains an element whose height does not coincide with one of the above heights, then it is easy to find the required proper element. Now let every element from $x+A$ have one of the heights $h(p^n a)$, $n \geq 0$. We assume that there exist elements $r_n \in R\,(n \geq 1)$ with $h(x) < h(x+r_1a) < h(x+r_2a) < \cdots$. It follows from the triangle inequality that

$$h((r_{n+1}-r_n)a) = h(x+r_na).$$

This implies

$$h(r_2 - r_1) < h(r_3 - r_2) < \cdots < h(r_{n+1} - r_n) < \cdots,$$

i.e., $\{r_n\}_{n\geq 1}$ is a Cauchy sequence in the p-adic topology of the ring R. Since R is complete, the limit r of this sequence exists. Since the heights of the elements $r - r_n$ increase with increasing n, the heights of elements $(r - r_n)a$ also increase with increasing n. Since $x + ra = (r - r_n)a + (x + r_n a)$, we obtain that $h(x + r_n a) < h(x + ra)$ for all n. However, the existence of an element $x + ra$ with such the height is impossible. Now it is clear that $x + A$ contains a proper element.

(3) Similar to the proof of (2), we assume that A is a free module. Since $r(M) = 1$, we have that A is a cyclic module, $A = Ra$. We use Case (2) which has been proved. For this purpose, we use $\widehat{R}$-hulls which will appear in Section 35, where $\widehat{R}$ is the p-adic completion of the ring R. (We can also deal with the tensor product $\widehat{R} \otimes M$.) The *$\widehat{R}$-hull* $\widehat{R}M$ of the module M is an $\widehat{R}$-module such that $\widehat{R}M$ contains M, $\widehat{R}M/M$ is a divisible torsion-free module, M is an isotype submodule in $\widehat{R}M$ (this easily follows from the property that $\widehat{R}M/M$ is a torsion-free module), $t(\widehat{R}M) = t(M)$, and $r(\widehat{R}M) = 1$. We prove that the induced mapping $p^\sigma M \to p^\sigma(M/A)$ is an epimorphism for every ordinal number σ. Since $\widehat{R}A$ is a cyclic $\widehat{R}$-module, we have that by (2), it is a nice submodule in $\widehat{R}M$. Consequently, every element of $p^\sigma(M/A)$ is the image of some element of $p^\sigma(\widehat{R}M)$. If $p^\sigma(\widehat{R}M)$ is a primary module, then $p^\sigma(\widehat{R}M) = p^\sigma M$ (we consider that M is an isotype submodule). Consequently, every element of $p^\sigma(M/A)$ is the image of some element of $p^\sigma M$. We consider the case, where $p^\sigma(\widehat{R}M)$ is not a primary module. Since $r(\widehat{R}M) = 1$, we have that $p^n a \in p^\sigma(\widehat{R}M)$ for some $n \geq 1$. In addition, $\widehat{R}M = M + \widehat{R}(p^n a)$ (consider that $\widehat{R}(p^n a) \not\subseteq M$ and $r(M) = 1$). If $y \in p^\sigma(\widehat{R}M)$, then we have $y = x + rp^n a$, where $x \in M$ and $r \in \widehat{R}$. In fact, for the element x, we have $x \in p^\sigma(\widehat{R}M) \cap M = p^\sigma M$. We also note that $rp^n a \in \widehat{R}(Ra) = \widehat{R}A$. As a result, we have that every element of the factor module $\widehat{R}M/\widehat{R}A$, which is contained in the image of the submodule $p^\sigma(\widehat{R}M)$, is also contained in the image of the submodule $p^\sigma M$. □

Isotype nice submodules are very useful. A submodule A of the module M is said to be *balanced* if A is an isotype nice submodule. It follows from Lemma 27.1 and remarks after the definition of a nice submodule that A is balanced in M if and only if the induced sequence

$$0 \to p^\sigma A \to p^\sigma M \to p^\sigma(M/A) \to 0$$

is exact for every ordinal number σ. Balanced submodules have many natural properties of direct summands. Consequently, balanced submodules are close to direct summands. We consider some properties of balanced submodules.

(g) Direct summands always are balanced submodules.

(h) If $A \subseteq B \subseteq M$ and A is a balanced submodule of the module M, then A is a balanced submodule of the module B.

It is obvious that A is an isotype submodule in B. We show that A is a nice submodule in the module B. It is sufficient to prove that $p^\sigma(B/A) = (p^\sigma B + A)/A$ for every limit ordinal number σ. We assume that for all ordinal numbers $\tau < \sigma$, there exist relations which are similar to the above relation. In this case, we have

$$p^\sigma(B/A) = \bigcap_{\tau<\sigma} p^\tau(B/A) = \bigcap_{\tau<\sigma} (p^\tau B + A)/A.$$

Let $b + A \in p^\sigma(B/A)$. Then

$$b + A \in p^\sigma(M/A) = (p^\sigma M + A)/A.$$

This gives $b = b_1 + a$, $b_1 \in p^\sigma M$, $a \in A$, and $b + A = b_1 + A$. We can assume that $b \in p^\sigma M$. Since $b + A \in (p^\tau B + A)/A$ for every $\tau < \sigma$, we have $b = b_\tau + a_\tau$, where $b_\tau \in p^\tau B$, $a_\tau \in A$. Furthermore, we obtain

$$a_\tau = b - b_\tau \in p^\tau M \cap A = p^\tau A \quad \text{and} \quad b \in p^\tau B.$$

Consequently, $b \in \bigcap_{\tau<\sigma} p^\tau B = p^\sigma B$ and $b + A \in (p^\sigma B + A)/A$.

Two additional properties are gathered in the following lemma.

Lemma 28.5. *Let $A \subseteq B \subseteq M$.*

(1) *If A is a nice submodule in M and B is a balanced submodule in M, then B/A is a balanced submodule in M/A.*

(2) *If A is a balanced submodule in M and B/A is a balanced submodule in M/A, then B is a balanced submodule in M.*

Proof. It follows from Lemma 28.3 that it is sufficient to verify only isotypeness in the both cases.

(1) We prove that $p^\sigma(B/A) = B/A \cap p^\sigma(M/A)$, where σ is an arbitrary ordinal number. Let $b + A \in B/A \cap p^\sigma(M/A)$. Then we have

$$b + A \in p^\sigma(M/A) = (p^\sigma M + A)/A.$$

Therefore $b = b_1 + a$, where $b_1 \in p^\sigma M$ and $a \in A$. Consequently, $b_1 \in B \cap p^\sigma M = p^\sigma B$ and $b + A = b_1 + A \in p^\sigma(B/A)$.

In (2), we prove that $p^\sigma B = B \cap p^\sigma M$. We take some element $b \in B \cap p^\sigma M$. We have

$$b + A \in B/A \cap p^\sigma(M/A) = p^\sigma(B/A) = (p^\sigma B + A)/A$$

since the submodule A is nice in B by property (h). Therefore $b = b_1 + a$, where $b_1 \in p^\sigma B$ and $a \in A$. This implies that $a = b - b_1 \in A \cap p^\sigma M = p^\sigma A$. Therefore $b = b_1 + a \in p^\sigma B$. □

(i) Assume that $A \subseteq B \subseteq M$, where B/A is an isotype submodule in M/A, A is a balanced submodule in B, and B is a balanced submodule in M. Then A is a balanced submodule in M.

As for (b), it remains to verify that A is a nice submodule of the module M, i.e., $p^\sigma(M/A) = (p^\sigma M + A)/A$ for all ordinal numbers σ. Let $x + A \in p^\sigma(M/A)$. If $x \in B$, then

$$x + A \in B/A \cap p^\sigma(M/A) = p^\sigma(B/A) = (p^\sigma B + A)/A,$$

since A is a nice submodule in B. We have $x = y + a$, where $y \in p^\sigma B$ and $a \in A$. Therefore

$$x + A = y + A \in (p^\sigma B + A)/A \subseteq (p^\sigma M + A)/A.$$

If $x \notin B$, then $x + B \in p^\sigma(M/B) = (p^\sigma M + B)/B$ and $x = y + b$, where $y \in p^\sigma M$ and $b \in B$. Consequently,

$$b + A = (x - y) + A \in B/A \cap p^\sigma(M/A) = p^\sigma(B/A) = (p^\sigma B + A)/A.$$

Furthermore, we have $x - y = b_1 + a$ for some elements $b_1 \in p^\sigma B$ and $a \in A$. This gives $x = y + b_1 + a \in p^\sigma M + A$ and $x + a \in (p^\sigma M + A)/A$.

Many results on balanced subgroups of Abelian groups are obtained in the work [154] of Hunter.

The notion of a balanced submodule leads to the following definition. The exact sequence of modules

$$0 \to A \xrightarrow{f} B \xrightarrow{g} C \to 0 \tag{6.1}$$

is said to be *balanced exact* if the image fA is a balanced submodule in the module B. The following proposition contains characterizations of balanced exact sequences (see Exercise 4 in Section 10).

Proposition 28.6. *For the exact sequence* (6.1), *the following conditions are equivalent*:

(1) *the sequence* (6.1) *is a balanced exact sequence*;

(2) *the induced sequence*

$$0 \to p^\sigma A \to p^\sigma B \to p^\sigma C \to 0 \tag{6.2}$$

is exact for every ordinal number σ;

(3) $g(p^\sigma B) = p^\sigma C$ *and* $g((p^\sigma B)[p]) = (p^\sigma C)[p]$ *for every ordinal number* σ.

Proof. The equivalence of conditions (1) and (2) was already noted after the definition of a balanced submodule. (Clearly, the mappings in (6.2) are induced by the mappings f and g.) For convenience, we identify A with fA in the sequence (6.1).

(1) $\Longrightarrow$ (3). It follows from (2) that the first relation in (3) holds. We verify that the second relation holds. Homomorphisms do not decrease heights. Consequently, the left set is contained in the right set. Let $x \in (p^\sigma C)[p]$. Then $x \in p^\sigma C$ and by (2), there exists an element $y \in p^\sigma B$ with $g(y) = x$. Since $px = 0$, we have that $g(py) = 0$ and $py \in A \cap p^{\sigma+1} B = p^{\sigma+1} A$ (since A is an isotype submodule). Consequently, there exists an element $z \in p^\sigma A$ with $py = pz$. Therefore $y - z \in (p^\sigma B)[p]$ and $g(y - z) = g(y) = x$ $(g(z) = 0$ by (2)).

(3) $\Longrightarrow$ (1). The first relation in (3) means that the sequence (6.2) is exact at the place $p^\sigma C$. It follows from Lemma 28.1 that it is sufficient to verify that A is an isotype submodule in the module B, i.e., the relation $p^\sigma A = A \cap p^\sigma B$ holds for every ordinal number σ. We use the induction on σ. The case of limit ordinal numbers is trivial. Consequently, let $\sigma = \tau + 1$ for some $\tau \geq 0$. For an element x of $A \cap p^\sigma B$, we choose $y \in p^\tau B$ with $x = py$. Then $g(y) \in (p^\tau C)[p]$ (since $g(x) = pg(y) = 0$). Consequently, there is an element $z \in (p^\tau B)[p]$ with $g(z) = g(y)$. We have $g(y - z) = 0$ and $y - z \in A \cap p^\tau B = p^\tau A$ (by the induction hypothesis). Since $p(y - z) = 0$, we have $x \in p^\sigma A$, which is required. □

Using height sequences, we can obtain more restricted classes of submodules and exact sequences than the classes of nice submodules and balanced exact sequences. A submodule A of the module M is called an h-*nice* submodule if $(M/A)(u) = (M(u) + A)/A$ for every height sequence u. An isotype h-nice submodule is called an h-*balanced* submodule. The properties of these new submodules are similar to the corresponding properties of nice and balanced submodules. We recommend to the reader to verify this. An exact sequence (6.1) is said to be h-*balanced exact* if A is an h-balanced submodule of the module B. We have that h-nice submodules, h-balanced submodules, and h-balanced exact sequences are nice submodules, balanced submodules, and balanced exact sequences, respectively. This follows from the following property. If for a given ordinal number σ, we take the height sequence $u = (\sigma, \sigma + 1, \sigma + 2, \ldots)$, then $p^\sigma M = M(u)$ for every module M.

Proposition 28.7. *The following conditions are equivalent*:

(1) *the sequence* (6.1) *is h-balanced exact*;

(2) *the induced sequence*

$$0 \to A(u) \to B(u) \to C(u) \to 0 \tag{6.3}$$

is exact for every height sequence u;

(3) $g(B(u)) = C(u)$ *and* $g((p^\sigma B)[p]) = (p^\sigma C)[p]$ *for every height sequence* u *and an ordinal number* σ;

(4) $g(B(u)) = C(u)$ *and* $g(B(u)[p]) = C(u)[p]$ *for every height sequence* u.

The proof is similar to the proof of Proposition 28.6. We only clarify some details. Similar to the sequence (6.2), the mappings in (6.3) are induced by the mappings f and g; this is correct, since homomorphisms can only increase height sequences. The sequence (6.3) is always exact in $A(u)$. By Lemma 28.1, the sequence (6.3) is exact in $B(u)$ if and only if the submodule A is isotype in B; the sequence (6.3) is exact in $C(u)$ if and only if A is h-nice in B. □

As was already noted, an h-balanced exact sequence is a balanced exact sequence. The converse is not true and the counterexample is in the class of mixed modules. The height sequences of elements of torsion-free modules do not have jumps. Considering this property and the above relation $p^\sigma M = M(u)$, we obtain that h-balanced exact sequences of torsion-free modules coincide with balanced exact sequences. We also have a similar situation for primary modules.

Proposition 28.8. *An exact sequence* (6.1) *of primary modules is* h*-balanced exact if and only if the sequence is balanced exact.*

Proof. It is sufficient to verify that if sequence (6.1) is balanced exact, then it is h-balanced exact. Thus, let (6.1) be a balanced exact sequence. By Lemma 28.1, A is an isotype submodule in B. It remains to prove that A is an h-nice submodule in B, i.e., $(B/A)(u) = (B(u) + A)/A$ for the height sequence $u = (\tau_0, \tau_1, \tau_2, \ldots)$. Let $b + A \in (B/A)(u)$. We use the induction on the order of b. If $o(b) = p$, then $b + A \in p^{\tau_0}(B/A)$. However, $p^{\tau_0}(B/A) = (p^{\tau_0}B + A)/A$, since A is nice in B. It is clear that $b \in B(u)$. We assume that the induction hypothesis holds. Since $pb + A \in (B/A)(pu)$, we have

$$pb + A \in (B(pu) + A)/A.$$

We can assume again that $pb \in B(pu)$. In this case, there exists an element $b_1 \in B(u)$ with $pb = pb_1$. Now we obtain $p(b - b_1) = 0$ and $(b - b_1) + A \in (B/A)(u)$ by the first induction step. Furthermore,

$$(b - b_1) + A \in (B(u) + A)/A \quad \text{and} \quad b + A \in (B(u) + A)/A. \qquad \square$$

Exercise 1 (see Exercise 6 in Section 7). Let A be a submodule of a primary module M. If the heights of all elements of $A[p]$ coincide in A and M, then A is an isotype submodule in M.

Exercise 2. (a) If A is an isotype submodule in M, then $A/p^\tau A$ is an isotype submodule in $M/p^\tau A$ for every ordinal number τ.

(b) If $p^\tau A$ is an isotype submodule in $p^\tau M$ and $A/p^\tau A$ is an isotype submodule in $M/p^\tau A$ for some τ, then A is an isotype submodule in M.

Exercise 3. The submodule $p^\tau M$ always is a nice submodule in M for every ordinal number τ.

Exercise 4. A submodule A is nice in M if and only if $A \cap p^\sigma M$ is a nice submodule in $p^\sigma M$ and $(A + p^\sigma M)/p^\sigma M$ is a nice submodule in $M/p^\sigma M$ for every ordinal number σ.

Exercise 5. A pure submodule is not necessarily isotype or nice.

Exercise 6. (a) The property to be a nice submodule is not transitive.

(b) Construct an example of a nice submodule A of the module M which is not nice in some larger submodule of the module M.

Exercise 7. Assume that $A \subseteq B \subseteq M$, A is a nice submodule in M, and B/A is a finitely generated module. If B/A is a primary module or R is a complete domain, then B is also a nice submodule.

Exercise 8. If A is a balanced submodule of the module M, then $p^\sigma A$ is a balanced submodule of the module $p^\sigma M$ and $A/p^\sigma A$ is a balanced submodule of the module $M/p^\sigma M$ for every ordinal number σ.

Exercise 9. The exact sequence (6.1) is balanced exact if and only if the induced sequence

$$0 \to A/p^\sigma A \to B/p^\sigma B \to C/p^\sigma C \to 0$$

is exact for every ordinal number σ.

Exercise 10. Construct an example of a balanced exact sequence which is not an h-balanced exact sequence. (Such an example is also an example of a nice submodule which is not h-nice.)

Exercise 11. Construct an example of a balanced submodule which is not a direct summand.

Exercise 12. In a reduced torsion-free module, isotype submodules coincide with pure submodules and nice submodules coincide with submodules closed in the p-adic topology.

Exercise 13. If M is a reduced torsion-free module of finite rank and every pure closed submodule of M is a direct summand of M, then M is equal to the direct sum of purely indecomposable modules.

29 Categories Walk and Warf

The aim of the section is to present two new categories. Similar to the category of quasihomomorphisms from Section 24, they differ from "ordinary" category of modules by morphisms. The both categories are adapted for studies of mixed modules "up to primary modules." Primary modules are ignored by the another choice of morphisms. In fact, they become the sero objects in new categories.

The category Walk already appeared in Section 22. The objects of this category are modules. For two modules M and N, the corresponding set of morphisms $\mathrm{Hom}_{\overline{W}}(M, N)$ coincides with $\mathrm{Hom}(M, N)/\mathrm{Hom}(M, t(N))$. The morphisms $\bar{f}\colon M \to N$ coincide with the residue classes $f + \mathrm{Hom}(M, t(N))$ with representative $f \in \mathrm{Hom}(M, N)$. The composition of morphisms in Walk is induced by the composition of representatives, i.e., $\bar{f}\bar{g} = \overline{fg}$ for $\bar{f}\colon M \to N$ and $\bar{g}\colon N \to L$. It is clear that the composition is associative and the identity morphism of the object M is $\bar{1}_M$. Therefore, we have a category. The monomorphisms in Walk coincide with morphisms $\bar{f}\colon M \to N$ such that the inclusion $f(x) \in t(N)$ implies the inclusion $x \in t(M)$ (equivalently, f induces the (ordinary) monomorphism $M/t(M) \to N/t(N)$). If T is a primary module, then $0 \to T$ and $T \to 0$ are isomorphisms.

Proposition 29.1. Walk *is an additive category with kernels and infinite sums and it satisfies the weak Grothendieck condition.*

Proof. It follows from the definition that the set of morphisms $\mathrm{Hom}_{\overline{W}}(M, N)$ is an Abelian group. It is also clear that the composition of morphisms is bilinear. Assume that we have arbitrary modules A_i ($i \in I$, where I is a finite or infinite subscript set). We take the direct sum M of the modules A_i with corresponding embeddings $e_i : A_i \to M, i \in I$. Then M considered together with the morphisms $\bar{e}_i$ ($i \in I$) is the direct sum of the objects A_i in Walk. Thus, Walk is an additive category with infinite sums. Let $\bar{f}\colon M \to N$ be some morphism. We set

$$L = \{x \in M \mid f(x) \in t(N)\}.$$

Then the pair $(L, \bar{l})$, where l is an embedding $L \to M$, is the kernel of the morphism $\bar{f}$.

Now we assume that $\bar{f}\colon A \to \bigoplus_{i\in I} B_i$ is a monomorphism in Walk and A is not a primary module. There exist an element $a \in A$ and a finite subset F in I such that $f(a)$ has infinite order and it is contained in $\bigoplus_{i\in F} B_i$. We set

$$C = \{x \in A \mid f(x) \in \bigoplus_{i\in F} B_i\}.$$

We obtain the required commutative diagram from the definition of the weak Grothendieck condition presented in Section 27. □

For any two modules M and N, the kernel of the induced homomorphism

$$\mathrm{Hom}(M, N) \to \mathrm{Hom}(M, N/t(N))$$

coincides with $\mathrm{Hom}(M, t(N))$. Thus, the factor module

$$\mathrm{Hom}(M, N)/\mathrm{Hom}(M, t(N))$$

is isomorphic to some submodule of the module $\mathrm{Hom}(M, N/t(N))$. The module $\mathrm{Hom}(M, N/t(N))$ can be identified with $\mathrm{Hom}(M/t(M), N/t(N))$. Thus, the group $\mathrm{Hom}_{\overline{W}}(M, N)$ always is an torsion-free R-module (if the group is not equal to zero). Let $\mathrm{End}_{\overline{W}}(M)$ be the endomorphism ring of the module M in Walk. Then $\mathrm{End}_{\overline{W}}(M)$ is a torsion-free R-algebra.

It is easy to characterize isomorphisms in the category Walk.

Proposition 29.2. *Two modules M and N are isomorphic to each other in* Walk *if and only if there exist primary modules S and T such that $M \oplus T \cong N \oplus S$.*

Proof. We assume that there exist S and T with $M \oplus T \cong N \oplus S$. The embedding $M \to M \oplus T$ and the projection $M \oplus T \to M$ are mutually inverse isomorphisms in Walk. The same is true for $N \oplus S$. Consequently, we obtain $M \cong M \oplus T$ and $N \cong N \oplus S$ in Walk. Therefore $M \cong N$.

We assume that M is isomorphic to N in Walk and $\bar{f}\colon M \to N$, $\bar{g}\colon N \to M$ are mutually inverse isomorphisms. For some $e : M \to t(M)$ and $h\colon N \to t(N)$, we have $fg = 1_M + e$ and $gf = 1_N + h$. We set $S = e(M)$ and $T = h(N)$. We define homomorphisms $\alpha\colon M \oplus T \to N \oplus S$ and $\beta : N \oplus S \to M \oplus T$ by the relations

$$\alpha(m, t) = (f(m) - t,\ e(m) - g(t)),$$
$$\beta(n, s) = (g(n) - s,\ h(n) - f(s))$$

for all $m \in M$, $t \in T$, $n \in N$, and $s \in S$. By calculations using the relations $ef = fh$ and $hg = ge$, it can be verified that α and β are mutually inverse isomorphisms. □

To make possible the use of the theorems from Section 27 in the category Walk, we consider small objects and objects with local endomorphism rings.

Lemma 29.3. (a) *The module M is a small object in* Walk *if and only if the rank of the module M is finite.*

(b) *The endomorphism ring in* Walk *of a module of rank* 1 *is local.*

Proof. (a) Let M be a module of infinite rank. The divisible hull D of the module $M/t(M)$ is the direct sum of $r(M)$ copies of the module K (K is the field of fractions of the domain R). We denote by f the composition of the canonical homomorphism $M \to M/t(M)$ with embedding $M/t(M)$ in D. Then the morphism $\bar{f}\colon M \to \bigoplus_{r(M)} K$ cannot factor through a direct sum of a finite number of modules K. This means that M is not a small object.

Now we assume that M has finite rank and let $\bar{f}\colon M \to \bigoplus_{i\in I} A_i$ be a morphism in Walk. In M, we fix a maximal linearly independent system of elements $a_1, \dots, a_n$ of infinite order. Then we take a finite subset F of the set I such that $f(a_k) \in \bigoplus_{i\in F} A_i$ for all $k = 1, \dots, n$. For every element x of infinite order of the module M, there exist elements $r, r_1, \dots, r_n \in R$ such that $rx = r_1a_1 + \cdots + r_na_n$ and $r \neq 0$. This gives $rf(x) \in \bigoplus_{i\in F} A_i$. Therefore, it is clear that if $f = g + h$, where $g\colon M \to \bigoplus_{i\in F} A_i$ and $h\colon M \to \bigoplus_{i\in I\setminus F} A_i$, then

$$h(M) \subseteq t\Big(\bigoplus_{i\in I\setminus F} A_i\Big).$$

Consequently, $\bar{h} = 0$ and $\bar{f} = \bar{g}$, i.e., $\bar{f}$ factors through the finite sum $\bigoplus_{i\in F} A_i$.

(b) If $M \cong K \oplus T$, where T is a primary module, then $\mathrm{End}_{\overline{W}}(M) \cong K$. Otherwise, the maximal divisible submodule of the module M is a primary module. Then we prove that the ring $\mathrm{End}(M)$ is a split extension of the ring R by the ideal $\mathrm{Hom}(M, t(M))$. (The notion of a split extension was introduced before Theorem 19.1. In the beginning of Section 19, there is also a remark on the sense of the inclusion $R \subseteq \mathrm{End}(M)$.) It is clear that

$$R \cap \mathrm{Hom}(M, t(M)) = 0.$$

Let $\alpha \in \mathrm{End}(M)$. We choose an element x in M of infinite order. Let $\alpha(x) = y$. If $y \in t(M)$, then $\alpha \in \mathrm{Hom}(M, t(M))$. If $y \notin t(M)$, then $p^k ux = p^m vy$, where k and m are nonnegative integers and u, v are invertible elements of the ring R. It follows from the relation $\alpha(p^m vy) = p^k uy$ that $m \leq k$. If we set $a = p^m x$, then it is easy to verify that α coincides on the submodule Ra with the multiplication by the element $p^{k-m}w$, where $w = uv^{-1}$. Since

$$\alpha = p^{k-m}w + (\alpha - p^{k-m}w),$$

we have that $p^{k-m}w \in R$ and $\alpha - p^{k-m}w \in \operatorname{Hom}(M, t(M))$. Thus, we have the module direct sum

$$\operatorname{End}(M) = R \oplus \operatorname{Hom}(M, t(M)),$$

which implies the isomorphism $\operatorname{End}_{\overline{W}}(M) \cong R$. □

A module M is said to be *completely decomposable* if M is a direct sum of modules of rank 1. Thus, the completely decomposable module M is equal to the direct sum $\bigoplus_{i\in I} A_i$, where A_i is a module of rank 1 for every subscript i. We consider the module M as an object of the category Walk. The small modules A_i are finitely approximable and countably approximable and the rings $\operatorname{End}_{\overline{W}}(A_i)$ are local (Lemma 29.3). It follows from Proposition 29.1 that we can apply Theorem 27.8 and Theorem 27.10 to the decomposition $M = \bigoplus_{i\in I} A_i$. Using Proposition 29.2, we formulate the corresponding results in module-theoretic terms.

Corollary 29.4. (a) *If $M = \bigoplus_{j\in J} B_j$ is another decomposition of the module M, where the rank of every module B_j is equal to 1, then there exist a bijection $\theta : I \to J$ and primary modules T_i and S_i such that*

$$A_i \oplus T_i \cong B_{\theta(i)} \oplus S_i$$

for all $i \in I$.

(b) *Let N be some direct summand of the module M. Then there exist primary modules S, T and a subset J in I such that*

$$N \oplus T \cong \Big(\bigoplus_{i\in J} A_i\Big) \oplus S.$$

We define one more quite useful category and present the main properties of the category. The objects of a new category also are modules. For a submodule A in M such that M/A is a primary module, we consider homomorphisms $f\colon A \to N$ which do not decrease heights of elements. This means that $h_M(a) \le h_N(fa)$ for all $a \in A$. Now we take the set of all such "partial" homomorphisms f for all submodules A and we define an equivalence relation on the set as follows. We assume that two homomorphisms $f\colon A \to N$ and $g\colon B \to N$ are equivalent if they coincide on some submodule C such that $C \subseteq A \cap B$ and M/C is a primary module. Indeed, we obtain an equivalence relation. It is useful to keep in mind that $M/(A \cap B)$ is a primary module. Now we consider the corresponding equivalence classes as morphisms from M into N. We denote by $\operatorname{Hom}_{\underline{W}}(M, N)$ the set of morphisms from M into N. If $\bar{f}$ is a morphism, then we represent it by a homomorphism $f\colon A \to N$ which does not decrease heights, where M/A is a primary module and f is contained in the equivalence class $\bar{f}$. We define the operation of addition of morphisms. Let $\bar{f}, \bar{g}\colon M \to N$ be two morphisms and let $f\colon A \to N$ and $g\colon B \to N$ be their representatives. We assume that $\bar{f} + \bar{g} = \overline{(f+g)_{|A\cap B}}$ or $\bar{f} + \bar{g} = \overline{f+g}$, for brevity. Thus, $\bar{f} + \bar{g}$ is represented by a homomorphism $f + g\colon A \cap B \to N$. The set $\mathrm{Hom}_{\underline{W}}(M, N)$ forms an Abelian group with respect to the defined addition. We especially note that the zero element coincides with the class of $f\colon A \to N$ such that $f(A)$ is a primary module.

We define the composition of morphisms. Let we have $\bar{f}\colon M \to N$ and $\bar{g}\colon N \to L$ with representatives $f\colon A \to N$ and $g\colon B \to L$. We take the composition $f_{|C}\, g\colon C \to L$, where $C = \{a \in A \mid f(a) \in B\}$ and M/C is a primary module. We assume that $\bar{f}\bar{g} = \overline{f_{|C}\, g}$ or, conditionally, $\bar{f}\bar{g} = \overline{fg}$ (it is clear that $\bar{f} = \bar{f}_{|C}$). So defined composition is associative. The morphisms of the form $\bar{1}_M$ are the identity morphisms. Thus, we have defined the category of modules with morphisms $\bar{f}$; it is denoted by Warf (see also Exercise 2 in connection to this topic).

We differently define the group $\mathrm{Hom}_{\underline{W}}(M, N)$ with the use of direct spectra. This notion is circumstantially presented in [92, Section 11]. Considering that direct spectra also appear in other parts of the work, we present brief definitions and two properties of direct spectra.

Let $\{A_i\}_{i\in I}$ be a family of modules, where the subscript set I is partially ordered and for all $i, j \in I$, there exists a subscript $k \in I$ with $i, j \leq k$. We assume that for every subscript pair $i, j \in I$ with $i \leq j$, there exists a homomorphism $e_i^j : A_i \to A_j$ such that $e_i^i = 1_{A_i}$ for every $i \in I$ and $e_i^j e_j^k = e_i^k$ provided $i \leq j \leq k$. In this situation, the family of the modules A_i with homomorphisms e_i^j $(i, j \in I)$ is called the *direct spectrum*. We consider the direct sum $M = \bigoplus_{i\in I} A_i$ and denote by B the submodule in M generated by all elements of the form $a_i - e_i^j a_i$, where $a_i \in A_i$, $i \leq j$. The factor module M/B is called the *limit* of the given direct spectrum or the *direct limit*; it is denoted by $\varinjlim_I A_i$.

We have two properties of direct limits which can be useful in the proof of the equivalence of the new definition and the original definition of the group $\mathrm{Hom}_{\underline{W}}(M, N)$.

(a) Every element of M/B can be represented in the form $a_i + B$, where $a_i \in A_i$ for some $i \in I$.

(b) The submodule B consists of all sums $a_{i_1} + \cdots + a_{i_n}$ $(a_{i_s} \in A_{i_s})$ such that I contains a subscript $k \geq i_1, \ldots, i_n$ with $e_{i_1}^k a_{i_1} + \cdots + e_{i_n}^k a_{i_n} = 0$.

Let M and N be two modules and let A be a submodule in M. We denote by $H_A(M, N)$ the submodule of the module $\mathrm{Hom}(A, N)$ consisting of all homomorphisms $A \to N$ which do not decrease heights. For every submodule B of the module A, we have the induced "restriction" mapping $H_A(M, N) \to H_B(M, N)$, $f \to f_{|B}$, where $f\colon A \to N$. Let A_i $(i \in I)$ be a family of all submodules of the module M such that M/A_i is a primary module. Assuming $i \leq j \Longleftrightarrow A_i \supseteq A_j$, we obtain a partial order on I. Let

$$e_i^j : H_{A_i}(M,N) \to H_{A_j}(M,N)$$

be the induced mapping, where $i \leq j$. Then the system of modules and homomorphisms $\{H_{A_i}(M,N);\, e_i^j\ (i,j \in I)\}$ is a direct spectrum. Let G be the limit of it. We state that modules $\mathrm{Hom}_{\underline{W}}(M,N)$ and G are isomorphic to each other. For example, there exists an isomorphism such that every morphism $\bar{f} \colon M \to N$ is mapped to the residue class $f + B$, where $f \colon A_i \to N$ is some representative of $\bar{f}$.

It is useful to note that a morphism $\bar{f} \colon M \to N$ is a monomorphism in Warf if and only if for every representative $f \colon A \to N$, the inclusion $f(a) \in t(N)$ implies the inclusion $a \in t(A)$ for every $a \in A$. The morphisms $0 \to T$ and $T \to 0$ are isomorphisms for every primary module T.

Proposition 29.5. Warf *is an additive category with kernels and infinite sums and it satisfies the weak Grothendieck condition.*

Proof. We have already proved that $\mathrm{Hom}_{\underline{W}}(M,N)$ is an Abelian group for any two modules M and N. It is clear that for given morphisms $\bar{f}_1, \ldots, \bar{f}_k$ from M into N, we can assume that their representatives $f_i : A_i \to N$ $(i = 1, \ldots, k)$ are defined on the same submodule (for example, we can take the intersection of all A_i). This simplifies considerations. In particular, it is now obvious that the composition is bilinear. Let we have an arbitrary set of modules A_i, $i \in I$. We consider the (ordinary) module direct sum $M = \bigoplus_{i \in I} A_i$ with embeddings $e_i : A_i \to M$, $i \in I$. The module M considered together with the morphisms $\bar{e}_i$ is the direct sum of modules A_i $(i \in I)$ in Warf. Indeed, let we have morphisms $\bar{f}_i : A_i \to N$ in Warf with representatives $f_i : C_i \to N$ $(i \in I)$. There exists a unique homomorphism $f \colon \bigoplus_{i \in I} C_i \to N$ such that $e_i f = f_i$ for all i. We obtain a unique morphism $\bar{f} \colon M \to N$ such that $\bar{e}_i \bar{f} = \bar{f}_i$ for all i. Thus, Warf is an additive category with infinite sums.

We take an arbitrary morphism $\bar{f} : M \to N$ with representative $f \colon A \to N$. We set

$$L = \{x \in M \mid rx \in \mathrm{Ker}(f) \text{ for some nonzero } r \in R\}.$$

Let $l : L \to M$ be an embedding. Then the pair $(L, \bar{l})$ is the kernel of the morphism $\bar{f}$. The property that Warf satisfies the weak Grothendieck condition can be proved similar to the corresponding assertion from Proposition 29.1. $\square$

It follows from the definition of morphisms that either $\mathrm{Hom}_{\underline{W}}(M,N) = 0$ or $\mathrm{Hom}_{\underline{W}}(M,N)$ is an torsion-free R-module. Consequently, the endomorphism ring $\mathrm{End}_{\underline{W}}(M)$ in Warf of the module M is a torsion-free R-algebra.

Proposition 29.6. (a) *Two modules M and N are isomorphic to each other in* Warf *if and only if there exist free submodules A and B such that M/A and N/B are primary modules and there is a module isomorphism $f\colon A \to B$ which preserves heights of elements.*

(b) *Two modules M and N of rank 1 are isomorphic to each other in* Warf *if and only if $U(M) = U(N)$.*

Proof. (a) We say that the isomorphism $f\colon A \to B$ *preserves heights* if $h_M(a) = h_N(fa)$ for all $a \in A$.

Let M and N be isomorphic to each other in Warf and let $\bar{f}\colon M \to N$ and $\bar{g}\colon N \to M$ be mutually inverse isomorphisms. We choose representatives $f\colon X \to N$ and $g\colon Y \to M$. Without loss of generality, we can assume that X and Y are free modules. In X, there exists a submodule A such that $fg = 1_A$ and X/A is a primary module. Similarly, Y contains a submodule C such that $gf = 1_C$ and Y/C is a primary module. Therefore, we obtain that Y/fA is also a primary module. Setting $B = fA$, we obtain the height-preserving isomorphism $f\colon A \to B$.

Conversely, if $f\colon A \to B$ is the isomorphism considered in the statement of the proposition, then $ff^{-1} = 1_A$ and $f^{-1}f = 1_B$. In this case, $\bar{f}$ and $\bar{f}^{-1}$ are mutually inverse isomorphisms and $M \cong N$ in Warf.

(b) The invariant $U(M)$ of the module M was defined in the previous section; it is the equivalence class of the height sequences of elements of infinite order in the module M. If $M \cong N$ in Warf, then by (a), there exists a height-preserving isomorphism $A \to B$, for some submodules A and B which are isomorphic to R. It is clear that $U(M) = U(N)$.

Let $U(M) = U(N)$. We take some elements x in M and y in N of infinite order. Since $U(x) \sim U(y)$, we can multiply the elements x and y by some nonzero elements of R and pass to elements with equal height sequences. Therefore, we assume that $U(x) = U(y)$. We consider height-preserving homomorphisms $f\colon Rx \to Ry$ $(rx \to ry)$ and $g\colon Ry \to Rx$ $(ry \to rx)$. Here $\bar{f}$ and $\bar{g}$ are mutually inverse isomorphisms in Warf from M onto N and from N onto M, respectively. □

For any two modules M and N, there exists a homomorphism of R-modules

$$\operatorname{Hom}(M,N) \to \operatorname{Hom}_{\underline{W}}(M,N),$$

such that every $f\colon M \to N$ is mapped onto the equivalence class $\bar{f}$. The kernel of it coincides with $\operatorname{Hom}(M, t(M))$. Therefore, $\operatorname{Hom}_{\overline{W}}(M,N)$ can be canonically embedded in $\operatorname{Hom}_{\underline{W}}(M,N)$. Similarly, $\operatorname{End}_{\overline{W}}(M)$ is an R-subalgebra in $\operatorname{End}_{\underline{W}}(M)$. Therefore Walk is a subcategory of the category Warf. We have the following result (it is sufficient to use Lemma 24.1 in (b)).

Corollary 29.7. (a) *The modules, which are isomorphic to each other in* Walk, *are isomorphic to each other in* Warf.

(b) *If a module is indecomposable in* Warf, *then it is indecomposable in* Walk.

We have one more result which is similar to Lemma 29.3.

Lemma 29.8. (a) *A module M is a small object in* Warf *if and only if the rank of M is finite.*

(b) *The endomorphism ring in* Warf *of any module of rank* 1 *is local.*

Proof. The proof of assertion (a) is similar to Part (a) in Lemma 29.3. We need to only specialize some details; we remain this to the reader.

(b) If $M \cong K \oplus T$, where T is a primary module, then $\mathrm{End}_{\underline{W}}(M) \cong K$. In the remaining principal case, the divisible submodule of the module M is a primary module. We show that $\mathrm{End}_{\underline{W}}(M) \cong R$. The ring R can be embedded in $\mathrm{End}_{\underline{W}}(M)$ by the mapping $r \to \overline{r}$, $r \in R$. We assume that $\overline{f} \in \mathrm{End}_{\underline{W}}(M)$, $\overline{f} \neq 0$, and $f\colon Rx \to M$ represents $\overline{f}$. If $f(x) = y$, then $y \neq t(M)$. There exist invertible elements $u, v \in R$ and integers $k, m \geq 0$ with $p^k ux = p^m vy$. Then $m \leq k$, since otherwise

$$h(p^k x) < h(p^m x) \leq h(p^m y)$$

(since f does not decrease heights). Similar to Lemma 29.3, we obtain that the restriction of f to some submodule Ra coincides with the multiplication by the element $p^{k-m}w$. Therefore $\overline{f} = \overline{p^{k-m}w}$ and $\mathrm{End}_{\underline{W}}(M) = R$ (under the identification r with $\overline{r}$). □

What can we say about completely decomposable modules if we pass to the category Warf? Every module of rank 1 is a small object in Warf; therefore, it is finitely approximable and countably approximable and the endomorphism ring of the module is local. Warf satisfies the weak Grothendieck condition. Now Theorem 27.8 and Theorem 27.10 imply the following result.

Corollary 29.9. *Let M be a completely decomposable module and let $M = \bigoplus_{i\in I} A_i$ be some decomposition of M, where $r(A_i) = 1$ for all $i \in I$. Then*

(1) *if $M = \bigoplus_{j\in J} B_j$, where $r(B_j) = 1$ for all $j \in J$, then there exists a bijection $\theta : I \to J$ such that $A_i \cong B_{\theta(i)}$ in* Warf *for all $i \in I$;*

(2) *if N is some direct summand of the module M, then there exists a subset $J \subseteq I$ such that $N \cong \bigoplus_{i\in J} A_i$ in* Warf.

For a completely decomposable module $M = \bigoplus_{i\in I} A_i$, we define invariants which are cardinal numbers. For every equivalence class e of height sequences (see

Section 28), we denote by $g(e, M)$ the number of summands A_i with $U(A_i) = e$. Now let N be some direct summand of the module M. By Corollary 29.9, $N \cong \bigoplus_{i \in J} A_i$ in Warf, where J is some subset in I. We set $g(e, N) = g\left(e, \bigoplus_{i \in J} A_i\right)$ for every class e. Corollary 29.9 can be reformulated in another form (consider Proposition 29.6 (b)).

Corollary 29.10. *Let M and N be two direct summands of completely decomposable modules. Then $M \cong N$ in* Warf *if and only if $g(e, M) = g(e, N)$ for all equivalence classes e of height sequences.*

Let M be a completely decomposable module or a direct summand of some completely decomposable module. We can assert that the numbers $g(e, M)$ do not depend on a concrete decomposition of this completely decomposable module into a sum of modules of rank 1, i.e., they uniquely determined by the module M. Consequently, the set of integers $g(e, M)$ for all classes e is a complete system of invariants for the module M in the category Warf. It is also independent by Exercise 3 and Theorem 30.2. (See Sections 4 and 6 on invariants of modules.) The defined invariants are called *Warfield invariants*. We also consider Warfield invariants $g(e, M)$ in Sections 31 and 32. In Section 32, we will define some invariants for an arbitrary module; the invariants coincide with $g(e, M)$ for any direct summand M of a completely decomposable module.

We have a satisfactory classification of direct summands of completely decomposable modules in the category Warf. The problem of description of these modules up to ordinary isomorphism is far more difficult. In what follows, we will classify one quite large class of completely decomposable modules and their direct summands (Theorem 32.6).

The category Warf was introduced by Warfield (see [310] and [313]); the category Walk was first considered by E. Walker. Endomorphism rings of modules and Abelian groups in the category Walk were studied by Goldsmith–Zanardo [121] and Breaz [36]. There is an interesting paper of K. Walker [297], where the author studies properties of Abelian groups in the category which is similar (in a certain sense) to the category Walk.

Exercise 1. In Walk, two modules M and N are isomorphic to each other if and only if $M \oplus t(N) \cong N \oplus t(M)$.

Exercise 2. Prove that in the definition of the category Warf, we can consider only partial homomorphisms $f \colon A \to N$ for torsion-free submodules A.

Exercise 3. For every equivalence class e of height sequence, there exists a module M of rank 1 with $U(M) = e$ (see also Theorem 30.2).

Exercise 4. For two mixed modules M and N of rank 1, describe the groups $\mathrm{Hom}_{\overline{W}}(M,N)$ and $\mathrm{Hom}_{\underline{W}}(M,N)$ in terms of the equivalence classes $U(M)$ and $U(N)$.

Exercise 5 (Files [73]). (a) If M is a sharp module, then there exists a split extension $\mathrm{End}(M) = A \oplus \mathrm{Hom}(M, t(M))$, where A is an R-algebra which is isomorphic to some pure subalgebra in $\widehat{R}$ containing R. Therefore

$$\mathrm{End}_{\overline{W}}(M) \cong A.$$

(b) If the rank of the module M from (a) is finite, then $\mathrm{End}_{\overline{W}}(M)$ is a discrete valuation domain. (Sharp modules are defined in remarks to Chapter 7.)

30 Simply presented modules

In this section, we consider completely decomposable modules such that for direct summands of the modules, we will construct a structural theory in Section 32 (we mentioned this theory at the end of the previous section). Our classification method is based on the very special representation of the module by generators and defining relations. The method allows us to obtain a relatively simple proof of the main results, since it is closely related to the structural theory.

If X is a subset in the module M, then RX denotes the submodule in M generated by X.

It is useful to present a known general information about generator systems and defining relations. Let M be a module and let $\{a_i\}_{i\in I}$ be some generator system of M. The generators can satisfy (nontrivial) relations of the form $r_{i_1}a_{i_1} + \cdots + r_{i_n}a_{i_n} = 0$, where r_{i_s} are nonzero elements of the ring R. To better study this relations, we take a free module F with free basis $X = \{x_i\}_{i\in I}$ numbered by elements of the same subscript set I (free bases were defined in Section 5). We consider a homomorphism $\phi : F \to M$, where $\phi(r_{i_1}x_{i_1} + \cdots + r_{i_n}x_{i_n}) = r_{i_1}a_{i_1} + \cdots + r_{i_n}a_{i_n}$. Let $H = \mathrm{Ker}(\phi)$. Then $M \cong F/H$. The relation $r_{i_1}a_{i_1} + \cdots + r_{i_n}a_{i_n} = 0$ in the module M corresponds to an arbitrary element $r_{i_1}x_{i_1} + \cdots + r_{i_n}x_{i_n}$ in H; the relation is called the *relation connecting the elements* a_{i_s} in the module M. If we choose some subset Δ in H such that the submodule generated by Δ, coincides with H, then Δ and the corresponding to Δ set of relations in M are called a *system of defining relations* of the module M. All relations connecting elements a_i in the module M can be considered as corollaries of defining relations. The triple $\{X, \Delta, \phi\}$ is called a *representation* of the module M.

Every module is a factor module of a free module. Therefore, every module can be represented by a system of defining relations with respect to some symbol set. In addition, let two modules be presented by defining relations with respect to some generator systems such that there is an one-to-one correspondence between these systems such that defining relations of one module correspond to defining relations of another module and conversely; then the modules are isomorphic to each other.

Is it true that there always exists a module with given generator system and given system of defining relations? The answer is positive; this follows from the following argument. If we have an arbitrary free module $F = \bigoplus_{i \in I} Rx_i$ and an arbitrary (abstract) system of relations, in which some words consisting of the elements x_i are equal to zero, then it is easy to find a module, for which these relations form a system of defining relations. For this purpose, it is sufficient to take the submodule H in F generated by the left parts of given relations and pass to the factor module. The set $\{a_i\}_{i \in I}$, where $a_i = x_i + H$ $(i \in I)$ is a generator system for the factor module. We note that some of the generators a_i can be equal to zero and some of the generators a_i can be equal to each other.

We define the main notion of the section. A module M is said to be *simply presented* if it has a generator system connected only by defining relations of the form $px = 0$ or $px = y$ and their corollaries.

Let M be a simply presented module. We show that we can choose a generator system Y of M such that

(1) $0 \notin Y$;

(2) Y does not contain equal elements;

(3) if $y \in Y$ and $py \neq 0$, then $py \in Y$.

These conditions can be presented in terms of representations of the module M, as follows. A representation $\{X, \Delta, \phi\}$ of the module M is said to be *standard* provided

(1) the inclusion $x \in X$ implies the relation $\phi(x) \neq 0$;

(2) if $x, y \in X$ and $x \neq y$, then $\phi(x) \neq \phi(y)$;

(3) if $x \in X$ and $p\phi(x) \neq 0$, then $p\phi(x) = \phi(y)$ for some $y \in X$.

We recall that the homomorphism ϕ in the representation $\{X, \Delta, \phi\}$ is a homomorphism with kernel $R\Delta$ from a free module F with basis X into the module M, where Δ is a system of defining relations of the module M of the form px, $px - y$, $x, y \in X$.

Lemma 30.1. *A simply presented module M has the standard representation $\{X, \Delta, \phi\}$. In addition, the set $Y = \phi(X)$ has the following properties:*

(1) $0 \notin Y$;

(2) *if $y \in Y$ and $py \neq 0$, then $py \in Y$;*

(3) *for every $y \in Y$, if we set*

$$Z = \{z \in Y \mid p^n z \neq y \textit{ for every } n \geq 0\},$$

then $y \notin RZ$.

Conversely, if M is some module and Y is a generator system of M which satisfies conditions (1)–(3), *then M is a simply presented module and there exists the standard representation $\{X, \Delta, \phi\}$ of the module M with $Y = \phi(X)$.*

Proof. First, we verify that the second part of the lemma holds. Let M be a module and let Y be a generator system of M which satisfies (1)–(3). Furthermore, let F be a free module with basis Y and let G be the kernel of the canonical homomorphism $\phi : F \to M$. Consequently, $M \cong F/G$. We denote by H the submodule in F generated by all elements of the form px and $px - y$ $(x, y \in Y)$ such that $px, px - y \in G$. Then $H \subseteq G$ and F/H is a simply presented module. We assume that $H \neq G$. Among elements of the module G not contained in H, we choose an element g such that the expression $g = r_1 y_1 + \cdots + r_n y_n$ $(r_i \in R$, $y_i \in Y)$ has the minimal possible integer n. Using (2), we can assume that all r_i are invertible elements. It is clear also that $n \geq 2$. Multiplying by r_1^{-1}, we can assume that $r_1 = 1$. Furthermore, we assert that for all $i > 1$ and $n > 0$, we have $p^n y_i \neq y_1$ in the module M. Otherwise, we obtain a more short expression for some element of $G \setminus H$ (with less integer n) by replacing $r_i x_i$ by $(r_i + p^n)x_i$ and omitting x_1. However, in this case, the relation $x_1 = -(r_2 x_2 + \cdots + r_n x_n)$ in M contradicts to condition (3).

Now let M be a simply presented module and let $\{X, \Delta, \phi\}$ be some representation of M. We denote by Y the set of all nonzero elements of the form $p^n \phi(x)$, where $n \geq 1$, $x \in X$. We verify that the generator system Y satisfies conditions (1)–(3). By the previous argument, this implies that the standard representation exists. It is also clear that if the representation $\{X, \Delta, \phi\}$ will be chosen to be standard, then $Y = \phi(X)$ and the proof of the first part of the theorem will be completed. It is obvious that the set Y satisfies (1) and (2). We pass to (3). We take y and Z satisfying (3).

We take the quasicyclic module $R(p^\infty)$ with generators $c_1, c_2, \ldots, c_n, \ldots$ such that $pc_1 = 0$ and $pc_{n+1} = c_n$ for every $n \geq 1$. Furthermore, let F be a free module with basis X. We construct a homomorphism $f \colon F \to R(p^\infty)$ by defining it on the basis X. If $x \in X$ and $p^n \phi(x) = y$, then we set $f(x) = c_{n+1}$; if $x \in X$ and $p^n \phi(x) \neq y$, then we set $f(x) = 0$. Since $\phi(\Delta) = 0$, it follows from easy calculations that $f(\Delta) = 0$; therefore, $f(R\Delta) = 0$. Consequently, the mapping f induces the homomorphism $F/R\Delta \to R(p^\infty)$ such that $a + R\Delta \to f(a)$ for $a \in F$. Since the module $F/R\Delta$ is isomorphic to M under the correspondence $a + R\Delta \to \phi(a)$ $(a \in F)$, we have the homomorphism $g \colon M \to R(p^\infty)$ such that

$g(y) = c_1 \neq 0$ and $g(Z) = 0$, whence $y \notin RZ$. Thus, Y satisfies (3) and the proof is completed. □

In what follows, all representations of simply presented modules are assume to be standard (unless otherwise stated); equivalently, we assume that generator systems satisfy conditions (1)–(3).

We present examples of simply presented modules (with generator systems which satisfy conditions (1)–(3)).

1. The cyclic module $R(p^n)$ is simply presented, since it has a generator system $x_1, \dots, x_n$ such that $px_1 = 0, px_2 = x_1, \dots, px_n = x_{n-1}$.

2. The quasicyclic module $R(p^\infty)$ is simply presented. It has a generator system $x_1, x_2, \dots, x_n, \dots$ such that $px_1 = 0, px_2 = x_1, \dots, px_n = x_{n-1}, \dots$.

3. The modules R and K are simply presented. Indeed, the required generator system of R is formed by the elements $1, p, p^2, \dots$. In the module K, we need to add the elements $1/p, 1/p^2, \dots$ to the elements $1, p, p^2, \dots$.

4. Any direct sum of simply presented modules A_i $(i \in I)$ is simply presented.

If we take the union of generator systems of all modules A_i, then we obtain the required generator system for the direct sum.

We have that direct sums of cyclic modules and divisible modules are simply presented. The following theorem provides important examples of mixed simply presented modules of rank 1 (see Walker [300] and Warfield [313]).

Theorem 30.2. *If v is an arbitrary height sequence, then there exists a simply presented module A_v of rank 1 which contains an element x_v of infinite order with $U(x_v) = v$. If α is any ordinal number, then there exists a simply presented primary module P_α such that $p^\alpha P_\alpha$ is a cyclic module of order p.*

Proof. We use the above mentioned property that there always exists a module which has the given generator system and the given defining relations. First, we consider the primary case. All finite properly ascending sequences of ordinal numbers $(\alpha_1, \dots, \alpha_n)$ with $\alpha_n = \alpha$ are generators of the module P_α. The defining relations have the form $p(\alpha_1, \dots, \alpha_n) = (\alpha_2, \dots, \alpha_n)$ $(n > 1)$ and $p(\alpha) = 0$. The module P_α satisfies the definition of a simply presented module. The height of the element $p^k(\alpha_1, \dots, \alpha_n)$ in P_α is equal to α_{k+1}, $0 \leq k \leq n-1$. Therefore, it can be proved (for example, with the use of the induction on α) that $p^\alpha P_\alpha$ is a cyclic module generated by (α).

Let $v = \{\sigma_i\}_{i \geq 0}$. If $\sigma_0 = \infty$, then K is the required module. If $\sigma_{n+1} = \infty$ and $\sigma_n < \infty$, then we set $A_v = P_{\sigma_n} \oplus K$ and we can take $(\sigma_0, \sigma_1, \dots, \sigma_n) + 1$ as the required element x_v, where $(\sigma_0, \sigma_1, \dots, \sigma_n)$ is one of generators of the module P_{σ_n} and $1 \in K$.

Finally, we assume that all coordinates σ_i are ordinal numbers, i.e., they are not equal to ∞. As the module A_v, we take the module with generators consisting of finite properly ascending sequences $(\alpha_1, \ldots, \alpha_n)$ for different n such that α_n is one of the ordinal numbers σ_i, $i \geq 0$. The module A_v has the following defining relations:

$$p(\alpha_1, \ldots, \alpha_n) = (\alpha_2, \ldots, \alpha_n), \quad n > 1,$$

and

$$p(\sigma_i) = (\sigma_{i+1}).$$

We can take $x_v = (\sigma_0)$. □

The following result is easily proved, but it gives an essential information about the object of our study.

Theorem 30.3 (Warfield [313]). *Any simply presented module is the direct sum of a primary module and modules of rank* 1.

Proof. Let X be a generator system of a simply presented module M (we recall that by our agreement, X necessarily satisfies the above conditions (1)–(3)). There exists an equivalence relation on X such that two elements x and y of X are equivalent if and only if $p^m x = p^n y$ for some positive integers m and n. Let X_i ($i \in I$) be the set of all equivalence classes. We denote by M_i the submodule in M generated by all elements of X_i, i.e., $M_i = RX_i$. It follows from Lemma 30.1 (or directly from the nature of defining relations) that $M = \bigoplus_{i \in I} M_i$. Every M_i is a simply presented module with generator system X_i. If X_0 is the equivalence class consisting of elements of finite order of the set X, then M_0 is a primary module. For $X_i \neq X_0$, we have the module M_i of rank 1 with generator system X_i consisting only of elements of infinite order. □

We have that a simply presented mixed module is completely decomposable. Therefore, the module has Warfield invariants defined in the previous section.

Now we list several elementary useful results on simply presented modules and their generator systems. First, we note that the generator system X of a simply presented module M can be equipped by a partial order $<$ such that $y < x$ if and only if $p^n x = y$ for some $n > 0$. If M is a primary module, then X satisfies the minimum condition. We continue the direction considered in Theorem 30.3. Precisely, we show that a certain partition of the set X induces a direct decomposition of the module M. For an element $y \in X$, we set $X_y = \{x \in X \mid y \leq x\}$.

(a) If M is a simply presented primary module and L is the set of minimal elements in X, then $M = \bigoplus_{y \in L} RX_y$.

For distinct $y, z \in L$, the sets X_y and X_z have the empty intersection. By the minimum condition, they cover the set X. Therefore, the sets X_y ($y \in L$) form a partition of the set X. The defined earlier equivalence relation on X can be specialized as follows. We say that two elements $x, y \in X$ are equivalent if $p^m x = p^n y \neq 0$ for some $m, n > 0$. (This specialization touches on only elements of finite order in X.) Then the sets X_y coincide with equivalence classes for the defined new equivalence relation on X. Similar to Theorem 30.3, we obtain that M is the direct sum of modules RX_y, $y \in L$.

(b) If M is a simply presented module with generator system X and Y is some subset in X, then RY and M/RY are simply presented modules.

Indeed, the module RY is generated by the set Y, and every relation between elements of Y has the form $px = 0$ or $px = y$. The factor module M/RY can be defined by generators $\bar{x} = x + Ry$ $(x \in X)$, by relations of the form $p\bar{x} = 0$ and $p\bar{x} = \bar{y}$, and by additional relations of the form $\bar{x} = 0$.

(c) Every nonzero element a of a simply presented module M with generator system X can be uniquely represented in the form $a = v_1x_1 + \cdots + v_kx_k$, where $x_1, \ldots, x_k$ are distinct elements of X and $v_1, \ldots, v_k$ are invertible elements of the ring R.

First, we note that if $z_1, \ldots, z_n$ are some distinct elements in X and, for example, z_1 is a maximal element in this set with respect to the order indicated before (a), then $z_1 \notin Rz_2 + \cdots + Rz_n$.

By properties (1)–(3) of the set X, the existence of such a representation is obvious. We assume that there exists another representation of the element a. By convention, it can be presented in the form

$$a = r_1x_1 + \cdots + r_kx_k + r_{k+1}x_{k+1} + \cdots + r_lx_l,$$

where either $r_i = 0$ or r_i is an invertible element. It follows from the just presented property of maximal elements that each of the elements $x_{k+1}, \ldots, x_l$ cannot be maximal in the set $x_1, \ldots, x_l$. Consequently, we can assume that x_1 is maximal. Then $x_1 \notin Rx_2 + \cdots + Rx_l$. There exists a homomorphism φ from the module $Rx_1 + \cdots + Rx_l$ into the quasicyclic module $R(p^\infty)$ which maps x_1 onto an element of order p and maps $x_2, \ldots, x_l$ onto 0. Since $R(p^\infty)$ is an injective module, we can assume that φ is defined on M. We have $\varphi a = v_1(\varphi x_1) = r_1(\varphi x_1)$ that implies $r_1 = v_1$. The induction on k completes the proof.

(d) If we consider the unique representation from (c) and $a \in p^\sigma M$, then $x_i \in p^\sigma M$ for all $i = 1, \ldots, k$.

For $\sigma = 0$, the assertion is trivial. We assume that the assertion holds for all ordinal numbers which are less than σ. The case of the limit ordinal number σ is not difficult. Assume that $\sigma - 1$ exists and $a = pb$ for some $b = u_1y_1 + \cdots + u_ly_l \in p^{\sigma-1}M$, where $y_1, \ldots, y_l$ are distinct elements from X and $u_1, \ldots, u_l$ are invertible elements from R. By the induction hypothesis, $y_j \in p^{\sigma-1}M$ for $j = 1, \ldots, l$. The element a can be represented in the form $w_1z_1 + \cdots + w_mz_m$ with distinct elements $z_1, \ldots, z_m$ of X and invertible elements $w_1, \ldots, w_m$. Each of the elements $z_1, \ldots, z_m$ has the form p^ny_j for some j and $n \geq 1$, whence

$z_1, \dots, z_m \in p^\sigma M$. It follows from (c) that these elements z are equal to the elements x, i.e., $x_1, \dots, x_k \in p^\sigma M$.

(e) Let M be a reduced simply presented primary module such that the length of M is a limit ordinal number. Then M is the direct sum of simply presented modules whose lengths are less than the length of M. The assertion remains true provided we replace the length by the Ulm length (the notions of the lengths were introduced in Section 28).

The required decomposition follows from Part (a). It follows from (d) that the last nonzero Ulm submodule of the module RA_y is generated by the element y. Consequently, the lengths of the modules RA_y are nonlimit ordinal numbers. Consequently, they are less than the length of the module M.

(f) If M is a simply presented module, then for every ordinal number σ, the modules $p^\sigma M$ and $M/p^\sigma M$ also are simply presented. In addition, if the module M is a reduced primary module, then all Ulm factors of M are direct sums of cyclic modules.

Let X be a generator system modules M. We set

$$Y = \{y \in X \mid h(y) \geq \sigma\}.$$

It follows from (d) that $p^\sigma M = RY$ and the result follows from (b). Since $M^\sigma = p^{\omega\sigma} M$ for every σ, Ulm submodules and Ulm factors of the module M are simply presented. Ulm factors of a reduced primary module do not have elements of infinite height. It remains to prove that the simply presented primary module M without elements of infinite height is a direct sum of cyclic modules. Indeed, since the length of the module M does not exceed ω, we have that by (e), M is the direct sum of bounded modules; therefore M is a direct sum of cyclic modules.

(g) If Y is some subset of a generator system X of a simply presented module M, then RY is a nice submodule in M.

Similar to (c), the element $a \in M \setminus A$ can be represented in the form $a = u_1x_1 + \cdots + u_kx_k + v_1y_1 + \cdots + v_ly_l$, where x_i and y_j are elements of $X \setminus Y$ and Y, respectively, and u_i, v_j are invertible elements of the ring R. For simplicity, we assume that $x_1, \dots, x_s$ $(s \leq k)$ are not contained in A and all remaining elements x are contained in A. We assert that the element $b = u_1x_1 + \cdots + u_sx_s$ of the residue class $a + A$ is proper with respect to A. Indeed, we take an arbitrary element c of A and we represent the sum $u_{s+1}x_{s+1} + \cdots + u_kx_k + v_1y_1 + \cdots + v_ly_l + c$ in the form of the sum from Part (c): $w_1z_1 + \cdots + w_mz_m$. Then $u_1x_1 + \cdots + u_sx_s + w_1z_1 + \cdots + w_mz_m$ is the unique expression for the element $a + c$ indicated in (c). It follows from Part (d) that

$$\begin{aligned} h(a+c) &= \min\{h(x_i), h(z_j) \mid i = 1, \dots, s;\ j = 1, \dots, m\} \\ &\leq \min\{h(x_i) \mid i = 1, \dots, s\} = h(b). \end{aligned}$$

Reduced simply presented primary modules coincide with totally projective modules. A reduced primary module M is said to be *totally projective* if

$$p^\sigma \operatorname{Ext}(M/p^\sigma M, N) = 0$$

for every ordinal number σ and every module N. These notions first appeared in the theory of Abelian p-groups, (equivalently, primary groups, or primary $\mathbb{Z}_p$-modules, or primary $\widehat{\mathbb{Z}}_p$-modules; see Section 4). With corresponding exchanges, the theory of p-groups can be extended to primary modules over an arbitrary discrete valuation domain. In the more general context, we say about this in Section 4 and in the remarks to the presented chapter.

We can better understand the role and importance of the results of the central section 32 of the chapter if we cover several aspects of the theory of p-groups related to the famous Ulm's theorem. In 1933, Ulm [290] has proved that two countable reduced p-groups A and B are isomorphic to each other if and only if coincide the corresponding Ulm–Kaplansky invariants of the groups A and B. Ten years before, Prüfer [252] has proved that Ulm factors of countable p-groups are direct sums of cyclic groups. The theorems of Ulm and Prüfer complement the Zippin theorem [329] on the existence of a countable p-group with given Ulm sequence. These three theorems form a completed structural theory of countable p-groups. We can say that they give a complete classification of countable Abelian p-groups.

The problem on the existence of p-groups with arbitrary given Ulm sequence was solved by Kulikov [189] and Fuchs [90] (see [93, Theorem 76.1]). As for the classification of (noncountable) p-groups, Kolettis made the first serious step only in 1960 [172]. He has extended the Ulm–Zippin theory to direct sums of countable p-groups. These groups are also distinguished by their Ulm–Kaplansky invariants. It is natural to formulate the following problem: describe (sufficiently large) classes of p-groups with similar property. In the process of homological consideration, Nunke ([246, 247]) defined totally projective groups. He has characterized direct sums of countable p-groups as totally projective groups A such that $p^{\omega_1} A = 0$, where ω_1 is the first noncountable ordinal number. Later, Hill [142] has extended the Ulm's theorem to totally projective groups (see also Walker [299]). Crawley and Hales ([51, 52]) used another approach and discovered simply presented p-groups ("T-groups" in their terminology); they have proved the Ulm's theorem for such groups. Using the result of Hill, they have proved that these groups coincide with totally projective groups in the reduced case. By this reason, we identify the words "a reduced simply presented primary module" and "a totally projective module".

Totally projective modules can be characterized in terms of the existence of certain families of nice submodules. We present necessary definitions.

We say that a primary module M satisfies the *Hill condition* (or "has a nice system," or "satisfies the third countability axiom") if M has a system $\mathcal{N}$ of nice submodules such that

(a) $0 \in \mathcal{N}$;

(b) if $\{A_i \mid i \in I\}$ is an arbitrary set from $\mathcal{N}$, then $\sum_{i \in I} A_i \in \mathcal{N}$;

(c) for every submodule $A \in \mathcal{N}$ and any countable or finite subset X of the module M, there exists a submodule $C \in \mathcal{N}$ such that $A, X \subseteq C$ and C/A is a countably generated module. (The definition of a countably generated module is presented in Section 8.)

We present one more definition.

Let M be a primary module and let

$$0 = A_0 \subset A_1 \subset \cdots \subset A_\sigma \subset \cdots \subset A_\mu = M$$

be a well ordered properly ascending chain of nice submodules of the module M satisfying the following properties:

(a) $A_{\sigma+1}/A_\sigma$ is a cyclic module of order p for every $\sigma < \mu$;

(b) $A_\sigma = \bigcup_{\tau<\sigma} A_\tau$ if σ is a limit ordinal number.

Such a chain is called a *nice composition series* of the module M. (Loth [214] defines and study other composition series.)

We gather main characterizations of totally projective modules.

Theorem 30.4. *For a reduced primary module M, the following conditions are equivalent*:

(1) *M is a totally projective module*;

(2) *M is a simply presented module*;

(3) *M satisfies the Hill condition*;

(4) *M has a nice composition series*;

(5) *M is projective with respect to all balanced exact sequences of primary modules.*

Balanced exact sequences are defined in Section 28. The meaning of condition (5) is explained in the proof of Theorem 32.15.

The following corollary follows from condition (4).

Corollary 30.5. *Let M be a reduced primary module and let A be a nice submodule in M. If A and M/A are totally projective modules, then M is also a totally projective module.*

Finally, we note that a reduced countably generated primary module is a totally projective module.

Theorem 30.3 and the proof of it allow to represent every simply presented module M in the form $P \oplus N$, where P is a primary simply presented module and N is a mixed simply presented module with generator system which contains only elements of infinite order. In what follows, we concentrate our effort to describe mixed simply presented modules.

Exercise 1. A torsion-free module M is simply presented if and only if M is the direct sum of modules of rank 1.

Exercise 2 (Walker [299, 300]). Prove that Walker modules P_α constructed in Theorem 30.2 have the following properties:

(1) $l(P_\alpha) = \alpha + 1$;

(2) $f(\alpha, P_\alpha) = 1$ (the invariants $f(\alpha, P_\alpha)$ are defined in the next section);

(3) $P_\alpha / R(\alpha) \cong \bigoplus_{\beta<\alpha} P_\beta$.

Exercise 3. Every totally projective module is a direct summand of the direct sum of modules P_α for distinct α.

31 Decomposition bases and extension of homomorphisms

In this section, we make a difficult preliminary work for the remaining material of the chapter. Technically, it is sometimes preferable to use terms related to certain subsets of modules instead of the category Warf. We also touch on an extension problem for isomorphisms between submodules of two given modules to isomorphisms of the modules.

Unless otherwise stated, the symbol M denotes a reduced mixed module. It is convenient to call a maximal linearly independent system, consisting of elements of infinite order of the module M, a basis of the module M (linearly independent systems are considered in Section 4). Thus, a basis X is a free submodule in M such that M/RX is a primary module.

We define two central notions of the section. The first notion is due to Rotman (see [263] and [265]).

A basis X of a module M is called a *decomposition basis* of the module M if for all $x_1, \dots, x_n$ in X and any $r_1, \dots, r_n$ in R, we have

$$h(r_1x_1 + \dots + r_nx_n) = \min\{h(r_1x_1), \dots, h(r_nx_n)\}.$$

Therefore, the behavior of elements of a decomposition basis is similar to the behavior of elements of distinct direct summands of the module.

A decomposition basis X is called a *nice decomposition basis* if RX is a nice submodule in M.

Every element x of infinite order in a module M of rank 1 forms a nice decomposition basis for M by Proposition 28.4 (3). Now let M be a completely decomposable module and let $M = \bigoplus_{i \in I} M_i$, where M_i is a module of rank 1 for all $i \in I$. In every M_i, we choose an element x_i of infinite order. Then the set $X = \{x_i \mid i \in I\}$ is a nice decomposition basis of the module M (take into account property (f) in Section 28).

Modules with decomposition basis are closely related to completely decomposable modules.

Proposition 31.1. *A module M has a decomposition basis if and only if M is isomorphic in the category* Warf *to some completely decomposable module.*

Proof. Let M have a decomposition basis $X = \{x_i \mid i \in I\}$. For every $i \in I$, we take a module A_i of rank 1 such that $U(x_i) = U(y_i)$ for some element $y_i \in A_i$ of infinite order (such a module A_i exists by Theorem 30.2). We set $N = \bigoplus_{i \in I} A_i$ and $Y = \{y_i \mid i \in I\}$. The bijection $X \to Y$ ($x_i \to y_1, i \in I$) induces the height-preserving isomorphism $RX \to RY$. By Proposition 29.6, we have $M \cong N$ in Warf.

Conversely, we assume that the module M is isomorphic in Warf to a completely decomposable module $N = \bigoplus_{i \in I} A_i$. Let $Y = \{y_i \mid i \in I\}$ be the above decomposition basis for N. By Proposition 29.6, there exist free submodules F and G in M and N, respectively, such that M/F and N/G are primary and there is a height-preserving isomorphism $f\colon F \to G$. For every $i \in I$, there exists a nonzero element $r_i \in R$ with $r_i y_i \in G$. The set $\{r_i y_i \mid i \in I\}$ is also a decomposition basis of the module N. We set $x_i = f^{-1}(r_i y_i) \in F$. Then $\{x_i \mid i \in I\}$ is a decomposition basis of the module M. □

With the use Theorem 30.2, property 4 from Section 30, and Proposition 29.6(b), we can verify that every completely decomposable module is isomorphic in Warf to some simply presented module.

Corollary 31.2. *Any direct summand of a module M with decomposition basis has a decomposition basis.*

Proof. Let $M = A \oplus B$, where A is a mixed module. The module M is isomorphic in Warf to some completely decomposable module N. In Warf, we have $N \cong C \oplus D$, where $A \cong C$. There exists a completely decomposable module A' with $C \cong A'$ (Corollary 29.9 (2)). Thus, $A \cong A'$ in Warf and A has a decomposition basis by Proposition 31.1. □

Let M be a module with decomposition basis X. It is known that M is isomorphic in Warf to some completely decomposable module N. Consequently, the module M has Warfield invariants $g(e, M)$ (see Corollary 29.10 and the text which is adjacent to the corollary). It follows from the proof of Proposition 31.1 that $g(e, M)$ is equal to the number of elements x of X with $U(x) \in e$.

For the main structural Theorem 32.6 on direct summands of simply presented modules, we need some additional information about decomposition bases. The information is contained in assertions 31.3–31.5 and 31.7.

Let $X = \{x_i \mid i \in I\}$ be some decomposition basis of the module M. For every i, we take an arbitrary integer $n_i \geq 0$. It is clear that the set $Y = \{p^{n_i} x_i \mid i \in I\}$ is also a decomposition basis for M. We say that Y is a *subordinate* of the basis X. If RY is a nice submodule in M, then Y is called a *nice subordinate*.

Lemma 31.3 (Hunter–Richman [155]). *Let X and Y be two decomposition bases with $X \subset RY$. Then there exists a set of subsets in Y (these subsets are called closed subsets) such that*

(1) *the union of every chain of closed subsets is a closed subset;*

(2) *every infinite set is contained in a closed set of the same cardinality;*

(3) *if S is closed, then*

  (a) *$X \cap RS$ is a decomposition basis for RS;*
  (b) *$S \cup (X \setminus RS)$ is a decomposition basis for RY.*

Proof. We will extend a given subset $S \subset Y$ in a such a manner that it satisfies (3). For an element $x \in X$, let π_x be the projection from RX onto the cyclic summand Rx. We set

$$\alpha S = \{x \in X \mid \pi_x(p^n s) \neq 0 \text{ for some } s \in S \text{ and } n \geq 0 \text{ with } p^n s \in RX\}.$$

Then (3a) is equivalent to the inclusion $\alpha S \subset RS$. For every element $z \in RX$, we define the following subset in X:

$$\operatorname{supp}_X(z) = \{x \in X \mid \pi_x(z) \neq 0 \text{ and } h(\pi_x(z)) = h(z)\}.$$

Now we set

$$\beta S = \bigcup \{\operatorname{supp}_X(z) \mid z \in RX \text{ and } \operatorname{supp}_Y(z) \subset S\}.$$

To verify later that the inclusion $\beta S \subset RS$ implies 3b), we note that for $z \in RX$ and $y \in RS$ with

$$h(y + z) \neq h(y) = h(z),$$

we have $\operatorname{supp}_Y(z) \subset S$.

We called the subset S closed if $\alpha S, \beta S \subset RS$. Since α and β permute with the union of chains, the union of every chain of closed sets is a closed set. In fact, we have verified already that conditions (1) and (3) hold. To verify (2), we first show that $|\alpha S| \leq |RS|$ and $|\beta S| \leq |RS|$. The first inequality is obvious. Now we note that $\operatorname{supp}_X(z)$ is finite for all z from RX. Furthermore, if z_1 and z_2 have equal projections onto RS and $\operatorname{supp}_Y(z_i) \subset S$ for $i = 1, 2$, then $\operatorname{supp}_X(z_1) = \operatorname{supp}_X(z_2)$. Therefore $|\beta S| \leq |RS|$. For the infinite subset S, let $S(1)$ be the least subset in Y such that

$$RS(1) \supset S \cup \alpha S \cup \beta S.$$

We set $S(n) = (S(n-1))(1)$ for $n > 1$. Then $S(1) \cup S(2) \cup \cdots$ is a closed set and the cardinality of the set is equal to the cardinality of S. □

Lemma 31.4. *Every countable or finite decomposition basis has a nice subordinate.*

Proof. Let $x_1, x_2, \ldots$ be a decomposition basis and let $\sigma_1, \sigma_2, \ldots$ be some numeration of ordinal numbers $h(p^n x_k)$, where $n \geq 0$ and $k \geq 1$. For every $j \geq 1$, we choose a nonnegative integer $s(j)$ such that for every $i \leq j$, we have either

$$\sigma_i < h(p^{s(j)} x_j), \qquad \text{or} \qquad \sigma_i > h(p^n x_j)$$

for all n. Then the subordinate $\{p^{s(j)} x_j \mid j = 1, 2, \ldots\}$ is nice (see Exercise 2). □

In connection to the following proposition, we recall that totally projective modules are defined before Theorem 30.4. We also note that if A is a nice submodule in M, then M/A is a reduced module (we assume that M is a reduced module).

Proposition 31.5 (Hunter–Richman [155]). *Let X and Y be two decomposition bases with $X \subset RY$. Then there exists a subordinate Z in X such that RZ is a nice submodule in RY and RY/RZ is a totally projective module.*

Proof. Using Lemma 31.4 and the property that countably generated reduced primary module is a totally projective module, we can assume that Y is a noncountable set. We well order the set Y, i.e., we have $Y = \{y_\sigma \mid \sigma < \lambda\}$, where λ is the least ordinal number of cardinality $|Y|$. For every $\sigma < \lambda$, we define a set Y_σ by taking as $Y_{\sigma+1}$ the closure of the set $Y_\sigma \cup \{y_\sigma\}$ in the sense of Lemma 31.3 and by setting $Y_\sigma = \bigcup_{\tau<\sigma} Y_\tau$ for limit ordinal number σ. Now we set $X_\sigma = X \cap RY_\sigma$. It follows from Lemma 31.3 that the defined sets have the following five properties:

(1) $X_\lambda = X$ and $Y_\lambda = Y$;

(2) $X_\sigma = \bigcup_{\tau<\sigma} X_{\tau+1}$ and $Y_\sigma = \bigcup_{\tau<\sigma} Y_{\tau+1}$;

(3) X_σ is a decomposition basis for RY_σ;

(4) $|Y_\sigma| \leq |Y|$;

(5) $Y_\sigma \cup (X \setminus X_\sigma)$ is a decomposition basis.

Let $X^\sigma = X_{\sigma+1} \setminus X_\sigma$. The residue classes with representatives in X^σ form a decomposition basis of the factor module $RY_{\sigma+1}/RY_\sigma$. Using induction, we can obtain a subordinate Z^σ in X^σ such that $R(Y_\sigma \cup Z^\sigma)$ is a nice submodule in $RY_{\sigma+1}$ and $RY_{\sigma+1}/R(Y_\sigma \cup Z^\sigma)$ has a nice composition series (these series are defined before Theorem 30.4). The set $Z = \bigcup_{\sigma<\lambda} Z^\sigma$ is a decomposition basis for RY. We set $Z_\sigma = \bigcup_{\tau<\sigma} Z^\tau$. Then $Z_\sigma = Z \cap RY_\sigma$.

We prove by the induction on σ that the submodule $R(Y_\rho \cup Z_\sigma)$ is nice in RY_σ provided $\rho \leq \sigma$. This is trivial if $\rho = \sigma$. If $R(Y_\rho \cup Z_\sigma)$ is a nice submodule in RY_σ, then

$$R(Y_\rho \cup Z_{\sigma+1}) = R(Y_\rho \cup Z_\sigma \cup Z^\sigma)$$

is a nice submodule in the module $R(Y_\sigma \cup Z^\sigma)$ which is a nice submodule in $RY_{\sigma+1}$. We assume that σ is a limit ordinal number and $R(Y_\rho \cup Z_\tau)$ is a nice submodule in RY_τ for all $\rho \leq \tau < \sigma$. If $y \in RY_\sigma$, then $y \in RY_\tau$ for some $\tau < \sigma$. We can assume that y has the largest height in the residue class $y + R(Y_\rho \cup Z_\tau)$. Since $Y_\tau \cup (Z \setminus Z_\tau)$ is a decomposition basis, y has the largest height in $y + R(Y_\rho \cup Z)$. Taking $\sigma = \lambda$, we obtain that $R(Y_\rho \cup Z)$ is a nice submodule in RY. If $\rho = 0$, then RZ is a nice submodule in RY. Since

$$\begin{aligned} &R(Y_{\rho+1} \cup Z)/R(Y_\rho \cup Z) \\ &\quad = R(Y_{\rho+1} \cup (Z \setminus RY_{\rho+1}))/R(Y_\rho \cup Z^\rho \cup (Z \setminus RY_{\rho+1})) \\ &\quad \cong RY_{\rho+1}/R(Y_\rho \cup Z^\rho), \end{aligned}$$

the first module in these relations has a nice composition series. These series provide a nice composition series for RY/RZ. Consequently, RY/RZ is a totally projective module (Theorem 30.4). □

We define some invariants of modules which are cardinal numbers. For every ordinal number σ, we take the fully invariant submodule $(p^\sigma M)[p]$. We have

$$(p^\sigma M)[p] = p^\sigma M \cap M[p],$$

and $(p^\sigma M)[p]$ is a space over the residue field R_p of the ring R. The dimension of the R_p-space $(p^\sigma M)[p]/(p^{\sigma+1} M)[p]$ is called the σth *Ulm–Kaplansky invariant* of the module M; it is denoted by $f(\sigma, M)$. If $l(M)$ is the length of the module M, we have $f(\sigma, M) = 0$ for ordinal numbers $\sigma \geq l(M)$.

We generalize these well-known Ulm–Kaplansky invariants. For a submodule A of the module M, we denote by $A(\sigma)$ the intersection $(p^\sigma M)[p] \cap (p^{\sigma+1} M + A)$.

Here $A(\sigma)$ is a submodule of the module $(p^\sigma M)[p]$ containing $(p^{\sigma+1}M)[p]$. We also note that

$$A(\sigma) = \{x \in M \mid px = 0,\ h(x) \geq \sigma \text{ and } h(x+a) > \sigma \text{ for some } a \in A\};$$

in addition, an element x of order p and of the height σ is contained in $A(\sigma)$ if and only if x is not proper with respect to A. The cardinal number

$$f_\sigma(M, A) = \dim_{R_p}((p^\sigma M)[p]/A(\sigma))$$

is called the σth *Ulm–Kaplansky invariant* of the module M with respect to the submodule A. We have

$$f_\sigma(M, A) \leq f(\sigma, M) \quad \text{and} \quad f_\sigma(M, 0) = f(\sigma, M).$$

We find more exact interrelations between the defined invariants and "relative" invariants. We denote by $U_\sigma(M)$ the space $(p^\sigma M)[p]/(p^{\sigma+1}M)[p]$ over the residue field R_p. We have the isomorphism

$$U_\sigma(M) \cong A(\sigma)/(p^{\sigma+1}M)[p] \oplus (p^\sigma M)[p]/A(\sigma) \tag{6.4}$$

and the relation

$$f(\sigma, M) = \dim_{R_p}(A(\sigma)/(p^{\sigma+1}M)[p]) + f_\sigma(M, A). \tag{6.5}$$

Later, we will use $f(\sigma, M)$ to obtain certain conclusions about $f_\sigma(M, A)$. To facilitate this procedure, we consider these invariants from another side. We have the canonical mapping

$$p_\sigma : p^\sigma M/p^{\sigma+1}M \to p^{\sigma+1}M/p^{\sigma+2}M,$$

induced by the multiplication by p,

$$p_\sigma : x + p^{\sigma+1}M \to px + p^{\sigma+2}M.$$

We denote by $K_\sigma(M)$ the kernel of p_σ. Then

$$K_\sigma(M) = \{x \in p^\sigma M \mid px \in p^{\sigma+2}M\}/p^{\sigma+1}M.$$

For a submodule $A \subseteq M$, we set

$$A_\sigma = A \cap p^\sigma M, \qquad A^*_\sigma = \{x \in A_\sigma \mid px \in A_{\sigma+2}\}, \qquad K_\sigma(A) = A^*_\sigma/A_{\sigma+1}.$$

The embedding $\kappa \colon A \to M$ induces the monomorphism

$$\kappa_\sigma \colon K_\sigma(A) \to K_\sigma(M), \qquad x + A_{\sigma+1} \to x + p^{\sigma+1}M$$

and the image of the monomorphism is equal to

$$(\{x \in A(\sigma) \mid px \in A_{\sigma+2}\} + p^{\sigma+1}M)/p^{\sigma+1}M.$$

Lemma 31.6. *There exist the following canonical isomorphisms*:

(1) $K_\sigma(M) \cong U_\sigma(M)$;

(2) $K_\sigma(M)/\operatorname{Im}(\kappa_\sigma) \cong (p^\sigma M)[p]/A(\sigma)$;

(3) $\operatorname{Im}(\kappa_\sigma) \cong A(\sigma)/(p^{\sigma+1}M)[p]$.

Proof. (1) Let $x + p^{\sigma+1}M \in K_\sigma(M)$. Then $px \in p^{\sigma+2}M$ and $px = py$ for some $y \in p^{\sigma+1}M$. We obtain $p(x-y) = 0$ and $x - y \in p^\sigma M$. Now we map $x + p^{\sigma+1}M$ onto $x - y + (p^{\sigma+1}M)[p]$. The inverse mapping is the mapping $x + (p^{\sigma+1}M)[p] \to x + p^{\sigma+1}M$.

(2) and (3) We naturally identify $(p^\sigma M)[p]/A(\sigma)$ with

$$U_\sigma(M)/(A(\sigma)/(p^{\sigma+1}M)[p]).$$

The isomorphism in (2) is induced by the isomorphism constructed in (1). For this purpose, it is sufficient to verify that the isomorphism from (1) induces the isomorphism from (3). However, this is the case. We recall that

$$A(\sigma) = (p^\sigma M)[p] \cap (p^{\sigma+1}M + A).$$

If

$$x + p^{\sigma+1}M \in \operatorname{Im}(\kappa_\sigma)$$

and we take the element y from (1), then $x - y \in A(\sigma)$. If

$$x + (p^{\sigma+1}M)[p] \in A(\sigma)/(p^{\sigma+1}M)[p],$$

then $x + p^{\sigma+1}M \in \operatorname{Im}(\kappa_\sigma)$. □

We denote $\operatorname{Im}(\kappa_\sigma)$ by $I_\sigma(A)$. It follows from Lemma 31.6 that the isomorphism (6.4) and the above relation (6.5) can be represented in the form

$$K_\sigma(M) \cong I_\sigma(A) \oplus K_\sigma(M)/I_\sigma(A) \tag{6.6}$$

and

$$f(\sigma, M) = \dim_{R_p} I_\sigma(A) + f_\sigma(M, A). \tag{6.7}$$

The notion of relative Ulm–Kaplansky invariants was presented by Hill. The papers of Hunter, Richman, and Walker [155, 156, 157] contain a more complete information about the invariants $f(\sigma, M)$, $f_\sigma(M, A)$, and $g(e, M)$ and interrelations between these invariants.

We return to decomposition bases and present the following definition (Warfield [313]). A decomposition basis X of the module M is called a *lower decomposition basis* if for every ordinal number σ, either $I_\sigma(RX)$ is a finite-dimensional R_p-space or $f(\sigma, M) = f_\sigma(M, RX)$ (Exercise 3 is related to this definition).

We say that the height sequence $\{\sigma_i\}_{i\geq 0}$ has a jump at σ, where σ is some ordinal number if for some $i \geq 0$, we have $\sigma_i = \sigma$ and $\sigma_{i+1} > \sigma + 1$, i.e., there is a jump between σ_i and σ_{i+1}.

Lemma 31.7 (Warfield [313]). *If M is a module with decomposition basis X, then there exists a subordinate X' in X such that X' is a lower decomposition basis.*

Proof. By the transfinite induction, we can find pairwise disjoint countable subsets X_λ $(\lambda \in \Lambda)$ of the set X such that X is the union of all X_λ and for every $\lambda \in \Lambda$ and every ordinal number σ if $I_\sigma(RX)$ is an infinite-dimensional space, then $I_\sigma(RX_\lambda)$ is either 0 or an infinite-dimensional space. With each X_λ, we associate some new set X'_λ as follows. For every $\lambda \in \Lambda$, let $S(\lambda)$ be the set of ordinal numbers σ such that $I_\sigma(RX_\lambda)$ is infinite-dimensional. If $S(\lambda)$ is the empty set, then we set $X'_\lambda = X_\lambda$. Otherwise, we assert that we can choose pairwise disjoint infinite subsets X^σ_λ in X_λ (one for every $\sigma \in S(\lambda)$) such that if $x \in X^\sigma_\lambda$, then $U(x)$ has a jump at σ. We reserve the proof of this assertion to another paragraph and assume that the assertion is true.

If $x \in X^\sigma_\lambda$, then there exists an integer $m(x)$ such that the height of $p^{m(x)}x$ exceeds σ. Let X'_λ be the set of elements $p^{m(x)}x$ for all $x \in X^\sigma_\lambda$ $(\sigma \in S(\lambda))$ and the elements $x \in X_\lambda$ that are not contained in each of the sets $X^\sigma_\lambda, \sigma \in S(\lambda)$. Let X' be the union of all sets X'_λ, $\lambda \in \Lambda$. By construction, if σ is an arbitrary ordinal number such that $I_\sigma(RX)$ is infinite-dimensional, then the dimension of the factor space $I_\sigma(RX)/I_\sigma(RX')$ is not less than the dimension of $I_\sigma(RX')$. Therefore, we have that X' is a lower decomposition basis.

To prove that X_λ can be partitioned as indicated above, we set $X_\lambda = Y$ and

$$S(\lambda) = \{\sigma_i \mid i = 0, 1, 2, \ldots\}.$$

(The case, where the set $S(\lambda)$ is finite, is quite easy.) Let x_{11} be some element from Y_{11}. If Y_{11} has only a finite number of elements whose height sequences have a jump at σ_2, then we set $Y_{12} = Y_{11}$. Otherwise, let Y'_{12} be an infinite subset in Y_{11} such that $Y_{11} \setminus Y'_{12}$ is infinite and contains x_{11} and $U(x)$ has a jump at σ_2 provided $x \in Y'_{12}$. We set $Y_{12} = Y_{11} \setminus Y'_{12}$ and we choose an element x_{12} in Y_{12} with $x_{12} \neq x_{11}$. Continuing in such a manner, we construct $Y_{13}, Y_{14}, \ldots$. Let $Y_1 = \bigcap_{i=1}^{\infty} Y_{1i}$, where Y_1 is still infinite, since contains all elements x_{1i}. Now we use the ordinal number σ_2 to apply the same procedure to $Y \setminus Y_1$ and obtain the set Y_2. As a result, we obtain pairwise disjoint infinite sets Y_i such that for $x \in Y_i$, $U(x)$ has a jump at σ_i. □

As a rule, the existing methods of the proof of isomorphisms between direct summands of simply presented modules use the technique of successive extensions of isomorphisms between certain submodules. We prove the main theorem on this topic. The following lemma includes the most part of technical details of the proof of this Theorem 31.9. First, we recall that the notions of a height-preserving isomorphism and a homomorphism which does not decrease heights were already considered in Section 29.

Lemma 31.8. *Let A and B be nice submodules of the modules M and N, respectively, M/A and N/B be primary modules, $\varphi\colon A \to B$ be a height-preserving isomorphism, $f_\sigma(M, A) = f_\sigma(N, B)$, and let*

$$t_\sigma : (p^\sigma M)[p]/A(\sigma) \to (p^\sigma N)[p]/B(\sigma)$$

be an isomorphism for every σ. If $x \in M$, then there exist nice submodules A_1 and B_1 in M and N, respectively, and there is a height-preserving isomorphism $\varphi_1\colon A_1 \to B_1$ such that it extends φ and satisfies the following properties:

(1) $A_1 = A + Rx$;

(2) *A_1/A and B_1/B are finitely generated modules*;

(3) *for every σ, t_σ induces the isomorphism*

$$A_1(\sigma)/A(\sigma) \to B_1(\sigma)/B(\sigma);$$

(4) $f_\sigma(M, A_1) = f_\sigma(N, B_1)$.

Proof. We can assume that $x \notin A$, $px \in A$, and x proper with respect to A. Let $h(x) = \tau$.

Case 1. $h(px) > \tau + 1$. Let $px = py$ with $h(y) > \tau$. Then the order of the element $x - y$ is equal to p, the height of $x - y$ is equal to τ, and the element is proper with respect to A. Consequently, if

$$t_\tau(x - y + A(\tau)) = z + B(\tau),$$

then the order of the element z is equal to p, the height of the element z is equal to τ, and z is proper with respect to B (see the above definition of the submodule $A(\sigma)$). Let $\varphi(px) = pw$ with $h(w) > \tau$. There is an extension of φ such that x is mapped onto $w + z$. We set $A_1 = A + Rx$ and $B_1 = B + R(w + z)$. The submodules A_1 and B_1 are nice, since A and B are nice, and A_1/A, B_1/B are finitely generated by Lemma 28.3 and Proposition 28.4. If $\sigma < \tau$, then $x \in p^{\sigma+1}M$ and $A_1(\sigma) = A(\sigma)$. Similarly, we obtain $B_1(\sigma) = B(\sigma)$. If $\sigma > \tau$, then the same relations follow from the property that the element x is proper with respect to A and $w + z$ is proper with respect to B. If $\sigma = \tau$, then $x - y \in A_1(\sigma)$ and

$$0 \neq t_\sigma(x - y + A(\sigma)) = z + B(\sigma) \in B_1(\sigma)/B(\sigma).$$

The modules A_1/A and B_1/B are cyclic modules of order p. Since there exist the canonical epimorphism $A_1/A \to A_1(\sigma)/A(\sigma)$ and a similar epimorphism for B, we have that $A_1(\sigma)/A(\sigma)$ and $B_1(\sigma)/B(\sigma)$ also cyclic of order p. Consequently, t_σ induces an isomorphism, which is required. This proves (1), (2), and (3). The proof of (4) follows from (3).

Case 2. $h(px) = \tau + 1$. Let $\varphi(px) = pw$ with $h(w) \geq \tau$. Then $h(w) = \tau$, since $h(pw) = \tau + 1$. If $h(w + b) \geq \tau + 1$ for some $b \in B$, then $h(b) = \tau$ and $b = \varphi(a)$, where $h(a) = \tau$ and $h(x + a) = \tau$. Therefore $x + a$ is proper with respect to A, $h(p(x + a)) > \tau + 1$, and we obtain Case 1. Consequently, w is proper with respect to B and it is obvious that $w \notin B$. There is an extension of φ which maps x onto w. We set $A_1 = A + Rx$ and $B_1 = B + Rw$. If $\sigma \neq \tau$, then, similar to (1), we obtain $A_1(\sigma) = A(\sigma)$ and $B_1(\sigma) = B(\sigma)$. If $\sigma = \tau$, then the relation $p(m+c+x) = 0$ does not hold for any elements $c \in A$ and $m \in p^{\sigma+1}M$. We again have $A_1(\sigma) = A(\sigma)$ and $B_1(\sigma) = B(\sigma)$ which completes the proof. □

Let A and B be submodules of the modules M and N, respectively, and let $\varphi\colon A \to B$ be some isomorphism. We assume that φ can be extended to an isomorphism $\psi\colon M \to N$. Since $p^\sigma M$ and $(p^\sigma M)[p]$ are fully invariant submodules, ψ induces the isomorphisms $(p^\sigma M)[p] \to (p^\sigma N)[p]$ and $A(\sigma) \to B(\sigma)$. Therefore, we have the equality $f_\sigma(M, A) = f_\sigma(N, B)$ for relative invariants.

The following theorem was originally proved by Hill [142]; he additionally assumed that M and N are primary modules. The presented formulation and proof are due to Walker [299].

Theorem 31.9. *Let M/A and N/B be two totally projective modules, where A and B are nice submodules and $f_\sigma(M, A) = f_\sigma(N, B)$ for all σ. Then every height-preserving isomorphism $\varphi\colon A \to B$ can be extended to an isomorphism $M \to N$.*

Proof. Let

$$t_\sigma : (p^\sigma M)[p]/A(\sigma) \to (p^\sigma N)[p]/B(\sigma)$$

be an isomorphism for every ordinal number σ. The modules M/A and N/B satisfy the Hill condition (Theorem 30.4). Let $\mathcal{E}$ and $\mathcal{D}$ be two systems of nice submodules for the modules M/A and N/B from the definition of the Hill condition before Theorem 30.4. We consider the family $\mathcal{F}$ of all height-preserving isomorphisms $C \to D$ such that

(a) $C/A \in \mathcal{E}, D/B \in \mathcal{D}$, and

(b) for every σ, t_σ induces the isomorphism $C(\sigma)/A(\sigma) \to D(\sigma)/B(\sigma)$.

On the set of such isomorphisms, we standardly define a partial order. Precisely, if $\varphi_i : C_i \to D_i$ are isomorphisms ($i = 1, 2$), then $\varphi_1 \leq \varphi_2 \iff C_1 \subseteq C_2$ and the restriction of φ_2 to C_1 coincides with φ_1. It follows from the Zorn lemma that there exists a maximal such an isomorphism $\varphi_0 : C \to D$. Condition (b) implies the relation $f_\sigma(M, C) = f_\sigma(N, D)$ for all σ. The submodules C and D are nice by Lemma 28.3. For every σ, t_σ induces the isomorphism

$$(p^\sigma M)[p]/C(\sigma) \to (p^\sigma N)[p]/D(\sigma).$$

We assume that $M \neq C$. Let $z \in M \setminus C$. By Lemma 31.8, there exist nice submodules $A_1 = C + Rz$, B_1 and a height-preserving isomorphism $A_1 \to B_1$ such that the isomorphism extends φ_0 and t_σ induces the isomorphism

$$A_1(\sigma)/C(\sigma) \to B_1(\sigma)/D(\sigma).$$

Consequently, the isomorphism $A_1 \to B_1$ induces the isomorphism

$$A_1(\sigma)/A(\sigma) \to B_1(\sigma)/B(\sigma).$$

We have the system $\mathcal{E}$ and nice submodules A and A_1. Consequently, it follows from Lemma 28.3 that there exists a nice submodule C_1 such that $A_1 \subset C_1$, $C_1/A \in \mathcal{E}$, and $C_1 = A_1 + \sum_{i=1}^{\infty} Rx_{1i}$. There exists a height-preserving isomorphism $A_1 + Rx_{11} \to B_2$ which satisfies condition (b) and extends the isomorphism $A_1 \to B_1$. In a similar manner, we use the system $\mathcal{D}$ and nice submodules B and B_2 to obtain a nice submodule D_2 with properties $D_2 = B_2 + \sum_{i=1}^{\infty} Ry_{2i}$ and $D_2/B \in \mathcal{D}$. There exists a height-preserving isomorphism $A_2 \to B_2 + Ry_{21}$ which extends the previous isomorphism and satisfies (b). Furthermore, we have $C_2 = A_2 + \sum_{i=1}^{\infty} Rx_{2i}$ with $C_2/A \in \mathcal{E}$. Similarly, there exists a height-preserving isomorphism $A_2 + Rx_{12} + Rx_{21} \to B_3$ which extends the previous isomorphism and satisfies (b). Now we have $D_3 = B_3 + \sum_{i=1}^{\infty} Ry_{3i}$ with $D_3/B \in \mathcal{D}$. Thus, there exists a height-preserving isomorphism $A_3 \to B_3 + Ry_{22} + Ry_{31}$ which extends the previous isomorphism and satisfies (b). Continuing in such a manner, we obtain a height-preserving isomorphism $\bigcup C_i = \bigcup A_i \to \bigcup D_i$ which extends φ_0 and satisfies (b). However, $(\bigcup C_i)/A \in \mathcal{E}$ and $(\bigcup D_i)/B \in \mathcal{D}$. Therefore $C = M$ and $D = N$, which completes the proof. □

Corollary 31.10. *If A is a nice submodule of the module M such that M/A is a totally projective module, then every automorphism of the module A, which preserves heights in M, can be extended to an automorphism of the module M.*

Corollary 31.11. *Let M and N be two modules, A be a nice submodule in M, and let M/A be a totally projective module. If $\varphi\colon A \to N$ is a homomorphism which does not decrease heights, then φ can be extended to a homomorphism from M into N.*

Proof. We extend φ to a homomorphism $\psi\colon A\oplus N \to A\oplus N$ by setting $\psi(a,x) = (a, \varphi(a) + x)$. The homomorphism ψ is a height-preserving in $M \oplus N$ automorphism of the submodule $A \oplus N$. By the previous corollary, ψ can be extended to some automorphism α of the module $M \oplus N$. Let $\theta : M \to M \oplus N$ be the corresponding embedding and let $\pi\colon M \oplus N \to N$ be the corresponding projection. Then $\theta\alpha\pi$ is the required extension of φ. □

Exercise 1. Prove that every subordinate of a nice decomposition basis is nice.

Exercise 2 (Hunter–Richman–Walker [156]). We assume that a decomposition basis X of the module M satisfies the following condition. For a given ordinal number σ, there exist at most a finite number of ordinal numbers $\tau < \sigma$ such that $h(x) = \tau$ and $h(px) \geq \sigma$ for some $x \in X$. Prove that RX is a nice submodule in M.

Exercise 3. Let X be a decomposition basis of the module M. Verify that $\dim_{R_p}(I_\sigma(RX))$ coincides with the number of elements in X such that Ulm sequences of the elements have a jump at σ in the sense indicated before Lemma 31.7.

The following three exercises are taken from the paper of Hunter and Richman [155].

Exercise 4. Let M be a module with nice decomposition basis. Then every decomposition basis of the module M has a nice subordinate.

Exercise 5. Let Z be a finite subset of some decomposition basis. Then RZ is a nice submodule.

Exercise 6. Any two finite decomposition bases contain subordinates which generate equal submodules.

Exercise 7 (Files [73]). If M is a sharp module of rank > 1, then M does not have a nice free submodule.

32 Warfield modules

We define Warfield modules and prove that they have complete systems of invariants consisting of cardinal numbers. The independence of this system is considered in remarks at the end of the chapter.

As earlier, we assume that M is a reduced mixed module (unless otherwise stated).

First of all, we formulate the following result on decomposition bases.

Proposition 32.1. *For a module M, let Y be a nice decomposition basis of M and let X be a decomposition basis of M. Then there exists a nice subordinate Z in X with $RZ \subseteq RY$. In addition, if M/RY is a totally projective module, then there exists a nice subordinate Z in X such that $RZ \subseteq RY$ and M/RZ is a totally projective module.*

Proof. Taking the subordinate if necessary, we can assume that $X \subset RY$. By Proposition 31.5, there exists a subordinate Z in X such that RZ is a nice submodule in RY and RY/RZ is a totally projective module. Since RY is a nice submodule in M, we have that RZ is also a nice submodule in M, i.e., Z is a nice subordinate. Let M/RY be a totally projective module. The module M/RZ can be considered as an extension of the nice submodule RY/RZ by the module M/RY. By Corollary 30.5 M/RZ is a totally projective module (see also Lemma 28.3). □

A mixed module M is called a *Warfield module* if M is isomorphic to a direct summand of some simply presented module. Every simply presented module is a Warfield module. There exist Warfield modules of rank 1 which are not simply presented (the example is given in [313]).

Theorem 32.2 (Warfield [313]). (1) *A Warfield module M contains a nice decomposition basis X such that every subordinate X' in X is nice and M/RX' is a totally projective module.*

(2) *If a module M contains a nice decomposition basis X such that M/RX is a totally projective module, then M is a Warfield module.*

Proof. (1) There exist a module N and a simply presented module L such that $M \oplus N = L$. By Corollary 31.2, M and N have decomposition bases W and Z, respectively. We set $Y = W \cup Z$. Let V be a decomposition basis for L such that it is a subset of some generator system of the module L which satisfies conditions (1)–(3) before Lemma 30.1. We can always do this as follows. By Theorem 30.3, the module L is a direct sum of simply presented modules L_i of rank 1 $(i \in I)$. In every L_i, we choose one element x_i of infinite order from some generator system of the modules L_i. We set $V = \{x_i \mid i \in I\}$. Since every subordinate in V is a subset of the same generator system, it follows from Lemma 31.7 that we can assume that V is a lower decomposition basis. By properties (g) and (b) from Section 30, V is a nice basis and L/RV is a totally projective module (see Theorem 30.4, the text before the theorem, and the remark before Proposition 31.5).

We can choose a subordinate Y' in Y which satisfies the following properties:

(1) $Y' \subseteq RV$;

(2) Y' is a lower decomposition basis;

(3) there exists a bijection $t : Y' \to V$ such that $U(y)$ and $U(t(y))$ are equivalent and $U(y) \geq U(t(y))$ for every $y \in Y'$.

Using Proposition 32.1 and passing to further subordinates (which does not change conditions (1)–(3)), we can assume that Y' is a nice subordinate and L/RY' is a totally projective module.

We can also replace V by some subordinate V' such that $U(y) = U(p^n t(y))$ provided $p^n t(y) \in V'$. In this case, we obtain a height-preserving isomorphism $\varphi\colon RV' \to RY'$ (it extends the bijection which is inverse to t). This implies the isomorphism of R_p-spaces

$$I_\sigma(RV') \cong I_\sigma(RY')$$

for all ordinal numbers σ; this follows from the definition of the monomorphism κ_σ before Lemma 31.6. Consequently, if these spaces are finite-dimensional, then

$$f_\sigma(L, RV') = f_\sigma(L, RY')$$

by the relation (6.7) from Section 31. If they are infinite-dimensional, then we take into account the property that V' and Y' are lower decomposition bases and obtain

$$f_\sigma(L, RV') = f(\sigma, L) = f_\sigma(L, RY').$$

The conditions of Theorem 31.9 hold. Therefore φ can be extended to an automorphism of the module L. In addition, Y' is turned out to be a subset of some generator system modules L which satisfies conditions (1)–(3) before Lemma 30.1.

We set $X = Y' \cap M$. Since Y' is a subordinate for Y, we have that X is a subordinate for W. Consequently, X is a decomposition basis for M. Since X is a subset of some generator system for L, every subordinate X' of X is a subset of the same generator system for L. Consequently, RX' is a nice submodule in M according to property (g) from Section 30. Let $Z' = Y' \cap N$. Then we have

$$L/RY' = L/R(X' \cup Z') = L/(RX' \oplus RZ') \cong M/RX' \oplus N/RZ'.$$

Consequently, M/RX' is a totally projective module as a direct summand of a totally projective module. This proves that the decomposition basis X of the module M has all the properties required in (1).

(2) Considering Proposition 31.1 and the remark after it, we can take a simply presented module N with nice decomposition basis Y and the totally projective

factor module N/RY (see the beginning of the proof of (1)). In addition, there exists a height-preserving isomorphism between RX and RY. (Therefore M and N are isomorphic to each other in Warf.) Now we note that the class of totally projective modules is closed with respect to direct sums. Consequently, it follows from Theorem 30.2 that there exist totally projective modules with arbitrarily large Ulm–Kaplansky invariants. More precisely, let T be a totally projective module such that $f(\sigma, T) = 0$ provided $f(\sigma, M) = f(\sigma, N) = 0$ for every σ. Otherwise, let $f(\sigma, T)$ be an infinite cardinal number with

$$f(\sigma, T) > \min(f(\sigma, M), f(\sigma, N)).$$

We consider the modules $M \oplus T$ and $N \oplus T$. They have the same Ulm–Kaplansky invariants. As for Ulm–Kaplansky invariants of the module $M \oplus T$ with respect to RX, it follows from relations (6.5) and (6.7) in Section 31 that the invariants are equal to Ulm–Kaplansky invariants of the module $M \oplus T$. The same is true for $N \oplus T$ and RY. Therefore, we can apply Theorem 31.9 to the modules $M \oplus T$ and $N \oplus T$ and their submodules RX and RY. As a result, we obtain $M \oplus T \cong N \oplus T$, where $N \oplus T$ is a simply presented module. □

In fact, Part (a) of the following corollary has been proved in Part (2) of Theorem 32.2; Part (b) follows from Proposition 32.1 and Part (1) of this theorem.

Corollary 32.3. (a) *For every Warfield module M, there exists a totally projective module T such that $M \oplus T$ is a simply presented module.*

(b) *Every decomposition basis X of the Warfield module M contains a nice subordinate Z such that M/RZ is a totally projective module.*

In connection to (a), we mention a deep result of Hunter, Richman, and Walker [157] which states that every Warfield module is a direct sum of a simply presented module and a Warfield module of at most countable rank.

The following two corollaries are related to category-theoretical properties of Warfield modules.

Corollary 32.4. *If M is a Warfield module and N is an arbitrary module, then we have the equality of morphism groups*

$$\mathrm{Hom}_{\overline{W}}(M, N) = \mathrm{Hom}_{\underline{W}}(M, N).$$

Proof. We always have the inclusion

$$\mathrm{Hom}_{\overline{W}}(M, N) \subseteq \mathrm{Hom}_{\underline{W}}(M, N);$$

the sense of the inclusion is explained before Corollary 29.7. We take a decomposition basis X of the module M which exists by Part (1) of Theorem 32.2. We assume that $\bar{\varphi} \in \mathrm{Hom}_{\underline{W}}(M, N)$ and a homomorphism $\varphi_1 \colon A \to N$ represents $\bar{\varphi}$. This means that A is a torsion-free submodule in M such that M/A is a primary module and φ_1 does not decrease heights of elements. For every $x \in X$, there exists an integer $n \geq 0$ (depending on x) with $p^n x \in A$. We denote by X' the set of such elements $p^n x$ for all x. Let φ be the restriction of φ_1 to X'. Then RX' is a nice submodule and M/RX' is a totally projective module. By Corollary 31.11, φ can be extended to a homomorphism $\psi \colon M \to N$. We obtain $\bar{\psi} = \bar{\varphi}$. Therefore $\bar{\varphi} \in \mathrm{Hom}_{\overline{W}}(M, N)$. □

The Warfield module M is a direct summand of a simply presented module; therefore, it is a direct summand of a completely decomposable module. Consequently, M has Warfield invariants $g(e, M)$ (see Sections 29–31 on these invariants).

Corollary 32.5. *For two Warfield modules M and N, the following conditions are equivalent*:

(1) $g(e, M) = g(e, N)$ *for every equivalence class e of height sequences*;

(2) $M \cong N$ *in the category Warf*;

(3) $M \cong N$ *in the category Walk*;

(4) *there exist primary modules S and T such that $M \oplus T \cong N \oplus S$*;

(5) *there exists a totally projective module T such that $M \oplus T \cong N \oplus T$.*

Proof. Conditions (1) and (2) are equivalent by Corollary 29.10. Conditions (2) and (3) are equivalent by Corollary 32.4. Conditions (3) and (4) are equivalent by Proposition 29.2. The implication (1) $\Longrightarrow$ (5) follows from the proof of Part (2) of Theorem 32.2. In addition, we need also consider the paragraph after the proof of Corollary 31.2. Finally, the implication (5) $\Longrightarrow$ (4) is trivial. □

If φ is an isomorphism between some modules M and N, then the restriction of φ is an isomorphism from $(p^\sigma M)[p]$ onto $(p^\sigma N)[p]$. Consequently, φ induces the isomorphism of the corresponding factor modules; this implies the equality of Ulm–Kaplansky invariants $f(\sigma, M) = f(\sigma, N)$ for every σ. It follows from Corollary 32.5 that the invariants $g(e, M)$ of Warfield modules satisfy an analogous property. It is remarkable that it is sufficient to consider the cardinal numbers $f(\sigma, M)$ and $g(e, M)$ to distinguish Warfield modules. We are ready to present the main classification theorem of this chapter.

Theorem 32.6 (Warfield [307, 311, 313]). *Two Warfield modules M and N are isomorphic to each other if and only if $f(\sigma, M) = f(\sigma, N)$ for every ordinal number σ and $g(e, M) = g(e, N)$ for every equivalence class e of height sequences.*

Proof. It is sufficient to prove sufficiency. Let X and Y be the decomposition bases of the modules M and N, respectively, considered in Part (1) of Theorem 32.2. By passing to subordinates if necessarily, we can assume that X and Y are lower bases by Lemma 31.7. Since the invariants $g(e, M)$ coincide with $g(e, N)$, it follows from remarks after the proof of Corollary 31.2 that there exists a bijection $t : X \to Y$ such that $U(x) = U(t(x))$ for all $x \in X$. The bijection t can be naturally extended to a height-preserving isomorphism $\varphi\colon RX \to RY$. Then we use the argument which is similar to the argument used in the proof of Part (1) of Theorem 32.2. Precisely, we have the isomorphism of R_p-spaces $I_\sigma(RX) \cong I_\sigma(RY)$ for all ordinal numbers σ. If these spaces are finite-dimensional, then we have $f_\sigma(M, RX) = f_\sigma(N, RY)$. For infinite dimensions, we have

$$f_\sigma(M, RX) = f(\sigma, M) = f(\sigma, N) = f_\sigma(N, RY)$$

(X and Y are lower bases!). By Theorem 31.9, φ can be extended to an isomorphism from M onto N. □

We can say that two Warfield modules are isomorphic to each other if and only if their corresponding Ulm–Kaplansky invariants coincide and their corresponding Warfield invariants coincide.

We present other formulations of Theorem 32.6.

Corollary 32.7. *For two Warfield modules M and N, the following conditions are equivalent*:

(1) $M \cong N$;

(2) *M and N have equal Ulm–Kaplansky invariants and are isomorphic to each other in Warf*;

(3) *$t(M) \cong t(N)$ and Warfield invariants of modules M and N coincide.*

Proof. The equivalence of (1) and (2) follows from Corollary 32.5 and Theorem 32.6. The equivalence of (1) and (3) is contained in Theorem 32.6 if we consider that Ulm–Kaplansky invariants of every module are equal to the title invariants of the torsion submodule of the module. □

We apply the obtained results to modules of rank 1; then we apply them to countably generated modules which were already considered before Theorem 8.2.

Corollary 32.8. (a) *Assume that M is a Warfield module of rank 1, x is an element of infinite order from M, and a homomorphism $\varphi\colon Rx \to N$ does not decrease heights. Then φ can be extended to a homomorphism $M \to N$.*

(b) *A module M of rank 1 is a Warfield module if and only if there exists an element x in M of infinite order such that M/Rx is a totally projective module.*

(c) *Two Warfield modules M and N of rank 1 are isomorphic to each other if and only if they have the same Ulm–Kaplansky invariants and $U(M) = U(N)$.*

Proof. (a) By Proposition 28.4, the submodule Rx is nice in M. Now we use Corollary 31.11.

Property (b) follows from Proposition 28.4 and Theorem 32.2. As for to (c), it follows from Proposition 29.6 that $M \cong N$ in Warf $\iff U(M) = U(N)$. Therefore, the result follows from Corollary 32.7. □

Let M be a module of rank 1 such that $t(M)$ is a totally projective module. If x is an element of infinite order, then M/Rx is a totally projective module, since it is a countably generated extension of a totally projective module. Consequently, M is a Warfield module and Corollary 32.8 can be applied to M (see the paper of Wallace [301]).

Corollary 32.9. *A countably generated module M is a Warfield module if and only if M has a decomposition basis.*

Proof. Any Warfield module always has a decomposition basis. Conversely, if M has a decomposition basis X, then X is a countable or finite set. It follows from Lemma 31.4 that X has a nice subordinate X'. Since M/RX' is a reduced countably generated module, it is a totally projective module (this was mentioned at the end of Section 30). By Theorem 32.2, M is a Warfield module. □

Thus, countably generated modules with decomposition bases are determined by their Ulm–Kaplansky invariants and Warfield invariants. Any module of rank 1 necessarily has a decomposition basis. Consequently, the property of determinity by invariants holds for countably generated modules of rank 1. This has been proved by Kaplansky–Mackey [169], Rotman [263], and Megibben [232]. The paper of Rotman and Yen [265] contains the existence theorems for such modules.

In Section 36, we will need the following result of Files [74].

Lemma 32.10. *If M is a Warfield module and $\lambda = l(t(M))$, then M has a nice decomposition basis X such that M/RX is a totally projective module and $p^\lambda M \subseteq RX$.*

Proof. There exists a totally projective module T such that $M \oplus T$ is a simply presented module (Corollary 32.3). In addition, we can choose T with $p^\lambda T = 0$. Indeed, it follows from the proof of Theorem 32.2 (2) that the role of T can be played by the module $T/p^\lambda T$ which is also simply presented (Section 30, property (f)). Thus, $M \oplus T = \bigoplus_{i\in I} M_i$, where all M_i are simply presented modules of rank 1. We have

$$p^\lambda(M \oplus T) = p^\lambda M = \bigoplus_{i\in I} p^\lambda M_i,$$

where for every $i \in I$, we have either $p^\lambda M_i = 0$ or $p^\lambda M_i \cong R$. If the first case is realized, then we take an element x_i of infinite order in M_i such that $\{x_i\}$ is a nice basis (see Theorem 32.2 (1)). In addition, we can assume that $x_i \in M$. In the second case, let $p^\lambda M_i = Rx_i$. The factor module M_i/Rx_i is also totally projective by property (f) from Section 30. As a result, we have the nice decomposition basis $X = \{x_i \mid i \in I\}$ for $\bigoplus M_i$, where $\bigoplus M_i/RX$ is a totally projective module. However, $X \subset M$. Consequently, X is a nice basis of the module M and M/RX is a totally projective module, since it is a direct summand of a totally projective module. By construction, $p^\lambda M \subseteq RX$. □

Warfield invariants $g(e, M)$ have two defects. They are defined for a restricted class of modules. In addition, it is not easy to prove that they are really "invariants" (the proof uses serious category-theoretical arguments). Stanton [278] defined new invariants for every module. These invariants coincide with $g(e, M)$ if $g(e, M)$ are defined; the invariants do not have the both above defects.

Let $u = \{\sigma_i\}_{i \geq 0}$ be a height sequence such that $\sigma_i \neq \infty$ for all i. We recall that if M is a module, then

$$M(u) = \{x \in M \mid U(x) \geq u\}$$

(see Section 28). Now we denote by $M^*(u)$ the submodule generated by all elements $x \in M(u)$ such that $h(p^i x) > \sigma_i$ for an infinite number of subscripts i. The modules $M(u)$ and $M^*(u)$ are fully invariant submodules of the module M. It follows from the definition that $pM(u) \subseteq M^*(u)$ and, therefore $M(u)/M^*(u)$ is a space over the residue field R_p.

In Section 28, we have defined the height sequence $p^i u = (\sigma_i, \sigma_{i+1}, \ldots)$ for every $i \geq 0$. The sequences u and $p^i u$ are contained in the same equivalence class of height sequences.

Lemma 32.11. *The mapping*

$$\varphi_i : M(u)/M^*(u) \to M(p^i u)/M^*(p^i u),$$

defined by the relation

$$\varphi_i(x + M^*(u)) = p^i x + M^*(p^i u),$$

is a monomorphism for every $i \geq 0$.

Proof. Since $x \in M^*(u)$ implies $p^i x \in M^*(p^i u)$, we have that φ_i is well defined; it is clear that φ_i is a homomorphism. We show that φ_1 is a monomorphism. In this case, every φ_i is also a monomorphism, since it is the composition of monomorphisms. Let $x + M^*(u) \in \operatorname{Ker}(\varphi_1)$. Then $px = \sum_{k=1}^{n} r_k x_k$, where all x_k are contained in the generator system for $M^*(pu)$ indicated in the definition. Let

$$U(x_k) = (\tau_{k0}, \tau_{k1}, \dots).$$

Then $\tau_{kj} \geq \sigma_{j+1}$ for every j. For every k, there exists an infinite number of subscripts j with $\tau_{kj} > \sigma_{j+1}$. We have $x_k = py_k$, where the element y_k has the height sequence $(\rho_k, \tau_{k0}, \tau_{k1}, \dots)$ with $\rho_k \geq \sigma_0$. Therefore $y_k \in M^*(u)$ and $p(x - \sum r_k y_k) = 0$. The elements of finite order in $M(u)$ are contained in $M^*(u)$. Consequently, $x - \sum r_k y_k \in M^*(u)$ and $x \in M^*(u)$, since $\sum r_k y_k \in M^*(u)$. □

Assuming for convenience that the monomorphisms φ_i are embeddings, we obtain an ascending chain of R_p-spaces

$$M(u)/M^*(u) \subseteq M(pu)/M^*(pu) \subseteq \cdots.$$

For a more accurate presentation, we can consider the direct spectrum of R_p-spaces:

$$\{M(p^i u)/M^*(p^i u);\ \varphi_{ij}\ (i, j \geq 0)\},$$

where the monomorphisms φ_{ij} have an obvious sense. (Direct spectra and their limits are defined in Section 29.) We take the union $\bigcup_{i \geq 0} M(p^i u)/M^*(p^i u)$ or, equivalently, the limit of our direct spectrum. We call attention to the following detail. Assume that the height sequence v is equivalent to u. Then $p^m v = p^n u$ for some integers $m, n \geq 0$. Therefore, it is clear that we can substitute v instead of u in the above union and then nothing changes. We conclude that the union depends only on the equivalence class e containing the sequences u, v, but the union does not depend on the sequences $u, v, \dots$ themselves. We denote by $S_e(M)$ the considered union. Let $h(e, M)$ be the dimension of $S_e(M)$ over the field R_p. Thus, we have the invariant $h(e, M)$ of the module M. The cardinal numbers $h(e, M)$ for all possible classes e are called *Stanton invariants* of the module M.

Lemma 32.12. (a) *Let M be a module of rank 1 and let e be an equivalence class of height sequences. Then $h(e, M) = 1$ for $e = U(M)$, and $h(e, M) = 0$ for $e \neq U(M)$.*

(b) *If $M = \bigoplus_{i \in I} M_i$, then $h(e, M) = \sum_{i \in I} h(e, M_i)$ for every class e.*

Proof. (a) We choose an element $x \in M$ of infinite order. Let

$$U(x) = u = (\sigma_0, \sigma_1, \dots) \in e;$$

therefore $e = U(M)$. Let $y \in M(u) \setminus M^*(u)$ and let $U(y) = (\tau_0, \tau_1, \dots)$. There exists a subscript $j \geq 0$ such that $\tau_i = \sigma_i$ for all $i \geq j$. Therefore, there exist an invertible element $s \in R$ and an integer $n \geq 0$ such that $p^n sx = p^n y$ (since M has rank 1). Consequently, the order of the element $y - sx$ is finite. Therefore $y - sx \in M^*(u)$ and $y + M^*(u) = sx + M^*(u)$. This proves that $M(u)/M^*(u)$ is an one-dimensional R_p-space. In other words, $h(e, M) = 1$.

Now we assume that $e \neq U(M)$ and $v = (\rho_0, \rho_1, \ldots) \in e$. Let $z \in M(v) \setminus M^*(v)$. Then almost all coordinates of the sequence $U(z)$ coincide with the corresponding ρ_i. Consequently, $U(z)$ and v are equivalent to each other, which is impossible. Thus, in fact, we have $M(v) = M^*(v)$ and $h(e, M) = 0$.

(b) We take some sequence $u \in e$. The relations

$$M(u) = \bigoplus_{i\in I} M_i(u) \quad \text{and} \quad M^*(u) = \bigoplus_{i\in I} M_i^*(u)$$

imply the canonical isomorphism

$$M(u)/M^*(u) \cong \bigoplus_{i\in I} M_i(u)/M_i^*(u).$$

It follows from such isomorphisms that

$$S_e(M) \cong \bigoplus_{i\in I} S_e(M_i)$$

which implies the required equality for invariants. □

Corollary 32.13 (Stanton [278]). *Let M be a direct summand of a completely decomposable module. Then $g(e, M) = h(e, M)$ for every equivalence class e of height sequences.*

Proof. First, we consider the following circumstance. If T is a primary module, then $T(u) = T^*(u)$ for every sequence $u \in e$. Therefore $h(e, T) = 0$ and $h(e, M \oplus T) = h(e, M)$, where M is an arbitrary module. Consequently, Stanton invariants are calculated up to primary direct summands.

By assumption, we have $M \oplus N = \bigoplus_{i\in I} A_i$, where N is some module, and all A_i is a modules of rank 1. It follows from Corollary 29.4 that for some primary modules S and T, we obtain

$$M \oplus T \cong \Big(\bigoplus_{i\in J} A_i\Big) \oplus S,$$

where $J \subseteq I$. Furthermore, we consider Lemma 32.12 (b) and obtain $h(e, M) = \sum_{i\in J} h(e, A_i)$. On the other hand, it follows from the definition of Warfield invariants that

$$g(e, M) = g\Big(e, \bigoplus_{i\in J} A_i\Big) = \sum_{i\in J} g(e, A_i).$$

Lemma 32.12 (a) completes the proof. □

Finally, we show that Warfield modules have some natural projectivity property. We recommend compare the following Lemma with Lemma 10.1 (1). We recall that h-balanced exact sequences were introduced in Section 28.

Lemma 32.14. *For every module M, there exists an h-balanced exact sequence*

$$0 \to L \to P \to M \to 0,$$

where P is a simply presented module.

Proof. Let X be a set such that the cardinality of X is equal to the cardinality of M and let $\theta : X \to M$ be some bijection. Let P be a module with a generator system X and defining relations of the form $px = 0$ and $px = y$ $(x, y \in X)$. In addition, we assume that $px = 0$ (resp., $px = y$) if and only if the relation $p\theta(x) = 0$ (resp., $p\theta(x) = \theta(y)$) holds in M. (See the beginning of Section 30 on the existence of such a module P.) The module M satisfies the definition of a simply presented module. The mapping θ can be standardly extended to an epimorphism $\varphi\colon P \to M$. We set $L = \operatorname{Ker}(\varphi)$. It remains to prove that the obtained sequence is an h-balanced exact sequence. We use Proposition 28.7. By an easy induction, we can verify that the height in P of every element x of X is equal to the height of the element $\theta(x)$ in M. By this property, we can assume that $U(x) = U(\theta(x))$ for every $x \in X$. Therefore, we obtain that $\varphi(P(u)) = M(u)$ for every height sequence u. We also need to prove that

$$\varphi((p^\sigma P)[p]) = (p^\sigma M)[p]$$

for all ordinal numbers σ. If $x \in X$ and the order of the element $\varphi(x)$ is equal to p, then the same is true for x. Consequently, $U(x) = U(\varphi(x))$. Now it is clear that the required relation is a corollary of the definition of the module P. □

Theorem 32.15 (Warfield [307, 313]). *A module M is projective with respect to any h-balanced exact sequence if and only if M is a Warfield module.*

Proof. Let the module M be projective with respect to every h-balanced exact sequence. This means that for every epimorphism $\pi\colon B \to C$ with h-balanced submodule $\operatorname{Ker}(\pi)$ and a homomorphism $\varphi\colon M \to C$, there exists a homomorphism $\psi\colon M \to B$ with $\varphi = \psi\pi$ (for comparison, see the end of Section 10 on the definition of purely projective modules).

By Lemma 32.14, there exists an epimorphism $\pi\colon P \to M$, where P is a simply presented module and $\operatorname{Ker}(\pi)$ is an h-balanced submodule in P. Since the module M is projective with respect to any h-balanced exact sequence, there exists a homomorphism $\chi : M \to P$ with $\chi\pi = 1_M$. Therefore $P = \operatorname{Ker}(\pi) \oplus M'$, where $M' \cong M$ (see the proof of (1) $\Longrightarrow$ (3) of Theorem 5.4). Thus, M is a Warfield module.

Now we assume that M is a Warfield module and

$$0 \to A \to B \xrightarrow{\pi} C \to 0 \tag{6.8}$$

is an h-balanced exact sequence. The module M is a direct summand of some simply presented module. Clearly, we can assume that M is simply presented. Such a module is the direct sum of a totally projective module and simply presented modules of rank 1 (Theorem 30.3). Using the straightforward argument, it can be verified that any direct sum of projective in our sense modules is projective. Consequently, it is sufficient to prove the assertion, where M is a totally projective module or a simply presented module of rank 1. Any totally projective module is projective with respect to every balanced exact sequence of primary modules (Theorem 30.4). Let we have an h-balanced exact sequence (6.8). It can be verified with the use of Proposition 28.7 that the induced sequence of torsion submodules

$$0 \to t(A) \to t(B) \to t(C) \to 0$$

is an h-balanced exact sequence; therefore, it is a balanced exact sequence (see the end of Section 28 and Proposition 28.8). Now it is clear that a totally projective module is projective with respect to the sequence (6.8).

Finally, let M be a mixed simply presented module of rank 1. There exists an element $x \in M$ of infinite order such that M/Rx is a totally projective module (Corollary 32.8 (b)). We assume that $\varphi\colon M \to C$ is some homomorphism. Since $\pi(B(U(x))) = C(U(x))$ by Proposition 28.7, there exists a $y \in B(U(x))$ with properties $\pi(y) = \varphi(x)$ and $U(x) \leq U(y)$. The mapping $\psi\colon Rx \to Ry$, which maps x onto y, is a homomorphism which does not decrease heights. By Corollary 32.8 (a), ψ can be extended to a homomorphism $M \to B$ which is also denoted by the symbol ψ. We take the homomorphism $\varphi - \psi\pi\colon M \to C$. Since the submodule Rx is contained in the kernel of it, we can consider $\varphi - \psi\pi$ as a homomorphism $M/Rx \to C$. We noted earlier that a totally projective module is projective with respect to the sequence (6.8). This means the existence of $\xi : M \to B$ such that $Rx \subseteq \operatorname{Ker}(\xi)$ and $\xi\pi = \varphi - \psi\pi$. Therefore $\varphi = (\psi + \xi)\pi$, which is required. □

In the proof of the theorem, it is mentioned that the class of balanced exact sequences is larger than the class of h-balanced exact sequences. The modules, which are projective with respect to every balanced exact sequence, form a very interesting subclass of the class of Warfield modules which is rich in content. The theory of such modules is developed by Warfield in [310].

Exercise 1 (Hunter–Richman–Walker [157]). If M is a Warfield module, then any direct sum of an infinite number of copies of the module M is a simply presented module.

Exercise 2 (Jarisch–Mutzbauer–Toubassi [164]). Let M be a module such that the torsion submodule of M is the direct sum of cyclic modules. Then M is a Warfield module if and only if $p^n M$ is a Warfield module for some $n \geq 1$.

Exercise 3. Transfer Corollary 32.9 to direct sums of countably generated modules.

Exercise 4. A countably generated module M is a Warfield module if and only if there exists a countably generated primary module T such that $M \oplus T$ is a simply presented module.

Exercise 5 (Files [73]). (a) A sharp module of rank > 1 is not a Warfield module.

(b) If two sharp modules have the same Ulm–Kaplansky invariants and the modules are isomorphic to each other in Warf, then they are isomorphic to each other.

Remarks. There exist different approaches to the study of mixed modules and Abelian groups. They can be considered as extensions of torsion modules by torsion-free modules (for example, by divisible modules). The diametrically opposed approach is to study these objects as extensions of torsion-free modules (for example, free) by using of torsion modules. Other ways combine these two approaches and represent mixed modules as couniversal squares (see Schultz [274, 275]) or construct them by the use of torsion-free modules (see Section 22).

A review of studies related to the Ulm's theorem is presented in Section 30. Furthermore, the paragraph before Lemma 32.10 contains some historical remarks on generalizations of the Ulm theorem to countably generated modules of rank 1. Mackey, a student of Kaplansky, has obtained a short proof of the Ulm's theorem; the argument of Mackey were published in [166] and were essential for subsequent generalizations of the Ulm's theorem.

Rotman and Yen (see [263, 265]) and Bang (see [29, 30, 31]) have extended the mentioned results on countably generated modules of rank 1 to countably generated modules of finite rank and their direct sums. These papers also contain other results on such modules (see also Stratton [283]); in particular, the paper [265] contains the existence theorem (similar theorems are discussed below).

As for the theory presented here, it is necessary to note that activity in this field proceeded in parallel for modules and for p-local Abelian groups (i.e., $\mathbb{Z}_p$-modules, see Section 4). Formally, the activity frequently proceeded for $\mathbb{Z}_p$-modules. However, the results on $\mathbb{Z}_p$-modules usually did not require changes (only the corresponding specifications) for the extension to modules over an arbitrary discrete valuation domain. This property was already mentioned in the book of Kaplansky [166]. In [307], Warfield says on possibility of some generalization of the main results on totally projective Abelian groups to primary modules.

In general, we follow the ideas of Warfield. His papers [307, 308, 310, 311], and [313] have given a powerful impetus to the studies of mixed modules and Abelian groups. Some alternative method of the proof of old and new (at that time) results was developed by Hunter, Richman and Walker in [156] and [157]. They used ideas from the theory of groups with valuations. Hill and Megibben [148] presented an absolutely different method, which can be called combinatorial. They do not use any category-theoretical methods in their proofs. Their methods also have other advantages. Jarisch, Mutzbauer, and Toubassi ([162, 163, 164]) obtained several results on simply presented modules and Warfield modules M such that $t(M)$ is the direct sum of cyclic modules and $M/t(M)$ is a divisible module.

If we need to deal with Warfield $\mathbb{Z}_p$-modules, we also speak about the local case and *local Warfield groups*. The theory of local Warfield groups was extended to Abelian groups, which are not necessarily $\mathbb{Z}_p$-modules. Assuming this common theory, we speak about the global case and the *global Warfield groups*.

The theory of global Warfield groups has roots in the Ulm–Zippin theory of countable p-groups and the Baer theory of completely decomposable torsion-free groups. These two theories were united by the notion of a simply presented group which was introduced for p-groups by Crawley and Hales in [51]; the notion was transferred to arbitrary groups in [308]. Global Warfield groups were originally defined as direct summands of simply presented groups. The study of such groups was initiated by the work [308] of Warfield. Lateron, the papers of Stanton [279, 282], Hunter–Richman [155], Hill–Megibben [150], and Hunter–Richman–Walker [157] appeared. For some aspects of this theory, it is very useful the theorem of Azumaya type for an additive category proved by Arnold, Hunter, and Richman in [15]. Several topics related to local and global Warfield groups are presented in the book of Loth [214].

We did not touch on the independence of systems of invariants $f(\sigma, M)$ and $g(e, M)$ for Warfield modules (this notion of independence was introduced in Section 6). For countable Abelian p-groups, there is the Zippin existence theorem (the theorem is mentioned at the end of Section 30). In general, an existence theorem asserts that there exists a certain module with given invariants (in our situation, the theorem asserts that there exists a Warfield module with given Ulm–Kaplansky invariants and Warfield invariants). In the papers of Hunter–Richman–Walker [155, 157], the existence theorems for local and global Warfield groups are proved.

The torsion submodule of a Warfield module is not necessarily a totally projective module. Torsion subgroups of local Warfield groups are called S-*groups*. Their description is contained in papers of Warfield [309], Hunter [154], Stanton (preprint), and Hunter–Walker [158]. In particular, [309] contains invariants for S-groups.

Returning to the beginning of these remarks, it is possible to say that we prefer the second approach to the studies of mixed modules. In connection to the mentioned first approach to mixed modules, we emphasize the very interesting paper of Mutzbauer and Toubassi [243] on mixed modules M such that $t(M)$ is the direct sum of cyclic modules and $M/t(M)$ is a divisible module (see also [162, 163, 164]).

Problem 13. (a) Which modules M satisfy the property that every pure closed submodule of M is a direct summand of M? (The paper [43] is related to this topic.)

(b) Which modules M satisfy the property that every balanced submodule of M is a direct summand of M?

Problem 14. (a) Describe modules M and N such that

$$\mathrm{Hom}_{\overline{W}}(M, N) = \mathrm{Hom}_{\underline{W}}(M, N)$$

(see Corollary 32.4).

(b) Find conditions implying that modules isomorphic to each other in Warf are isomorphic in Walk.

(c) Find conditions which imply that the indecomposability of the module M in Walk is equivalent to the indecomposability of M in Warf? (Corollary 32.4 is related to (b) and (c).)

The following two problems are contained in the list of 27 problems formulated by Warfield in [311]; some of the problems are still open.

Problem 15. Which R-algebras are endomorphism algebras in Walk or Warf of mixed modules? In particular, answer the question in the case, where R is a complete domain or the modules are of finite rank (see Corollary 22.3). More precisely, obtain characterization theorems or realization theorems in the sense of Chapter 4 for endomorphism rings (algebras) in Walk or Warf of mixed modules.

In the papers of Bang [30, 31], it has been proved in fact that for direct sums of countably generated modules over a complete discrete valuation domain, $M \cong N$ if and only if M and N have equal Ulm–Kaplansky invariants and they are isomorphic to each other in Warf. (Stratton [283] proved that the completeness is essential in this result.) Similar results hold for Warfield modules and sharp modules (Corollary 32.7 and Exercise 6 in Section 32).

Problem 16. Find classes of modules satisfying the following property. For any two modules M and N in such a class, if M and N have equal Ulm–Kaplansky invariants and they are isomorphic to each other in the category Warf (or Walk), then M and N are isomorphic.

Problem 17. In the sense of the book [183, Sections 32, 34], study relations between some subcategories of the category Walk or Warf and some subcategories of left modules over the endomorphism ring of the mixed module A in Walk or Warf. For example, prove the equivalence between the categories. Taking A as a module of rank 1, try to prove Corollary 29.4 and Corollary 29.9 using these equivalences and without use of Section 27. Probably, it is possible to use [310, Theorem 3.11]. ([311, Problem 14] is related to this problem.)

Problem 18. Construct the theory of mixed modules that are extensions of totally projective modules by divisible torsion-free modules (of finite rank). (It should be noted that there is an important paper [243].)

Chapter 7

Determinity of modules by their endomorphism rings

In Chapter 7, we consider the following topics:

- theorems of Kaplansky and Wolfson (Section 33);
- theorems of a topological isomorphism (Section 34);
- modules over completions (Section 35);
- endomorphisms of Warfield modules (Section 36).

In Chapter 4, we were interested in the question: what can we say about endomorphism rings (or endomorphism algebras) of modules? In this chapter, we consider another fundamental problem on endomorphism rings. The solution of this problem can answer to the naturally appeared in Chapter 4 question on uniqueness of the realization of the ring by the endomorphism ring of some module. This is the following problem: how much would the endomorphism ring (or the endomorphism algebra) determine the original module? In the classical formulation, the following results decide this problem. If $\mathrm{End}(M) \cong \mathrm{End}(N)$, then $M \cong N$, or, more generally, there exists a semilinear isomorphism of the modules M and N. Such theorems are called *isomorphism theorems* (in the weak sense). The isomorphism theorem in the strong sense usually means that the given isomorphism of endomorphism rings $\psi\colon \mathrm{End}(M) \to \mathrm{End}(N)$ is induced by an isomorphism (or a semilinear isomorphism) $\varphi\colon M \to N$. The last property means that $\psi(\alpha) = \varphi^{-1}\alpha\varphi$ for every $\alpha \in \mathrm{End}(M)$ (see (c) in Section 2). This type of isomorphism theorems is related to the following problem: determine modules such that all automorphisms of their endomorphism rings are internal. If we assume that endomorphism rings are equipped by the finite topology, then we can consider continuous isomorphisms of endomorphism rings. They contain more information about the original modules. In this case, we speak about *theorems of a topological isomorphism.*

In Section 33, we will see that the situation with isomorphism theorems for primary modules and torsion-free modules over a complete domain is very good. In the class of torsion-free modules over an incomplete domain, isomorphism theorems are quite rare. Here "nonisomorphism theorems" are more common (see Theorem 19.10). In contrast to the case of primary modules, the situation for mixed modules is more complicated (even in the case of mixed modules over a complete domain). It is not absolutely clear which classes of mixed modules can provide isomorphism theorems and which methods are necessary for the proof of such theorems. However, this is not surprising if we consider Theorem 19.10 and Corollary 22.3.

In Section 34, we examine the role of the finite topology in isomorphism theorems for mixed modules. In Sections 34 and 35, we obtain quite large classes of mixed modules, where isomorphism theorems hold. We also present the feasibility bounds for such theorems. In Section 36, we give the positive answer to the isomorphism problem in the weak sense for endomorphism algebras of Warfield modules.

Since our endomorphism rings are R-algebras, it is natural to consider their algebra isomorphisms. This does not imply significant loss of generality.

In the presented chapter, R is a commutative discrete valuation domain and $\widehat{R}$ is the p-adic completion of R.

33 Theorems of Kaplansky and Wolfson

We obtain isomorphism theorems for endomorphism rings of primary modules and torsion-free modules over a complete discrete valuation domain.

Let A be an algebra over some commutative ring S (algebras are defined in Section 19). The mapping $s \to s \cdot 1_A$ ($s \in S$) is a homomorphism from the ring S into the center of the algebra A. If the homomorphism is an embedding, i.e., A is a faithful S-module, then we identify the elements s and $s \cdot 1_A$ and we assume that S is a subring of the center of the algebra A. A mapping $A \to B$ which is a ring isomorphism and an isomorphism of S-modules is called an *isomorphism* from the algebra A onto the S-algebra B. Consequently, a ring isomorphism $\psi\colon A \to B$ is an isomorphism of algebras provided

$$\psi(sa) = s\psi(a) = (s \cdot 1_B)\psi(a)$$

for all $s \in S$ and $a \in A$. We also have

$$\psi(sa) = \psi(s \cdot 1_A \cdot a) = \psi(s \cdot 1_A)\psi(a),$$

whence $\psi(s \cdot 1_A) = s \cdot 1_B$. If A and B are faithful algebras, then we obtain $\psi(s) = s$. Therefore, isomorphisms of faithful S-algebras coincide on elements of S with the identity mapping.

Now we take some R-module M. We have the endomorphism R-algebra $\mathrm{End}(M)$ (see the beginning of Section 19). The canonical mapping $R \to \mathrm{End}(M)$ is not an embedding only if M is a bounded module. Let M be a bounded module and let p^k be the least upper bound of the orders of elements of M. Then M is an R_{p^k}-module and $\mathrm{End}(M)$ is an R_{p^k}-algebra (the conversion of an R-module into an R_{p^k}-module is discussed in Section 4). According to the above, we identify R_{p^k} with the image of the embedding $R_{p^k} \to \mathrm{End}(M)$. If M is a nonbounded module (i.e., is not a bounded module), then we identify R with the image of the embedding $R \to \mathrm{End}(M)$. Thus, if M and N are R-modules, then we can consider ring isomorphisms as well as "stronger" isomorphisms of R-algebras between $\mathrm{End}(M)$ and $\mathrm{End}(N)$.

In the study of isomorphism theorems, it is not necessarily to assume that considered modules are modules over the same ring. In such a situation, semilinear isomorphisms of modules are used. Let S and T be two rings, M be an S-module, and let N be a T-module. An additive isomorphism $\varphi\colon M \to N$ is called a *semilinear isomorphism* of the modules ${}_SM$ and ${}_TN$ if there exists a ring isomorphism $\tau : S \to T$ such that $\varphi(sa) = \tau(s)\varphi(a)$ for all $s \in S$ and $a \in A$. We say that an isomorphism of endomorphism rings (or endomorphism algebras)

$$\psi\colon \mathrm{End}_S(M) \to \mathrm{End}_T(N)$$

is induced by the semilinear isomorphism $\varphi\colon {}_SM \to {}_TN$ provided $\psi(\alpha) = \varphi^{-1}\alpha\varphi$ for every $\alpha \in \mathrm{End}(M)$. It should be noted that even for $S = T$, we frequently can prove only that ψ is induced by some semilinear isomorphism $\varphi\colon {}_SM \to {}_SN$ (but not by an ordinary isomorphism). In this situation, τ is some automorphism of the ring S.

We always assume that M and N are modules over the same ring R. Excluding Theorem 33.2 and several exercises, we consider R-algebra isomorphisms between $\mathrm{End}(M)$ and $\mathrm{End}(N)$. These restrictions do not lead to a considerable loss of generality, but release from technical difficulties. Considered isomorphisms of endomorphism algebras are their R-algebra isomorphisms.

We describe the center of the endomorphism ring of a primary module. This is interesting itself and this will be used in the study of isomorphisms of endomorphism rings. We note that direct summands of the module are invariant with respect to central endomorphisms. Indeed, let M be a module, A be a direct summand of M, $\gamma \in Z(\mathrm{End}(M))$, and let $\varepsilon : M \to A$ be the projection. Then

$$\gamma A = \gamma(\varepsilon M) = \varepsilon(\gamma M) \subseteq A.$$

Furthermore, if Ra and Rb are cyclic modules of orders p^m and p^n, then the existence of a homomorphism $\alpha\colon Ra \to Rb$ with $\alpha a = b$ is equivalent to the inequality $n \le m$. It is convenient to consider primary modules as modules over a complete domain.

Theorem 33.1 (Kaplansky [166]). *Let M be a primary module over a complete discrete valuation domain R. Then $Z(\text{End}(M)) = R_{p^k}$ provided the module M is bounded and p^k is the least upper bound of the orders of elements of M; otherwise, $Z(\text{End}(M)) = R$.*

Proof. It is sufficient to prove that an arbitrary element $\gamma \in Z(\text{End}(M))$ is contained in the corresponding ring.

(a) We assume that the module M is bounded. By Theorem 4.8, M is a direct sum of cyclic modules whose orders do not exceed p^k, where p^k is the element from the theorem. The module M has a cyclic direct summand Ra of order p^k. Considering Example 12.2, we obtain $\gamma a = \bar{s}a$, where $\bar{s} \in R_{p^k}$. For an arbitrary element $x \in M$ there exists a $\alpha \in \text{End}(M)$ such that $\alpha a = x$. We have

$$\gamma x = \gamma(\alpha a) = \alpha(\gamma a) = \alpha(\bar{s}a) = \bar{s}x,$$

whence $\gamma = \bar{s}$.

(b) For a nonbounded module M, we separately consider the case, where the module has the form $M = A \oplus D$, where A is a bounded module and D is a nonzero divisible module. Let Ra be a cyclic direct summand of the module A which is maximal among the elements of the module A of order p^k. It follows from (a) that the restriction of the endomorphism γ to A coincides with the multiplication by some element $\bar{s} \in R_{p^k}$. We denote by E some direct summand of the module D which is isomorphic to $R(p^\infty)$. By Example 12.5, the restriction of the endomorphism γ to E coincides with the multiplication by some element $r \in R$. For an arbitrary element $y \in D$, there exist $e \in E$ and $\beta \in \text{End}(M)$ with $y = \beta e$. Therefore

$$\gamma y = \gamma(\beta e) = \beta(\gamma e) = \beta(re) = r(\beta e) = ry.$$

This relation means that γ coincides with r on the whole module D. Now we choose an element $e \in E$ of order p^k and the endomorphism $\xi \in \text{End}(M)$ with $\xi a = e$. We have

$$re = \gamma(\xi a) = \xi(\gamma a) = \xi(\bar{s}a) = \bar{s}e.$$

Consequently, $(r - \bar{s})e = 0$ and $\bar{s} = \bar{r}$, where $\bar{r} = r + p^k R$. We can state that γ is the multiplication by r on the whole module M. Thus, $\gamma = r$.

It remains to consider the case, where $M = A \oplus D$ and the module A is not a bounded module. Basic submodules of the module M also are not bounded (see Theorem 7.2). By Corollary 9.3, there exist decompositions

$$M = Ra_1 \oplus \cdots \oplus Ra_i \oplus M_i, \quad i = 1, 2, \ldots,$$

such that $M_i = Ra_{i+1} \oplus M_{i+1}$ and the orders p^{n_i} of elements a_i satisfy $1 \leq n_1 \leq \cdots \leq n_i \leq \cdots$. For every pair of subscripts i and j with $i < j$, we take some endomorphism ε_{ji} of the module M which maps a_j onto a_i. Then $\gamma a_i = \bar{s}_i a_i = s_i a_i$ for all i, where $\bar{s}_i = s_i + p^{n_i} R$. If $i < j$, then it follows from

$$s_i a_i = \gamma a_i = (\varepsilon_{ji}\gamma) a_j = (\gamma \varepsilon_{ji}) a_j = \varepsilon_{ji} s_j a_j = s_j a_i$$

that $s_j - s_i \in p^{n_i} R$. Consequently, $\{s_i\}_{i \geq 1}$ is a Cauchy sequence in R. Taking the limit s of the sequence, we obtain $s a_i = s_i a_i$ for every i. Let x be some element of the module M. There are an element a_i and an endomorphism $\alpha \in \mathrm{End}(M)$ such that $\alpha a_i = x$. Furthermore,

$$\gamma x = \gamma \alpha a_i = \alpha \gamma a_i = \alpha s a_i = s x \quad \text{and} \quad \gamma = s. \qquad \square$$

Our first isomorphism theorem is related to primary modules. Under a ring isomorphism, the center is mapped into the center. Therefore, it follows from Theorem 33.1 that we can assume that the modules M and N from Theorem 33.2 are modules over the same ring. Similar to Theorem 33.1, it is convenient to distinguish the case of the bounded module M and the case of the nonbounded module M. Let M be a bounded module and let p^k be the least upper bound of the orders of elements of M. If N is some R-module and $\mathrm{End}(M) \cong \mathrm{End}(N)$, then it follows from $p^k M = 0$ that $p^k \,\mathrm{End}(M) = 0 = p^k \,\mathrm{End}(N)$ and $p^k N = 0$. Consequently, N is a bounded module. It is clear that p^k is the least upper bound of the orders of elements of the module N. This is sufficient to consider only two cases in the theorem.

Theorem 33.2 (Kaplansky [166]). *Let R be a complete discrete valuation domain and let M and N be either R_{p^k}-modules or nonbounded primary R-modules. Then every ring isomorphism between* $\mathrm{End}(M)$ *and* $\mathrm{End}(N)$ *is induced by some semilinear isomorphism between M and N.*

Proof. First, we note that the following property holds. Let M be a cyclic primary module, N be some primary module, and let $\mathrm{End}(M) \cong \mathrm{End}(N)$. Then N is an indecomposable module (property (a) from Section 2); therefore N is a cyclic or quasicyclic module (Corollary 7.4). Using Examples 12.2 and 12.5, we obtain that $M \cong N$.

We begin work with some ring isomorphism $\psi\colon \mathrm{End}(M) \to \mathrm{End}(N)$. For $\alpha \in \mathrm{End}(M)$, we write $\psi(\alpha) = \alpha^*$. We use some notations and results from the proof of Theorem 33.1.

Let M and N be two R_{p^k}-modules. It follows from Theorem 33.1 that ψ induces the automorphism $\bar{r} \to \bar{r}^*$ ($\bar{r} \in R_{p^k}$) of the ring R_{p^k}. The modules M and N are direct sums of cyclic modules. Let a be a generator of a cyclic direct summand module M of maximal order p^k. If $\varepsilon : M \to Ra$ is the projection, then ε is an idempotent of the ring $\mathrm{End}(M)$ and ε^* is an idempotent of the ring $\mathrm{End}(N)$. Consequently, $\varepsilon^* N$ is a direct summand of the module N. By property (d) from Section 2, ψ induces the ring isomorphism

$$\mathrm{End}(Ra) \to \mathrm{End}(\varepsilon^* N).$$

By the above remark, $\varepsilon^* N$ is a cyclic module Rb of order p^k. For an arbitrary element $x \in M$, we choose an endomorphism α of the module M with $x = \alpha a$. We define $\varphi\colon M \to N$ by the relation $\varphi x = \alpha^* b$. This definition is correct, i.e., it does not depend on the choice of the endomorphism α. If $x = \alpha_1 a$, where $\alpha_1 \in \mathrm{End}(M)$, then $(\alpha - \alpha_1)a = 0$ and $\varepsilon(\alpha - \alpha_1) = 0$. Therefore

$$(\varepsilon(\alpha - \alpha_1))^* = \varepsilon^*(\alpha^* - \alpha_1^*) = 0 \quad \text{and} \quad (\alpha^* - \alpha_1^*)b = 0.$$

We take one more element $y \in M$ and we choose $\beta \in \mathrm{End}(M)$ with $y = \beta a$. Then $x + y = (\alpha + \beta)a$ and

$$\varphi(x + y) = (\alpha + \beta)^* b = \alpha^* b + \beta^* b = \varphi x + \varphi y,$$

i.e., φ preserves the addition operation. For every element $r \in R$, we have $rx = r(\alpha a) = (r\alpha)a$. Therefore

$$\varphi(rx) = (r\alpha)^* b = r^* \alpha^* b = r^* \varphi(x).$$

If $\varphi x = \alpha^* b = 0$, then $(\varepsilon\alpha)^* = \varepsilon^* \alpha^* = 0$. Therefore $\varepsilon\alpha = 0$, $x = \varepsilon\alpha a = 0$, and $\mathrm{Ker}(\varphi) = 0$. The module Rb is a direct summand in N of maximal order. Consequently, for every element $z \in N$, there exists an endomorphism $\delta \in \mathrm{End}(N)$ with $z = \delta b$. We have $\delta = \alpha^*$ for some $\alpha \in \mathrm{End}(M)$. Then $z = \delta b = \alpha^* b = \varphi x$, where $x = \alpha a$. We obtain that φ is a bijection; in other words, φ is a semilinear isomorphism.

For $\mu \in \mathrm{End}(M)$, we represent the element $z = \varphi x$ in the form $z = \alpha^* b$ for some $\alpha \in \mathrm{End}(M)$. Now we have

$$\mu^* z = (\alpha\mu)^* b = \varphi((\alpha\mu)a) = \varphi(\mu x) = \varphi(\mu(\varphi^{-1} z)) = (\varphi^{-1}\mu\varphi)z,$$

i.e., $\psi(\mu) = \mu^* = \varphi^{-1}\mu\varphi$. Therefore φ induces ψ.

Now let M and N be nonbounded primary R-modules. The isomorphism ψ induces the automorphism $r \to r^*$ ($r \in R$) of the ring R (Theorem 33.1). If M is a quasicyclic module, then (similar to the beginning of the proof) we can prove that N is a quasicyclic module and $M \cong N$.

We partition the remaining part of the proof into two cases.

1. The module M has the form $M = A \oplus D$, where A is a bounded module and D is a nonzero divisible module. Let Ra be a cyclic direct summand of maximal

order p^k of the module A, E be a quasicyclic direct summand of the module D, and let $c_1, \dots, c_n, \dots$ be a generator system of the module E such that $pc_1 = 0$ and $pc_{n+1} = c_n$ for $n \geq 1$. We denote by $\varepsilon : M \to Ra$ and $\pi : M \to E$ the corresponding projections. Similar to the first part of the proof, we obtain that $\varepsilon^* N$ is a cyclic direct summand of order p^k of the module N and by the above, $\pi^* N$ is a quasicyclic direct summand of the module N. Let $\varepsilon^* N = Rb$ and let $d_1, \dots, d_n, \dots$ be a generator system of the module $\pi^* N$ such that $pd_1 = 0$ and $pd_{n+1} = d_n$ for $n \geq 1$. We represent an arbitrary element $x \in M$ in the form $x = x_1 + x_2$, where $x_1 \in A$ and $x_2 \in D$, and we take an endomorphism α of the module M such that $\alpha a = x_1$ and $\alpha c_n = x_2$ for some n. Then we set $\varphi x = \alpha^*(b + d_n)$. To prove that φx does not depend on α and n, we take $\alpha_1 \in \operatorname{End}(M)$ such that $\alpha_1 a = x_1$, $\alpha_1 c_m = x_2$, and $m \geq n$. Then $\varepsilon(\alpha - \alpha_1) = 0$ and $(p^{m-n}\alpha - \alpha_1)c_m = 0$. Consequently, the endomorphism $\pi(p^{m-n}\alpha - \alpha_1)$ annihilates $E[p^m]$. It follows from Example 12.5 that the endomorphism is divisible by p^m. Then the endomorphism $(\pi(p^{m-n}\alpha - \alpha_1))^*$ is also divisible by p^m. Consequently, it annihilates the element d_m. We obtain that $\alpha^* b = \alpha_1^* b$, $\alpha^* d_n = p^{m-n}\alpha^* d_m = \alpha_1^* d_m$, and $\alpha^*(b + d_n) = \alpha_1^*(b + d_m)$. Similar to the case of bounded modules, we can verify that φ is a semilinear isomorphism inducing ψ.

2. Let $M = A \oplus D$, where the module A is not a bounded module. We take direct decompositions of the module M presented in the proof of Theorem 33.1. Let ε_i be the projection $M \to Ra_i$. For subscripts $i < j$, we define the endomorphism ε_{ji} of the module M which maps a_j onto a_i and annihilates the complement summand for Ra_j; also we define the endomorphism ε_{ij} which maps a_i onto $p^{n_j - n_i} a_j$ and annihilates the complement to Ra_i. In this case, we have:

(1) ε_i is a pairwise orthogonal idempotents;

(2) $\varepsilon_i \varepsilon_{ij} = \varepsilon_{ij} \varepsilon_j = \varepsilon_{ij}$;

(3) $\varepsilon_{ij} \varepsilon_{ji} = p^{|n_j - n_i|} \varepsilon_i$;

(4) $\varepsilon_{ij} \varepsilon_{jk} = \varepsilon_{ik}$ if $i < j < k$ or $i > j > k$.

The submodules $\varepsilon_i^* N$ are cyclic direct summands of the module N. It follows from (2) that $\varepsilon_{i+1,i}^*$ maps from $\varepsilon_{i+1}^* N$ into $\varepsilon_i^* N$. We set $\varepsilon^* N = Rb_i$ and show that we can choose generators b_i such that $\varepsilon_{i+1,i}^* b_{i+1} = b_i$ for all i. If $b_1, \dots, b_i$ are already chosen and the element e_{i+1} generates the submodule $\varepsilon_{i+1}^* N$, then $\varepsilon_{i+1,i}^* e_i = sb_i$ for some $s \in R$. By (3), we obtain $\varepsilon_{i,i+1}^* sb_i = p^{n_{i+1} - n_i} e_{i+1}$. Considering the orders of elements, we obtain that s is an invertible element of the ring R. We take the element $b_{i+1} = s^{-1} e_{i+1}$. Then $\varepsilon_{i+1,i}^* b_{i+1} = b_i$. By property (4), $\varepsilon_{ji}^* b_j = b_i$ for all $i < j$. For $x \in M$, we choose an endomorphism $\alpha \in \operatorname{End}(M)$ such that $\alpha a_i = x$ for some i. Let $\varphi : x \to \alpha^* b_i$. If $\alpha_1 a_j = x$ and $j \geq i$, then $\varepsilon_j(\varepsilon_{ji}\alpha - \alpha_1) = 0$. Therefore $\varepsilon_j^*(\varepsilon_{ji}^* \alpha^* - \alpha_1^*) = 0$ which means

$\alpha^* b_i = \alpha_1^* b_j$. Therefore, the mapping φ is well defined. We can verify again that φ is a semilinear isomorphism which induces ψ. □

Let $\psi\colon \operatorname{End}(M) \to \operatorname{End}(N)$ be a ring isomorphism for the primary R-modules M and N. In the beginning of the section, it is noted that ψ is an isomorphism of R-algebras provided $\psi(r) = r$ for all $r \in R$. It follows from Theorem 33.1 that isomorphisms between the algebras $\operatorname{End}(M)$ and $\operatorname{End}(N)$ coincide with ring isomorphisms that do not move elements of the center.

To formulate one interesting corollary of Theorem 33.2, we set $M = N$. Let ψ be some automorphism of the ring $\operatorname{End}(M)$. We assume that ψ is induced by some automorphism φ of the module M, i.e., $\psi(\alpha) = \varphi^{-1}\alpha\varphi$ for every $\alpha \in \operatorname{End}(M)$. Since automorphisms of M are invertible elements of the ring $\operatorname{End}(M)$, the last relation means that ψ is an internal automorphism of the ring and the algebra $\operatorname{End}(M)$.

Corollary 33.3. *Let M and N be two primary modules. Then every isomorphism of the algebras $\operatorname{End}(M) \to \operatorname{End}(N)$ is induced by some isomorphism $M \to N$. Every automorphism of the endomorphism algebra of a primary module is an internal automorphism.*

We call the method used in the proof of Theorem 33.2 the *Kaplansky method.* It has the following sense. Primitive idempotents of the endomorphism ring of a primary module correspond to direct summands of the module which are isomorphic to $R(p^k)$ or $R(p^\infty)$. To construct an isomorphism from the module M onto the module N, Kaplansky used the transfer of properties of such summands by endomorphisms to obtain the required elements of the module N.

For torsion-free modules, quite completed results are also obtained. (We will use the property of such modules indicated in Corollary 11.7.) We consider the main case of reduced modules (see Exercise 6). For a torsion-free module M, the canonical mapping $R \to \operatorname{End}(M)$ is an embedding and we assume that R is a subring in $\operatorname{End}(M)$. Exercises 4, 5, and 7 are related to the following result.

Lemma 33.4. *Let M be a reduced torsion-free module over a complete discrete valuation domain R. Then the center of the ring $\operatorname{End}(M)$ is equal to R.*

Proof. Let $\gamma \in Z(\operatorname{End}(M))$. We choose some direct summand A of the module M which is isomorphic to R. Then $\gamma A \subseteq A$ and γ coincides on A with the multiplication by some element $s \in R$ (Example 12.1). For an element $x \in M$, there exist $a \in A$ and $\alpha \in \operatorname{End}(M)$ with $\alpha a = x$. We have

$$\gamma x = \gamma(\alpha a) = \alpha(\gamma a) = \alpha(sa) = sx \quad \text{and} \quad \gamma = s. \qquad \square$$

For torsion-free modules, we consider only isomorphisms of endomorphism algebras. Similar to the case of primary modules, difficulties appeared under the study ring isomorphisms are not very interesting (see Exercise 6).

Theorem 33.5 (Wolfson [319]). *Let M and N be two reduced torsion-free modules over a complete discrete valuation domain R. Then every isomorphism of the algebras* $\operatorname{End}(M)$ *and* $\operatorname{End}(N)$ *is induced by some isomorphism between the modules M and N. Every automorphism of the algebra* $\operatorname{End}(M)$ *is an internal automorphism.*

Proof. Let $\psi\colon \operatorname{End}(M) \to \operatorname{End}(N)$ be an isomorphism of algebras and $\psi(\alpha) = \alpha^*$ for $\alpha \in \operatorname{End}(M)$. We use the Kaplansky method. We fix some direct summand Ra of the module M which is isomorphic to R. Let $\varepsilon : M \to Ra$ be the projection. Then $\varepsilon^* : N \to \varepsilon^* N$ is a projection and

$$\operatorname{End}(\varepsilon^* N) \cong \operatorname{End}(\varepsilon N) \cong R.$$

Consequently, $\varepsilon^* N$ is an indecomposable module and $\varepsilon^* N \cong R$. We have $\varepsilon^* N = Rb$. We define $\varphi\colon M \to N$ as follows. Let $x \in M$ and let $x = \alpha a$, where α is an endomorphism of the module M. We set $\varphi x = \alpha^* b$. Similar to Theorem 33.2, we obtain that φ is an isomorphism of modules M and N which induces ψ. □

Theorem 19.10 leads to the idea that for torsion-free modules over a incomplete domain R, isomorphism theorems are possible only in some quite restricted classes of modules (for example, see Exercise 8 in Section 37).

By Theorem 16.1, $\operatorname{End}(M)$ is a complete topological ring with respect to the finite topology. If the ring R is equipped by the discrete topology, then $\operatorname{End}(M)$ is a complete topological R-algebra. This circumstance is very useful in the search for isomorphism theorems. Precisely, we can consider continuous in two directions (with respect to finite topologies) isomorphisms of endomorphism rings (or endomorphism algebras). Such isomorphisms are called *topological isomorphisms.* Thus, we say that a ring (resp., algebra) isomorphism $\psi\colon \operatorname{End}(M) \to \operatorname{End}(N)$ is a topological isomorphism of endomorphism rings (resp., endomorphism algebras) if ψ and ψ^{-1} are continuous with respect to finite topologies on $\operatorname{End}(M)$ and $\operatorname{End}(N)$. Every isomorphism $\varphi\colon M \to N$ induces the topological isomorphism

$$\psi\colon \operatorname{End}(M) \to \operatorname{End}(N), \quad \psi(\alpha) = \varphi^{-1}\alpha\varphi, \quad \alpha \in \operatorname{End}(M),$$

of endomorphism algebras; ψ is an isomorphism of algebras (see property (c) from Section 2). Now we take an arbitrary neighborhood of zero U_Y in the ring $\operatorname{End}(N)$, where Y is some finite subset of the module N, and

$$U_Y = \{\beta \in \operatorname{End}(N) \mid \beta Y = 0\}$$

(see Section 16). Then $\psi^{-1}U_Y = U_{\varphi^{-1}Y}$, where

$$U_{\varphi^{-1}Y} = \{\alpha \in \mathrm{End}(M) \mid \alpha(\varphi^{-1}Y) = 0\}$$

is a neighborhood of zero in the ring $\mathrm{End}(M)$. Conversely, ψ maps neighborhoods of zero of the ring $\mathrm{End}(M)$ onto neighborhoods of zero of the ring $\mathrm{End}(N)$. Consequently, ψ is a topological isomorphism.

Topological isomorphisms reflect the structure of a module more completely. The examples of the modules $R(p^\infty)$ and $\widehat{R}$ show that primary modules cannot be distinguished from torsion-free modules by ordinary isomorphisms of endomorphism rings in the general case (see also the paragraph after the proof of Theorem 33.6). In Theorem 33.2 and Theorem 33.5, the finite topology is not explicitly present. However, it follows from Proposition 16.2 that every isomorphism $\mathrm{End}(M) \to \mathrm{End}(N)$ is continuous for reduced primary modules or torsion-free modules M and N.

There are topological variants of Theorem 33.2 and Theorem 33.5 which extend these assertions. In the following theorem, we extend Theorem 33.2 to mixed modules and present a unified formulation of the both theorems. First, the following useful result holds. Let ε be some idempotent of the algebra $\mathrm{End}(M)$. The canonical isomorphism $\mathrm{End}(\varepsilon M) \cong \varepsilon \,\mathrm{End}(M)\varepsilon$, indicated in Section 2, property (b), is a topological isomorphism if we assume that $\mathrm{End}(\varepsilon M)$ is equipped by the finite topology and $\varepsilon \,\mathrm{End}(M)\varepsilon$ is equipped by the topology induced by the finite topology of the algebra $\mathrm{End}(M)$.

Theorem 33.6. (1) *Let M and N be two modules over a discrete valuation domain R and let $\psi\colon \mathrm{End}(M) \to \mathrm{End}(N)$ be a topological isomorphism of algebras. Then there exists an isomorphism $\varphi\colon t(M) \to t(N)$ such that $\psi(\alpha)$ and $\varphi^{-1}\alpha\varphi$ coincide on $t(N)$ for every $\alpha \in \mathrm{End}(M)$.*

(2) *Let M be either a primary module or a reduced torsion-free module over a complete discrete valuation domain R and let N be an arbitrary R-module. Then every topological isomorphism of algebras $\mathrm{End}(M) \cong \mathrm{End}(N)$ is induced by some isomorphism $M \cong N$.*

Proof. (1) The required isomorphism φ can be constructed by the Kaplansky method (we can consider only the torsion submodules of the modules M and N). We need to only recall the following property. If ε is a primitive idempotent of the ring $\mathrm{End}(M)$ and $\varepsilon M \cong R(p^k)$, then it is clear that $\psi(\varepsilon)N \cong R(p^k)$. Let $\varepsilon M \cong R(p^\infty)$. Then $\mathrm{End}(\varepsilon M) \cong \widehat{R}$ (Example 12.5). Consequently, $\mathrm{End}(\psi(\varepsilon)N) \cong \widehat{R}$. We assume that $\psi(\varepsilon)N$ is a torsion-free module. Since $\psi(\varepsilon)N$ can be considered as a module over the endomorphism ring of it, $\psi(\varepsilon)N$ is an $\widehat{R}$-module. Consequently, $\psi(\varepsilon)N \cong \widehat{R}$ (Example 12.1). By the property mentioned before the theorem, the rings $\mathrm{End}(\varepsilon M)$ and $\mathrm{End}(\psi(\varepsilon)N)$ are topologically isomorphic to each other. However, this is impossible, since the first topology is the p-adic topology and the second topology is discrete. Consequently, $\psi(\varepsilon)N$ is a primary module which is isomorphic to $R(p^\infty)$. Now we can use the Kaplansky method.

(2) We take some topological isomorphism $\psi\colon \operatorname{End}(M) \to \operatorname{End}(N)$. First, we assume that M be a primary module and y is an arbitrary element of the module N. Since ψ is continuous, there exist elements $x_1, \ldots, x_n \in M$ such that $\psi(\alpha)y = 0$ provided $\alpha x_1 = \cdots = \alpha x_n = 0$ for some α. There exists a nonzero element $r \in R$ such that $rx_i = 0$ for all i. Therefore $0 = \psi(r)y = ry$. Consequently, N is a primary module and Corollary 33.3 completes the proof.

Let M be a reduced torsion-free module. We assume that the module N has a direct summand which is isomorphic to $R(p^k)$ or $R(p^\infty)$. We consider the isomorphism ψ^{-1}. It follows from the proof of (1) that the module M also has a similar summand. This is a contradiction. Consequently, N is a reduced torsion-free module. It remains to refer to Theorem 33.5. □

Earlier, we considered two special situations, where endomorphism algebras are isomorphic to each other, but the corresponding modules are not isomorphic to each other. Precisely, if D is a divisible primary module, then the module $C = \operatorname{Hom}(R(p^\infty), D)$ is a complete torsion-free module and $\operatorname{End}(D) \cong \operatorname{End}(C)$ (Section 13). Let M be an adjusted cotorsion module. Using the Section 21, it can be proved that the restriction mapping $\alpha \to \alpha_{|t(M)}$ defines an isomorphism of algebras $\operatorname{End}(M)$ and $\operatorname{End}(t(M))$. Now we understand that the both isomorphisms of endomorphism algebras are not topological.

Exercise 1 (Baer [28]). Every isomorphism between rings of linear operators of two vector spaces over division rings is induced by a semilinear isomorphism of these spaces.

Exercise 2 (Baer [27], Kaplansky [166]). Prove that every isomorphism between endomorphism rings of two primary Abelian groups is induced by some isomorphism between these groups.

Exercise 3. Every isomorphism between endomorphism rings of two p-adic torsion-free modules is induced by some isomorphism of these modules.

Exercise 4. Let R be a discrete valuation domain and let M be an R-module which has a direct summand isomorphic to R. Then $Z(\operatorname{End}(M)) = R$.

Exercise 5. If D is a divisible torsion-free module over a discrete valuation domain R, then $Z(\operatorname{End}(D)) \cong K$, where K is the field of fractions of the domain R.

Exercise 6 (Wolfson [319]). Let M and N be two torsion-free modules over a complete discrete valuation domain R and let $\psi\colon \operatorname{End}(M) \to \operatorname{End}(N)$ be some ring isomorphism. Then the both modules M and N are either nondivisible or divisible. In the first case, there exists a semilinear isomorphism of R-modules $M \to N$ which induces ψ. In the second case, M and N are vector spaces over the field of fractions K of the domain R and ψ is induced by some semilinear isomorphism of these spaces.

Exercise 7. Let R be a discrete valuation domain, T be a reduced primary R-module, A be a reduced torsion-free R-module, and let $M = T \oplus A$.

(1) If T is a bounded module, then $Z(\operatorname{End}(M)) = R + p^k Z(\operatorname{End}(A))$, where p^k is the least upper bound of the orders of elements of the module T.

(2) If T is not a bounded module, then the center $Z(\operatorname{End}(M))$ is canonically isomorphic to the closure $\bar{R}$ of the subring R in the p-adic topology of the ring $Z(\operatorname{End}(A))$; more precisely, if $\bar{R} \cong P$, where P is some pure subring in $\widehat{R}$, then $Z(\operatorname{End}(M))$ consists of multiplications by elements of P.

34 Theorems of a topological isomorphism

We know that $\operatorname{End}(M)$ is a complete (in the finite topology) R-algebra; we simply say "the endomorphism algebra" (see Sections 19, 33). Till the end of the chapter, all considered isomorphisms between endomorphism algebras are isomorphisms of R-algebras.

We study the problem of a topological isomorphism for endomorphism algebras of mixed modules. We accept strongest assumptions. Precisely, we assume that the domain R is complete and isomorphisms of endomorphism algebras are topological. We will see that under such assumptions even for mixed modules of rank 1, two central questions also have the negative answer. We keep in mind the following questions. Whether the (topological) isomorphism $\operatorname{End}(M) \cong \operatorname{End}(N)$ implies the isomorphism $M \cong N$? Whether every (topological) automorphism of the algebra $\operatorname{End}(M)$ is an internal automorphism? We can only prove that for every mixed module M, there exists a unique (up to isomorphism over M) module $\widetilde{M}$ such that $M \subseteq \widetilde{M}$ and every topological isomorphism $\operatorname{End}(M) \cong \operatorname{End}(N)$ is induced by some isomorphism $\widetilde{M} \cong \tilde{N}$. Clearly, it follows from Theorem 33.6 (1) and properties of the cotorsion hull $M^\bullet$ that the new "hull" $\widetilde{M}$ is contained in $M^\bullet$. We present one quite general condition, where $\widetilde{M} = M$, which allows us to formulate results on the isomorphism problem in their usual form.

Let M be a reduced module. According to Section 21, we assume that $M \subseteq M^\bullet$. In addition, $t(M^\bullet) = t(M)$ and $M^\bullet/M$ is a divisible torsion-free module. We also use other properties of cotorsion hulls.

The group $\mathrm{Hom}(M, N)$ is an R-module, since R is a commutative ring (see Section 2). More specifically, if $\varphi \in \mathrm{Hom}(M, N)$ and $r \in R$, then the relation $(r\varphi)(x) = \varphi(rx)$ $(x \in M)$ defines a homomorphism $r\varphi \in \mathrm{Hom}(M, N)$ which gives a module multiplication on $\mathrm{Hom}(M, N)$.

We extend the notion of the finite topology to homomorphism groups. For a finite subset $X \subseteq M$, we set

$$U_X = \{\varphi \in \mathrm{Hom}(M, N) \mid \varphi X = 0\}.$$

Here U_X is a submodule of the R-module $\mathrm{Hom}(M, N)$. The module $\mathrm{Hom}(M, N)$ is turned into a topological R-module whose basis of neighborhoods of zero consists of the submodules U_X for all finite subsets X (we consider the discrete topology on the ring R). The topological R-module $\mathrm{Hom}(M, N)$ is a topological Abelian group such that a module multiplication is continuous with respect to the discrete topology and the finite topology on R and $\mathrm{Hom}(M, N)$, respectively (see also Section 11 on topological modules). The mapping

$$\psi\colon \mathrm{Hom}(M, N) \to \mathrm{Hom}(M', N')$$

is called a *topological isomorphism* if ψ and ψ^{-1} are continuous with respect to finite topologies isomorphisms of R-modules (we considered topological isomorphisms of algebras in the previous section).

If $M \subseteq M'$, then we have the induced homomorphism of R-modules $\mathrm{Hom}(M', N) \to \mathrm{Hom}(M, N)$, $\varphi \to \varphi_{|M}$. We call it the restriction homomorphism.

Lemma 34.1 (May [227]). *Let M be a reduced module and let $t(M) = T \neq 0$.*

(1) *There exists a maximal reduced module $\widetilde{M}$ with properties $M \subseteq \widetilde{M}$, $t(\widetilde{M}) = T$, and the induced mapping $\mathrm{Hom}(\widetilde{M}, T) \to \mathrm{Hom}(M, T)$ is a topological isomorphism. Between any two such maximal modules, there exists a unique isomorphism which coincides on M with the identity mapping.*

(2) *$\widetilde{M}$ is an $\widehat{R}$-module, $\widetilde{M}/M$ is a divisible torsion-free module, and $\widetilde{M}$ can be considered as the uniquely defined submodule in $M^\bullet$.*

(3) *Every $\alpha \in \mathrm{End}(M)$ can be uniquely extended to $\tilde{\alpha} \in \mathrm{End}(\widetilde{M})$; in fact, $\tilde{\alpha}$ is the restriction to $\widetilde{M}$ of the unique extension α to $\alpha^\bullet \in \mathrm{End}(M^\bullet)$.*

Proof. Let M' be some reduced module such that $M \subseteq M'$ and $t(M') = T$. We define two (possible) properties of the module M'.

$(*M')$ The induced mapping $\mathrm{Hom}(M', T) \to \mathrm{Hom}(M, T)$ is a topological isomorphism.

$(**M')$ For every $x \in M'$, there exist elements $y_1, \dots, y_n \in M$ such that $\beta'(x) = 0$ for every $\beta' \in \mathrm{Hom}(M', T^\bullet)$ such that $\beta'(M) \subseteq T$ and $\beta'(y_i) = 0$ for all $i = 1, \dots, n$.

We assert that $(*M')$ holds if and only if M'/M is a divisible torsion-free module and $(**M')$ holds.

We note that the following property holds. For an arbitrary nondivisible module N, there exists a nonzero homomorphism $N \to R(p)$. Since T is a primary module, there always exists a nonzero homomorphism $N \to T$. Now we assume that M'/M is a nondivisible module. There exists a nonzero homomorphism $\bar{\beta} : M'/M \to T$. Let β be the composition of the canonical homomorphism $M' \to M'/T$ with $\bar{\beta}$. Then $\beta \in \mathrm{Hom}(M', T)$ and $\beta \neq 0$, but the restriction of β to M is equal to zero. This contradicts to $(*M')$. Consequently, M'/M is a divisible module. Let $px \in M$, where $x \in M' \setminus M$. Then the height of the element px in M is equal to zero. Otherwise, $M' \setminus M$ has torsion elements, which is impossible by $t(M') = t(M)$. Consequently, there exists a homomorphism $\beta : M \to T$ such that the height of $\beta(px)$ is equal to zero. This β cannot be extended to a homomorphism $M' \to T$. This is a contradiction. Consequently, M'/M is a torsion-free module. Condition $(**M')$ immediately follows from the property that the isomorphism in $(*M')$ is a topological isomorphism. We only recall one property. The restriction of the homomorphism β' to M can be extended to some homomorphism from M' in T. This extension coincides with β', since M'/M is a divisible module and $T^\bullet$ is a reduced module.

Now we assume that M'/M is a divisible torsion-free module with property $(**M')$. We know that there exists a unique embedding M' in $M^\bullet$ which is the identity mapping on M (property (f) from Section 21). Consequently, we can assume that $M \subseteq M' \subseteq M^\bullet$. The mapping in $(*M')$ is injective, since the module M'/M is divisible and T is a reduced module. To verify that it is surjective, we take any homomorphism $\beta : M \to T$. It can be extended to a homomorphism $\beta^\bullet : M^\bullet \to T^\bullet$ (property (d) in Section 21). Let $x \in M'$ and let $y_1, \dots, y_n$ be the elements of M whose existence is stated in $(**M')$. We choose a positive integer k such that $p^k(\beta y_i) = 0$ for all $i = 1, \dots, n$. Let β' be the restriction of $p^k\beta^\bullet$ to M'. Then $(**M')$ implies that $\beta'(x) = 0$. Therefore $\beta^\bullet(x) \in T$. The restriction of $\beta^\bullet$ to M' is the required extension of β. Thus, the considered mapping is a bijection. It is always continuous. It follows from $(**M')$ that the inverse mapping is continuous. Consequently, property $(*M')$ holds.

We define $\widetilde{M}$ as the sum of all submodules M' in $M^\bullet$ such that $M \subseteq M'$ and $(*M')$ holds. Then $\widetilde{M}/M$ is a divisible torsion-free module, since $M^\bullet/M$ is torsion-free. For all such M', property $(**M')$ implies $(**\widetilde{M})$. By the proved properties, $(*\widetilde{M})$ holds. We have obtained a unique maximal submodule $\widetilde{M}$ in $M^\bullet$. If $\widetilde{M}_1$ is one more maximal submodule, then the uniqueness property of the embedding in $M^\bullet$ (property (f) from Section 21) proves the existence of an isomorphism $\widetilde{M_1} \to \widetilde{M}$ which coincides on M with the identity mapping.

It follows from (1) and (2) that it remains to prove the property that $\widetilde{M}$ is an $\widehat{R}$-module. However, $M^\bullet$ is an $\widehat{R}$-module by property (7) from Section 21. Therefore, there exists an $\widehat{R}$-submodule M' in $M^\bullet$ generated by $\widetilde{M}$. Then M'/M is a divisible torsion-free module, since $\widehat{R}/R$ is such a module (see Section 11). Why does property $(**M')$ hold? Every homomorphism $\beta' : M' \to T^\bullet$ is an $\widehat{R}$-homomorphism (Section 4). The elements of M' are combinations of elements of $\widetilde{M}$ with coefficients in $\widehat{R}$. Since $(**\widetilde{M})$ holds, it is clear that $(**M')$ holds. Since $\widetilde{M}$ is maximal, we have $M' = \widetilde{M}$.

Let us prove (3). Let $\alpha \in \mathrm{End}(M)$. It is sufficient to verify that $\alpha^\bullet(\widetilde{M}) \subseteq \widetilde{M}$. We set $M' = \alpha^\bullet(\widetilde{M}) + M$. Then M'/M is a divisible torsion-free module. In addition, $(**M')$ holds. We need to only verify $(**M')$ for elements of the form $\alpha^\bullet(x)$, where $x \in \widetilde{M}$. However, this easily follows from the application of $(**\widetilde{M})$ to $\alpha^\bullet\beta'$, where $\beta' \in \mathrm{Hom}(M', T^\bullet)$. Consequently, $(*M')$ holds. Therefore

$$M' \subseteq \widetilde{M} \quad \text{and} \quad \alpha^\bullet(\widetilde{M}) \subseteq \widetilde{M}. \qquad \square$$

Now we formulate some quite general theorem of a topological isomorphism. In this theorem, we keep in mind that there exists an isomorphism $\tilde{\varphi}\colon \widetilde{M} \to \tilde{N}$ such that $\psi(\alpha)$ and $\tilde{\varphi}^{-1}\tilde{\alpha}\tilde{\varphi}$ coincide on N for every $\alpha \in \mathrm{End}(M)$, where ψ is some topological isomorphism $\mathrm{End}(M) \to \mathrm{End}(N)$.

Theorem 34.2 (May [227]). *Let M be a reduced module with nonzero torsion submodule and let N be an arbitrary module. We assume that $M/t(M)$ is a divisible module. Then every topological isomorphism of algebras $\mathrm{End}(M) \to \mathrm{End}(N)$ is induced by some module isomorphism $\widetilde{M} \to \tilde{N}$. If R is a complete domain, then the assumptions on $M/t(M)$ can be omitted.*

Proof. We denote by T the torsion submodule $t(M)$. Let $\psi\colon \mathrm{End}(M) \to \mathrm{End}(N)$ be some topological isomorphism. First, we assume that R is a complete domain and M/T is a nondivisible module. Then M/T has a direct summand which is isomorphic to R (Corollary 11.8). Using Theorem 5.4, it can be proved that M also has a direct summand which is isomorphic to R. Similar to the proof of Part (1) of Theorem 33.6, it can be proved that the corresponding summand of the module N, obtained with the use of ψ, is isomorphic to R. Thus, he modules M and N have direct summands which are isomorphic to R. Similar to the corresponding part of the proof of Theorem 33.5, we use the method of Kaplansky to prove the existence of an isomorphism $\varphi\colon M \to N$ inducing ψ. It is clear that the isomorphism $\tilde{\varphi}\colon \widetilde{M} \to \tilde{N}$ also induces ψ.

Now we assume that M/T is a divisible module and the domain R is not necessarily complete. Similar to Theorem 33.6, it can be verified that the module N does not have direct summands of the form $R(p^\infty)$ and K, i.e., N is a reduced module. By this theorem, there exists an isomorphism $\varphi\colon T \to t(N)$ such that $\psi(\alpha)$ and $\varphi^{-1}\alpha\varphi$ coincide on $t(N)$ for every $\alpha \in \mathrm{End}(M)$. In particular, if $\alpha T = 0$, then $\psi(\alpha)t(N) = 0$. Since M/T is a divisible module, $\mathrm{Hom}(M/T, M) = 0$, where we identify $\mathrm{Hom}(M/T, M)$ with the set of all $\alpha \in \mathrm{End}(M)$ annihilating T. Therefore $\mathrm{Hom}(N/t(N), N) = 0$. Therefore $N/t(N)$ is a divisible module, since $t(N) \neq 0$. Consequently, $M^\bullet = T^\bullet$, $N^\bullet = t(N)^\bullet$ by property (f) in Section 21, and we can assume that $\widetilde{M} \subseteq T^\bullet$ and $\tilde{N} \subseteq t(N)^\bullet$. In addition, φ can be extended to an isomorphism $\varphi^\bullet : T^\bullet \to t(N)^\bullet$ by property (d) from Section 21. Endomorphisms of modules $T^\bullet$ and $t(N)^\bullet$ are completely determined by their action on torsion submodules. Consequently, we obtain that

$$\psi(\alpha)^\bullet = (\varphi^{-1}\alpha\varphi)^\bullet = (\varphi^\bullet)^{-1}\alpha^\bullet\varphi^\bullet$$

for every $\alpha \in \mathrm{End}(M)$. Furthermore, the isomorphism $\varphi^\bullet$ induces the isomorphism

$$\psi^\bullet : \mathrm{End}(T^\bullet) \to \mathrm{End}(t(N)^\bullet)$$

which extends ψ. We have $\psi(\alpha)^\bullet = \psi^\bullet(\alpha^\bullet)$. We take the isomorphism ψ^{-1} and use similar arguments to obtain the relation

$$\psi^{-1}(\beta)^\bullet = \varphi^\bullet\beta^\bullet(\varphi^\bullet)^{-1}$$

for every $\beta \in \mathrm{End}(N)$. We show that $\varphi^\bullet(\widetilde{M}) \subseteq \tilde{N}$. By symmetry, we also have $(\varphi^\bullet)^{-1}(\tilde{N}) \subseteq \widetilde{M}$. As a result, we obtain the isomorphism $\tilde{\varphi}\colon \widetilde{M} \to \tilde{N}$ which induces ψ.

Thus, we verify that $\varphi^\bullet(\widetilde{M}) \subseteq \tilde{N}$. We set $N' = \varphi^\bullet(\widetilde{M}) + N$. Since $\widetilde{M}/M$ and M/T are divisible modules, $\widetilde{M}/T$ is also a divisible module. Consequently, $p\widetilde{M} + T = \widetilde{M}$. Furthermore, we have

$$N' = \varphi^\bullet(p\widetilde{M} + T) + N = p\varphi^\bullet(\widetilde{M}) + pN + N = pN' + N.$$

Therefore N'/N is a divisible torsion-free module, since $N^\bullet/N$ is a torsion-free module. We verify that property $(**N')$ from Lemma 34.1 holds. By this lemma, $(*N')$ holds. By the definition of $\tilde{N}$, we obtain

$$N' \subseteq \tilde{N} \quad \text{and} \quad \varphi^\bullet(\widetilde{M}) \subseteq \tilde{N}.$$

First, we assert that

$$\psi(\mathrm{Hom}(M, T)) = \mathrm{Hom}(N, t(N)).$$

Since ψ is a topological isomorphism, the assertion is true by the following property. The set $\operatorname{Hom}(M,T)$ consists of all $\alpha \in \operatorname{End}(M)$ such that the sequence $\{p^k\alpha\}_{k\geq 0}$ converges to zero in the finite topology. We also present another proof. Let $\alpha \in \operatorname{Hom}(M,T)$. For every $y \in N$, it follows from the continuity of ψ that there exist elements $x_1,\dots,x_n \in M$ such that $\psi(\beta)y = 0$ if $\beta x_1 = \cdots = \beta x_n = 0$, where $\beta \in \operatorname{End}(M)$. There exists a positive integer k such that $p^k\alpha(x_i) = 0$ for all $i = 1,\dots,n$. Consequently, $p^k\psi(\alpha)y = 0$, i.e., $\psi(\alpha)y \in t(N)$ and $\psi(\alpha) \in \operatorname{Hom}(N,t(N))$. We obtain

$$\psi \operatorname{Hom}(M,T) \subseteq \operatorname{Hom}(N,t(N)).$$

The converse inclusion is proved similarly.

We need only to verify that for every $x \in \widetilde{M}$, property $(**\,N')$ holds for $\varphi^\bullet(x)$. By property $(**\,\widetilde{M})$, we can choose elements $y_1,\dots,y_n \in M$ such that $\beta^\bullet(x) = 0$ for every $\beta \in \operatorname{Hom}(M,T)$, where $\beta(y_i) = 0$ for all i. It follows from the topological isomorphism that there exist elements $y'_1,\dots,y'_m \in N$ such that if $\beta' \in \operatorname{Hom}(N,t(N))$ and $\beta'(y'_j) = 0$ for all j, then $\beta' = \psi(\beta)$ for some $\beta \in \operatorname{Hom}(M,T)$, where $\beta(y_i) = 0$ for all i. However, then $\beta^\bullet(x) = 0$. Consequently,

$$(\beta')^\bullet(\varphi^\bullet(x)) = \psi(\beta)^\bullet(\varphi^\bullet(x)) = (\varphi^\bullet\psi(\beta)^\bullet)(x) = (\beta^\bullet\varphi^\bullet)(x) = \varphi^\bullet(\beta^\bullet(x)) = 0.$$

Therefore $(**\,N')$ holds and the proof is completed. □

We need one condition which implies the relation $\widetilde{M} = M$. As it usually is, M^1 denotes the first Ulm submodule of the module M.

Lemma 34.3. *Let M be a reduced module with nonbounded torsion submodule over a complete domain R. If M^1 is a cotorsion module, then $\widetilde{M} = M$.*

Proof. We assume that there exists an element $x \in \widetilde{M} \setminus (\widetilde{M}^1 + M)$. Let $y_1,\dots,y_n \in M$ be the elements from property $(**\,\widetilde{M})$ of Lemma 34.1. We set

$$A = \widetilde{M}^1 + Ry_1 + \cdots + Ry_n \subseteq \widetilde{M}.$$

Since R is a complete domain, A is a nice submodule in $\widetilde{M}$ by Proposition 28.4 (2). It is clear that $(x + A) \cap \widetilde{M}^1 = \varnothing$. Consequently, the residue class $x + A$ has in $\widetilde{M}/A$ the finite height, say k. There exists the composition of the mappings

$$\beta\colon \widetilde{M} \to \widetilde{M}/A \to (\widetilde{M}/A)/p^{k+1}(\widetilde{M}/A) \to t(M)$$

such that $\beta(y_i) = 0$ for every i, but $\beta(x) \neq 0$. This contradicts to $(**\,\widetilde{M})$. Consequently, $\widetilde{M} = \widetilde{M}^1 + M$. Since $\widetilde{M}/M$ is a torsion-free module, we have $\widetilde{M}^1 \cap M = M^1$. Now we have

$$\widetilde{M}/M = (\widetilde{M}^1 + M)/M \cong \widetilde{M}^1/(\widetilde{M}^1 \cap M) = \widetilde{M}^1/M^1.$$

Consequently, $\widetilde{M^1}/M^1$ is a divisible torsion-free module or the module is equal to zero. If $\widetilde{M^1} \neq M^1$, then the cotorsion module M^1 is a direct summand in $\widetilde{M^1}$, which is impossible, since $\widetilde{M^1}$ is a reduced module. Consequently, $\widetilde{M^1} = M^1$ and $\widetilde{M} = M$. □

Theorem 34.4 (May [227]). *Let R be a complete domain, M be a reduced module, and let $\psi\colon \operatorname{End}(M) \to \operatorname{End}(N)$ be a topological isomorphism.*

(1) *If M^1 is a cotorsion module, then ψ is induced by some embedding from N in M.*

(2) *If M^1 is a cotorsion module, then every topological automorphism of the algebra $\operatorname{End}(M)$ is an internal automorphism.*

(3) *If M^1 is the direct sum of a bounded module and a torsion-free module of finite rank, then ψ is induced by some isomorphism from M onto N. In particular, this is true provided $M^1 = 0$.*

Proof. If $t(M)$ is a bounded module, then $M = t(M) \oplus X$, where $X = 0$ or X is some torsion-free module. If $X = 0$, then ψ is induced by some isomorphism from M onto N by Theorem 33.6. If $X \neq 0$, then the module M has a direct summand which is isomorphic to R. In this case, N has a similar summand and ψ is induced by some isomorphism from M onto N (see the beginning of the proof of Theorem 34.2). Consequently, we can assume that $t(M)$ is a nonbounded module. By Theorem 34.2, ψ is induced by some isomorphism $\tilde{\varphi}\colon \widetilde{M} \to \tilde{N}$. However, it follows from Lemma 34.3 that $\widetilde{M} = M$. Consequently, $\tilde{\varphi}^{-1}_{|N}$ is the injection $N \to M$ which induces ψ in the sense that $\psi(\alpha)$ and $\tilde{\varphi}^{-1}\alpha\tilde{\varphi}$ coincide on N for every $\alpha \in \operatorname{End}(M)$.

In (2), we have $\tilde{\varphi}\colon M \to M$ and $\tilde{\varphi}$ induces ψ.

In (3) $M^1 = t(M) \oplus F$, where $t(M)$ is a bounded module and F is a free module of finite rank. In particular, M^1 is a complete module; therefore, M^1 is a cotorsion module of finite rank. By (1), there exists an embedding $N \to M$. Therefore N^1 is a cotorsion module, since it is isomorphic to some submodule in M^1. Therefore $\widetilde{M} = M$, $\tilde{N} = N$ and $\tilde{\varphi}\colon M \to N$ induces ψ. □

In general, Parts (2) and (3) are not true without the corresponding condition for the first Ulm submodule. We present examples which support this.

We need the notion of a small homomorphism (see remarks to Chapter 4). Let M and N be two primary modules. A homomorphism $\varphi\colon M \to N$ is said to be *small* if for every positive integer e, there exists a positive integer n with $\varphi((p^n M)[p^e]) = 0$. All small homomorphisms from M in N form the R-module $\operatorname{Small}(M, N)$. All small endomorphisms of the module M form the ideal $\operatorname{Small}(M)$ of the algebra $\operatorname{End}(M)$.

Let G be a module with torsion submodule T. Then T and T^1 is a fully invariant submodules in G. Consequently, every endomorphism β of the module G induces the endomorphism $\bar{\beta}$ of the factor module T/T^1 by the relation $\bar{\beta}(x+T^1) = \beta(x)+T^1$, $x \in T$. The endomorphism β is said to be *small with respect to* the module T^1 if $\bar{\beta}$ is a small endomorphism of the module T/T^1. All small with respect to T^1 endomorphisms of the module G form an ideal in $\mathrm{End}(G)$.

An automorphism of a ring or an algebra, which is not an internal automorphism, is called an *external* automorphism.

Proposition 34.5 (May [227]). *Let R be a complete domain. There exists a mixed module M of rank* 1 *such that the algebra* $\mathrm{End}(M)$ *has a topological external automorphism.*

Proof. We use a known result on the existence of a primary module with given Ulm sequence (Fuchs [93, Theorem 76.1]). We also apply theorems from [49] and [59] on a split realization to endomorphism rings of primary modules (see also remarks at the end of Chapter 4). It follows from these results that we can choose primary modules T_1 and T_2 such that T_i^1 is a nonbounded direct sum of cyclic modules,

$$\mathrm{End}(T_i/T_i^1) = R \oplus \mathrm{Small}(T_i/T_i^1)$$

(the direct sum of R-modules), and $\mathrm{Hom}(T_i/T_i^1, T_j/T_j^1)$ consists of small homomorphisms for $i,j=1,2$ and $i \neq j$. We set $T = T_1 \oplus T_2$. We have $T^\bullet = T_1^\bullet \oplus T_2^\bullet$ and we denote by ε_i the projection $T^\bullet \to T_i^\bullet$, $i=1,2$. We prove that

$$\mathrm{End}(T^\bullet) = R\varepsilon_1 \oplus R\varepsilon_2 \oplus I$$

(the direct sum of R-modules), where I is the ideal of small with respect to the module T^1 endomorphisms of the module $T^\bullet$. First, we verify that $\mathrm{End}(T_1) = R \oplus I_1$, where I_1 is the ideal of small with respect to T_1^1 endomorphisms of the module T_1. It is clear that $R \cap I_1 = 0$. Let $\alpha \in \mathrm{End}(T_1)$ and let $\bar{\alpha}$ be the endomorphism of the module T_1/T_1^1 induced by α. Then $\bar{\alpha} = r + \bar{\beta}$, where $r \in R$ and $\bar{\beta} \in \mathrm{Small}(T_1/T_1^1)$. Since $\alpha = r + (\alpha - r)$, we have that $\overline{\alpha - r} = \bar{\alpha} - \bar{r} = \bar{\alpha} - r = \bar{\beta}$ and $\alpha - r \in I_1$. We obtain $\mathrm{End}(T_1) = R \oplus I_1$. Similarly, we have $\mathrm{End}(T_2) = R \oplus I_2$, where I_2 is the ideal of small with respect to the module T_2^1 endomorphisms of the module T_2. Furthermore, we have

$$\mathrm{End}(T) = \mathrm{End}(T_1) \oplus \mathrm{End}(T_2) \oplus \mathrm{Hom}(T_1,T_2) \oplus \mathrm{Hom}(T_2,T_1) = R\varepsilon_1 \oplus R\varepsilon_2 \oplus I.$$

Now we note that the ring $\mathrm{End}(T^\bullet)$ can be identified with the ring $\mathrm{End}(T)$ by the restriction mapping (see property (d) from Section 21).

Using Ext, the submodule $T_i^{1\bullet}$ (respectively, $T^{1\bullet}$) can be naturally embedded in $T_i^\bullet$ (respectively, $T^\bullet$). Since T_i^1 is a nonbounded direct sum of cyclic modules,

$(T_i^{1\bullet})^1$ is a torsion-free module by Proposition 21.5. We choose nonzero elements $x_i \in (T_i^{1\bullet})^1$, $i = 1, 2$. We set $x = x_1 + x_2$. Let M be a pure submodule in $T^\bullet$ such that $T \subseteq M$ and $M/T = K(x+T)$, where K is the field of fractions of the domain R. (The divisible torsion-free module $T^\bullet/T$ is a K-space.) Let $\beta \in I$. It is easy to verify that $\beta T^1 = 0$ (see Exercise 1). Consequently, $\beta T^{1\bullet} = 0$ and $\beta x = 0$. Therefore, we have $R \oplus I \subseteq \mathrm{End}(M)$. If $\alpha \in \mathrm{End}(M)$ (say, $\alpha = r_1\varepsilon_1 + r_2\varepsilon_2 + \beta$, where $r_1, r_2 \in R$ and $\beta \in I$) then $\alpha(x) = r_1x_1 + r_2x_2$. Then $\alpha(x) \in M$ if and only if $r_1 = r = r_2$. Consequently, $\alpha = r + \beta \in R \oplus I$ and $\mathrm{End}(M) = R \oplus I$. In particular, $\mathrm{End}(M) \subset \mathrm{End}(T^\bullet)$.

We choose some invertible element $u \in R$ which is not equal to 1 and we set $\gamma = \varepsilon_1 + u\varepsilon_2$. Here γ is an automorphism of the module $T^\bullet$ (indeed, $\gamma^{-1} = \varepsilon_1 + u^{-1}\varepsilon_2$). By the relation $\psi(\alpha) = \gamma^{-1}\alpha\gamma$ with $\alpha \in \mathrm{End}(T^\bullet)$, an automorphism of the algebra $\mathrm{End}(T^\bullet)$ is defined. Since I is an ideal, the restriction of ψ is an automorphism of the algebra $\mathrm{End}(M)$. We assume that ψ is an internal automorphism of the algebra $\mathrm{End}(M)$. Then there exists an automorphism φ of the module M such that $\psi(\alpha) = \varphi^{-1}\alpha\varphi$ for every $\alpha \in \mathrm{End}(M)$. Consequently, we have $\alpha(\varphi\gamma^{-1}) = (\varphi\gamma^{-1})\alpha$ for every $\alpha \in \mathrm{End}(M)$. In particular, $\varphi\gamma^{-1}$ is permutable with any endomorphism which is small with respect to the module T^1. In the last part of the proof of Theorem 33.1, all considered endomorphisms, besides γ, can be chosen small with respect to the module T^1. As a result, we obtain that $\varphi\gamma^{-1}$ is contained in the center of the ring $\mathrm{End}(T)$, i.e., $\varphi\gamma^{-1}$ is an invertible element from R, whence $\gamma \in \mathrm{End}(M)$. However, $\gamma(x) = x_1 + ux_2 \notin M$; this is a contradiction. Consequently, ψ is an external automorphism.

For the proof of the property that ψ is continuous in the finite topology, we take an element $y \in M$. It is sufficient to verify that there exists an element $y_1 \in M$ such that if $\alpha \in \mathrm{End}(M)$ and $\alpha(y_1) = \alpha(x) = 0$, then $\psi(\alpha)(y) = 0$. Considering that γ is an automorphism, we have $\alpha(\gamma^{-1}y) = 0$. For some $k \geq 0$ and $a \in R$, we have $p^k y = a(x_1 + x_2)$. By the choice of x_2, we have $x_2 = p^k x_3$, where $x_3 \in T^{1\bullet}$. Then

$$p^k\gamma^{-1}(y) = a(x_1 + u^{-1}x_2) = p^k y + p^k a(u^{-1} - 1)x_3.$$

We set $y_1 = \gamma^{-1}(y) - a(u^{-1} - 1)x_3$ and we note that $y_1 \in M$, since $y_1 \in y + T$. Let $\alpha = r + \beta \in \mathrm{End}(M)$ and let $\alpha(y_1) = \alpha(x) = 0$. Since $\beta(x) = 0$, we have $r = 0$; therefore, $\alpha = \beta$. Since $\beta(x_3) = 0$, we have $\alpha(\gamma^{-1}y) = 0$, which is required. By symmetry, we obtain that ψ^{-1} is continuous and ψ is a topological automorphism. □

In Part (3) of Theorem 34.4, the condition for the submodule M^1 is the best (in some sense) possibility for such theorems of a topological isomorphism. Using constructions similar to ones used in Proposition 34.5, May [227] has obtained the following results. Let A be a reduced module over a complete domain R such that $t(A)^1 = 0$ and A is not a direct sum of a bounded module and a torsion-free module of finite rank. Then there exists a reduced module M such that $M^1 \cong A$ and M does not satisfy the theorem of a topological isomorphism in the weak sense. If A is a primary module, then we can take a mixed module of rank 1 as M. In addition, there exists a mixed module N of rank 1 such that the algebras $\operatorname{End}(M)$ and $\operatorname{End}(N)$ are topologically isomorphic to each other, but the modules M and N are not isomorphic to each other.

The following exercises are taken from the paper of May [227].

Exercise 1. Let M be a module and let $T = t(M)$. If β is a small with respect to T^1 endomorphism of the module M, then $\beta(T^1) \subseteq T^2$.

Let S be some R-algebra. If we take as a subbasis of neighborhoods of zero the set of right annihilators of elements of the right ideal in S generated by primitive idempotents of finite additive order in S, then we obtain the topology on S which is called the *intrinsic* topology.

Exercise 2. Prove that for a primary module T, which is not a direct sum of a bounded module and a nontrivial divisible module, the intrinsic topology on $\operatorname{End}(T)$ coincides with the finite topology. (See Proposition 16.2; it is also useful to compare the intrinsic topology with two finite topologies defined before Theorem 18.4 and Proposition 18.6.)

Exercise 3. Let M be a module with torsion submodule T. We assume that the ring $\operatorname{End}(M)$ is Hausdorff in the intrinsic topology.

(1) The restriction mapping $f\colon \operatorname{End}(M) \to \operatorname{End}(T)$ is injective.

(2) $\operatorname{End}(T)/f\operatorname{End}(M)$ is an torsion-free R-module.

(3) The intrinsic topology on $\operatorname{End}(M)$ is induced by the finite topology on $\operatorname{End}(T)$.

Exercise 4. The endomorphism ring of the module M from Proposition 34.5 is complete in the intrinsic topology, and $\operatorname{End}(M)$ is not isomorphic to $\operatorname{End}(P)$ for every primary module P.

35 Modules over completions

Let R be an incomplete discrete valuation domain and let $\widehat{R}$ be the p-adic completion of R. We show that every reduced R-module can be embedded in a minimal (in a certain sense) reduced $\widehat{R}$-module. In general, modules over complete domains have more simple structures. Therefore, such an embedding can be useful in the study of modules over incomplete domains. At the end of the section, we briefly present several results on the isomorphism problem for endomorphism rings of mixed modules; some of the results use embeddings from R-modules in $\widehat{R}$-modules. Other results on this topic are contained in the next section. In two remaining sections of the chapter, arbitrary (i.e., not necessarily topological) isomorphisms are considered. It is clear that nontopological isomorphisms weaker reflect the structure of original modules. In this connection, it is interesting that every isomorphism between endomorphism algebras of two Warfield modules is a topological isomorphism (Theorem 36.4).

In this section and in the next section, we denote by RX the submodule generated by a subset X of some module. In Chapter 6, we agreed that for a mixed module M, the rank of the factor module $M/t(M)$ is called the rank of M. If this factor module is a divisible module, then it is a K-space and the dimension of it is equal to the rank of the module M.

Our constructions are based on properties of cotorsion hulls and tensor products. Let M be a reduced R-module. Similar to the previous section, we consider M as a submodule of the cotorsion hull $M^\bullet$ such that $t(M^\bullet) = t(M)$ and $M^\bullet/M$ is a divisible torsion-free module. By property (7) from Section 21, $M^\bullet$ is an $\widehat{R}$-module. Consequently, we can take the $\widehat{R}$-submodule in $M^\bullet$ generated by the submodule M. We denote it by $\widehat{R}M$; it is called the *$\widehat{R}$-hull* of the module M. By construction, $\widehat{R}M$ is a reduced module. It was mentioned in Section 13 that the tensor product $\widehat{R}\otimes M$ is an $\widehat{R}$-module (all tensor products are considered over the ring R). The $\widehat{R}$-hull of the module M can be constructed by the use of this tensor product. For this purpose, we consider the exact sequence of R-modules

$$0 \to R\otimes M \to \widehat{R}\otimes M \to \widehat{R}/R\otimes M \to 0.$$

The module $R\otimes M$ can be identified with M under the correspondence $r\otimes m \to rm$ ($r \in R$, $m \in M$) and M can be identified with the image in $\widehat{R}\otimes M$. Furthermore, $\widehat{R}/R\otimes M$ is a divisible torsion-free module, since $\widehat{R}/R$ has an analogous structure. We have $\widehat{R}\otimes M = D \oplus A$, where D is the largest divisible submodule of the $\widehat{R}$-module $\widehat{R}\otimes M$ and A is a reduced $\widehat{R}$-module A such that $M \subseteq A$ (see the remark after Theorem 6.1).

Lemma 35.1. *The $\widehat{R}$-hull $\widehat{R}M$ is isomorphic to the factor module of the $\widehat{R}$-module $\widehat{R}\otimes M$ with respect to the largest divisible submodule of $\widehat{R}\otimes M$.*

Proof. A balanced mapping $\widehat{R}\times M \to \widehat{R}M$, $(r, m) \to rm$ induces the epimorphism of R-modules $g\colon \widehat{R}\otimes M \to \widehat{R}M$, $r\otimes m \to rm$. In addition, $gD = 0$, since $\widehat{R}M$ is a reduced module. Let H be the submodule in $\widehat{R}\otimes M$ generated by all elements of the form $\sum r_i\otimes m_i - 1\otimes\sum r_i m_i$, where $r_i \in \widehat{R}$, $m_i \in M$, and

$\sum r_i m_i \in M$. It easily follows from the relation $p\widehat{R} + R = \widehat{R}$ that H is divisible. If $\sum r_i \otimes m_i \in \operatorname{Ker}(g)$, then

$$\sum r_i \otimes m_i = \sum r_i \otimes m_i - 1 \otimes \sum r_i m_i \in H \subseteq D;$$

therefore $\operatorname{Ker}(g) = D$. Consequently, $\widehat{R}M \cong (\widehat{R} \otimes M)/D$.

In the second proof, the above exact sequence of tensor products is used. Since A/M is a divisible torsion-free module, it follows from property (f) of Section 21 that $M^\bullet = A^\bullet$, and we can assume that $M \subseteq A \subseteq M^\bullet$. Let h be the restriction of g to A and let $h^\bullet : M^\bullet \to M^\bullet$ be an extension of h. Since h coincides on M with the identity mapping, $h^\bullet$ is the identity automorphism and $A = \widehat{R}M$ (see assertion (d) in Section 21). □

We present several simple properties of $\widehat{R}$-hulls. They follow from the definition of the $\widehat{R}$-hull and Lemma 35.1.

(1) $\widehat{R}M/M$ is a divisible torsion-free module, $t(\widehat{R}M) = t(M)$, and M is an isotype submodule in $\widehat{R}M$ (such submodules were introduced in Section 27).

Since $M^\bullet/M$ is torsion-free, we have that $\widehat{R}M/M$ is torsion-free and $t(\widehat{R}M) = t(M)$. Since the module $\widehat{R}/R$ is divisible, we have that the module $\widehat{R}M/M$ is divisible (the proof of Lemma 34.1 has a similar point).

(2) The module $\widehat{R}M$ can be characterized as a reduced $\widehat{R}$-module such that $\widehat{R}M$ contains M as an R-module, $\widehat{R}M$ is generated by M as an $\widehat{R}$-module, and $\widehat{R}M/M$ is a divisible torsion-free module.

(3) The rank of the $\widehat{R}$-module $\widehat{R}M$ does not exceed the rank of the R-module M.

If $\{a_i\}_{i \in I}$ is a maximal linearly independent system of elements of infinite order of the module M, then $\{a_i + t(M)\}_{i \in I}$ is a basis of the K-space $K(M/t(M))$ (see Section 4). Every element of $\widehat{R}M/t(M)$ can be linearly expressed through this basis; this gives the required result.

(4) Let M_i $(i \in I)$ be the set of reduced R-modules. Then

$$\widehat{R}\Big(\bigoplus_{i \in I} M_i\Big) \cong \bigoplus_{i \in I} \widehat{R}M_i. \tag{7.1}$$

Since $\operatorname{Ext}(K/R, -)$ commutes with finite direct sums of modules, we assume that

$$\bigoplus_{i \in I} M_i^\bullet \subseteq \Big(\bigoplus_{i \in I} M_i\Big)^\bullet.$$

Under such an approach, the isomorphism in (7.1) become an equality. We can also use Lemma 35.1. The canonical isomorphism

$$\widehat{R} \otimes \Big(\bigoplus_{i \in I} M_i\Big) \cong \bigoplus_{i \in I} (\widehat{R} \otimes M_i)$$

implies the required isomorphism of the factor modules with respect to divisible submodules.

(5) We obtain a covariant functor F from the category of reduced R-modules into the category of reduced $\widehat{R}$-modules if we set $F(M) = \widehat{R}M$ and $F(f)$ is the induced homomorphism $\widehat{R}M \to \widehat{R}N$ for any two R-modules M, N and any homomorphism $f\colon M \to N$. Here $F(f)$ can be also defined directly. Since $M^\bullet$ and $N^\bullet$ are $\widehat{R}$-modules, $f^\bullet$ is an $\widehat{R}$-module homomorphism $M^\bullet \to N^\bullet$. Then $F(f)$ is the restriction of $f^\bullet$ to $\widehat{R}M$. We can also define $F(f)$ with the use of the induced homomorphism $1 \otimes f$ of tensor products.

(6) $\mathrm{End}(M)$ is a subalgebra of the algebra $\mathrm{End}(\widehat{R}M)$ (i.e., the R-endomorphisms of the module $\widehat{R}M$ coincide with the $\widehat{R}$-endomorphisms of $\widehat{R}M$).

For every endomorphism α of the module M, we take the corresponding $F(\alpha)$, as in (5). In other words, we extend α to an endomorphism $\alpha^\bullet$ of the module $M^\bullet$; then we take the restriction of $\alpha^\bullet$ to $\widehat{R}M$.

Isomorphic modules have isomorphic $\widehat{R}$-hulls. For Warfield modules, the converse assertion holds.

Proposition 35.2 (Files [74]). *If M and N are two Warfield modules and $\widehat{R}M \cong \widehat{R}N$, then $M \cong N$.*

Proof. Since every primary module M is an $\widehat{R}$-module, we have $\widehat{R}M = M$. We can assume that M and N are not primary modules. Let $\varphi\colon \widehat{R}M \to \widehat{R}N$ be some isomorphism. There exists a primary module T such that $M \oplus T = \bigoplus_{i\in I} M_i$ and $N \oplus T = \bigoplus_{j\in J} N_J$, where all modules M_i and N_j have rank 1 (Corollary 32.3 and Theorem 30.3). For every $i \in I$, we choose an element of infinite order $x_i \in M_i \cap M$. Then $X = \{x_i \mid i \in I\}$ is a decomposition basis for the $\widehat{R}$-modules M and $\widehat{R}M$. It should be considered the property that by (4) and (3), we have

$$\widehat{R}M \oplus T = \bigoplus_{i\in I} \widehat{R}M_i,$$

where every $\widehat{R}M_i$ is a module of rank 1. Similarly, we take an element of infinite order y_j in every $N_j \cap N$ and obtain a decomposition basis $Y = \{y_j \mid j \in J\}$ for N and $\widehat{R}N$. Since φX and Y are two decomposition bases for $\widehat{R}N$, there exist subordinates $X' = \{x_i' \mid i \in I\}$ in X and $Y' = \{y_j' \mid j \in J\}$ in Y, respectively, and the bijection $\tau : I \to J$ such that

$$h_{\widehat{R}M}(p^k\varphi(x_i')) = h_{\widehat{R}N}(p^k y'_{\tau(i)})$$

for all $i \in I$ and $k \geq 0$ (see the paragraph after the proof of Corollary 31.2). Since the submodules M and N are isotype, we have

$$h_M(p^k x_i') = h_N(p^k\varphi(x_i')) = h_N(p^k y'_{\tau(i)})$$

for all i and k. We obtain the isomorphism $\chi : RX' \to RY'$ which preserves heights in M and N if we set $\chi(x'_i) = y'_{\tau(i)}$ for every $i \in I$. In addition, we have

$$t(M) = t(\widehat{R}M) \cong t(\widehat{R}N) = t(N)$$

and $M \cong N$ by Corollary 32.7. □

Every R-module M has the largest $\widehat{R}$-submodule. We denote it by $C(M)$. This submodule will play some role in the next section.

Lemma 35.3. *If M is a reduced module containing a nice basis, then $C(M) = t(M)$.*

Proof. Let X be a nice basis for M. The torsion submodule $t(M)$ is an $\widehat{R}$-module; therefore, it is contained in $C(M)$. If $C(M) \neq t(M)$, then we can choose a nonzero element $x \in C(M) \cap RX$. Then $\widehat{R}x \subseteq M$. Therefore

$$\widehat{R}/R \cong \widehat{R}x/Rx \subseteq M/Rx$$

and M/Rx is a nonreduced module. Since

$$M/RX \cong (M/Rx)/(RX/Rx)$$

and RX/Rx is a reduced module, M/RX is a nonreduced module. However, this is impossible, since RX is a nice submodule. Consequently, $C(M) = t(M)$. □

As a rule, the Kaplansky method is not applicable to endomorphism rings of mixed modules. A formal obstruction is the absence of endomorphisms which map from elements of infinite order onto cyclic direct summands of the module. Practically all methods of solution of the isomorphism problem for endomorphism rings of mixed modules are based on the use of cotorsion hulls. This was partially shown in Section 34. We can assume that every reduced module is standardly embedded in the cotorsion hull of the module. Now we assume that the endomorphism algebras of reduced mixed modules M and N are isomorphic to each other. In this case, it can be proved usually that the torsion submodules $t(M)$ and $t(N)$ are isomorphic to each other (see Theorem 36.4). Furthermore, the general case can be frequently reduced to the case, where $M/t(M)$ and $N/t(N)$ are divisible modules. This is one of the common situations. For example, such a reduction is always possible if R is a complete domain. Then we have $M^\bullet = T^\bullet = N^\bullet$, where $T = t(M)$, and we can assume that the modules M and N are contained in the cotorsion hull $T^\bullet$. Furthermore, the algebras $\mathrm{End}(M)$ and $\mathrm{End}(N)$ can be considered as subalgebras in $\mathrm{End}(T^\bullet)$ if we identify every endomorphism $\alpha \in \mathrm{End}(M)$ with $\alpha^\bullet \in \mathrm{End}(T^\bullet)$ and do the same with $\mathrm{End}(N)$. May [223] has proved that $\mathrm{End}(M) = \mathrm{End}(N)$ in this case. In addition, the original isomorphism ψ of the endomorphism algebras is induced by an isomorphism from M onto N if and only if there exists an invertible element $u \in \widehat{R}$ with $uM = N$. If R is a complete domain, then ψ is induced by an isomorphism from M onto N if and only if $M = N$. Therefore, the problem is to find the element u. If ψ is not induced by any isomorphism from M onto N (for example, this case is possible for Warfield modules, as in Theorem 36.4), then we need to continue the search for conditions implying the existence of an isomorphism between M and N.

Almost all existing results on the isomorphism problem for mixed modules are related to modules with totally projective torsion submodule, Warfield modules, or some larger classes of modules related to these two classes. The following theorem of May and Toubassi [229] can be considered as the first result in this topic.

Theorem 35.4. *Let M be a mixed module of rank 1 with totally projective torsion submodule. If N is a module of rank 1, then every isomorphism $\mathrm{End}(M) \to \mathrm{End}(N)$ is induced by some isomorphism $M \to N$.*

In [225], May determined quite general conditions which imply that an isomorphism $\mathrm{End}(M) \cong \mathrm{End}(N)$ is induced by an isomorphism of $\widehat{R}$-hulls $\widehat{R}M \cong \widehat{R}N$, where M is some mixed module and N is a general module. He has obtained some interesting applications to modules M with totally projective torsion submodule and Warfield modules M. If R is a complete domain, then we obtain the isomorphism of the modules M and N. In another paper [223], May has proved a more strong theorem for modules over a complete domain.

Theorem 35.5. *Let M be a reduced mixed module over a complete domain R. We assume that every submodule G in M such that M/G is a primary module, contains a nice submodule A such that M/A is a totally projective module. Then every isomorphism from $\mathrm{End}(M)$ onto $\mathrm{End}(N)$ is induced by some isomorphism from M onto N.*

May has proved that the assertion of the theorem holds if the module M satisfies one of the following conditions:

(1) the rank of M does not exceed $\aleph_0$ and $t(M)$ is a totally projective module;

(2) M is a Warfield module.

May presented examples which imply that Theorem 35.5 does not hold provided either $t(M)$ is not totally projective, or the rank of M exceeds $\aleph_0$, or M is not reduced. In each of the three cases, the corresponding endomorphism algebras have external automorphisms. We present one of the examples in greater detail. The example demonstrates obstructions to the search for isomorphism theorems for mixed modules.

Example 35.6. There exist a primary module T without elements of infinite height and a continual set of mixed modules M_i of rank 1 such that $t(M_i) = T$, M_i/T is a divisible module, $\operatorname{End}(M_i) = R \oplus \operatorname{Small}(T)$, and $\operatorname{Hom}(M_i, M_j) = \operatorname{Small}(T)$ for all distinct $i, j < 2^{\aleph_0}$. Each of the algebras $\operatorname{End}(M_i)$ has an external automorphism.

In the following exercises, M is a reduced module.

Exercise 1 (May [225]). For every ordinal number σ, the inclusion $\widehat{R}(p^\sigma M) \subseteq p^\sigma(\widehat{R}M)$ holds. The relation $\widehat{R}(p^\sigma M) = p^\sigma(\widehat{R}M)$ holds if either σ is finite, or M has rank 1, or M is a Warfield module.

Exercise 2. The rank of the $\widehat{R}$-module $\widehat{R} \otimes M$ is equal to the rank of the module M (see property (3)).

Exercise 3. Prove that $C(M)$ is a fully invariant submodule in M and $M/C(M)$ is a torsion-free module.

36 Endomorphisms of Warfield modules

We prove that a Warfield module is determined by the endomorphism algebra (in the weak sense) of the module in the class of such modules. We are based on the paper of Files [74]. We widely use $\widehat{R}$-hulls introduced in the previous section. As earlier, our constructions are based on properties of cotorsion hulls. (The meaning of these constructions is presented at the end of Section 35.)

Let M be a reduced module such that $M/t(M)$ is a divisible module. We set $T = t(M)$. By property (f) from Section 21, $T^\bullet = M^\bullet$. We recall that $T^\bullet$ is an adjusted cotorsion module (Proposition 21.3). Every endomorphism α of the module M has a unique extension to an endomorphism $\alpha^\bullet$ of the module $T^\bullet$ (property (d) in Section 21). In such a manner, $\operatorname{End}(M)$ can be embedded in $\operatorname{End}(T^\bullet)$. Similar arguments are true for every pure submodule N in $T^\bullet$ containing T. Therefore, we consider $\operatorname{End}(M)$ and $\operatorname{End}(N)$ as subalgebras of the same algebra $\operatorname{End}(T^\bullet)$. Under such conditions, we have $\operatorname{End}(M) = \operatorname{End}(N)$ if and only if $\alpha^\bullet(N) \subseteq N$ for all $\alpha \in \operatorname{End}(M)$ and $\beta^\bullet(M) \subseteq M$ for all $\beta \in \operatorname{End}(N)$. We note that $\operatorname{End}(T) = \operatorname{End}(T^\bullet)$, since the submodule T is fully invariant.

The Ulm submodules M^σ and the factors M_σ were defined in Section 28.

Lemma 36.1. *Let T be a reduced primary module. We assume that M and N are pure submodules in $T^\bullet$ which contain T and have nice decomposition bases. If* $\operatorname{End}(M) = \operatorname{End}(N)$, *then* $\operatorname{Hom}(M, T) = \operatorname{Hom}(N, T)$.

Proof. By Lemma 35.3, $C(M) = T = C(N)$. It is sufficient to prove that $\operatorname{Hom}(M, C(M)) = \operatorname{Hom}(N, C(N))$. We take $\alpha \in \operatorname{Hom}(M, C(M))$. Then $r\alpha(M) \subseteq M$. Consequently, $r\alpha^\bullet(N) \subseteq N$ for all $r \in \widehat{R}$. Consequently, the $\widehat{R}$-submodule generated by $\alpha^\bullet(N)$ in N is contained in N. Therefore $\alpha^\bullet(N) \subseteq C(N)$ and $\alpha \in \operatorname{Hom}(N, C(N))$. We obtain one inclusion. Another inclusion follows from the symmetry. □

Quite delicate interrelations can exist between the heights of elements of infinite order of the cotorsion hull $T^\bullet$ and the length $l(T)$ of a reduced primary module T (the length is defined in Section 28). In particular, to construct certain endomorphisms, we need to know that cotorsion hulls of some primary modules contain elements of infinite order and sufficiently large height. Papers of May [223, 225] contain information about this topic. We need one result from [223]. In the proof of Lemma 36.2, the notion of the inverse limit is considered; the notion is presented in [92]. (The dual construction of the direct limit was introduced in Section 29.)

Lemma 36.2. *Let T be a nonbounded primary module such that $T \subseteq \bar{T}$, where $\bar{T}$ is a totally projective module of length $< l(T) + \omega$. Let σ be the least ordinal number such that T^σ is a bounded module. If $\tau < \sigma$, then the rank of the module $(T^\bullet)^\tau$ is not less than $2^{\aleph_0}$. The rank of the module $(T^\bullet)^\sigma$ is at least $2^{\aleph_0}$ excluding the case, where σ is a limit ordinal number of confinality $> \omega$; $(T^\bullet)^\sigma$ is a primary module in this case.*

Proof. It is clear that σ is also the least ordinal number such that $\bar{T}^\sigma$ is a bounded module. We choose n with $p^n(\bar{T}^\sigma) = 0$. If $\tau < \sigma$, then T_τ is a nonbounded module. It follows from [92, Theorem 56.7] that $(T^\bullet)_\tau$ has a summand which is isomorphic to the p-adic completion $\widehat{T_\tau}$ of the module T_τ (see also Proposition 21.5). Therefore, the rank of the module $(T^\bullet)^\tau$ is at least $2^{\aleph_0}$.

We assume that σ is a nonlimit ordinal number. Let $\sigma = \tau + 1$ for some τ. It follows from the above properties that the rank of the module $(T^\bullet)^\sigma$ is not less than the rank of the maximal divisible torsion submodule of the module $\widehat{T_\tau}/T_\tau$. The embedding $T \to \bar{T}$ induces the homomorphism $T_\tau \to \bar{T}_\tau$ such that the orders of elements of the kernel are bounded by the element p^n. The image G of this homomorphism should be a nonbounded module, since T_τ is a nonbounded module. Furthermore, $\bar{T}_\tau$ is the direct sum of cyclic modules, since the totally projective module $\bar{T}$ is simply presented (see the end of Section 30 and property (f)). We have an epimorphism $T_\tau \to G$ such that the kernel of it is bounded by the element p^n. The same also holds for the induced mappings $\widehat{T_\tau} \to \widehat{G}$ and $\widehat{T_\tau}/T_\tau \to \widehat{G}/G$. The rank of the maximal divisible torsion submodule of the module $\widehat{G}/G$ is at least $2^{\aleph_0}$. Consequently, the same holds for $\widehat{T_\tau}/T_\tau$. The assertion has been proved for $\sigma = \tau + 1$.

Now we assume that σ is a limit ordinal number. We take inverse spectra of modules $\{T/T^{\tau}, \tau < \sigma; \pi_{\rho}^{\tau}\}$ and $\{\bar{T}/\bar{T}^{\tau}, \tau < \sigma; \bar{\pi}_{\rho}^{\tau}\}$, where π_{ρ}^{τ} $(\rho \leq \tau)$ is the canonical homomorphism $T/T^{\tau} \to T/T^{\rho}$ and $\bar{\pi}_{\rho}^{\tau}$ has the similar sense. Let L and $\bar{L}$ be the limits of these spectra. As earlier, it is sufficient to prove the corresponding result on the rank of the maximal divisible torsion submodule of the module $L/(T/T^{\sigma})$. Let $\varphi\colon T \to \bar{T}/\bar{T}^{\sigma}$ be the restriction of the canonical homomorphism. Then the set of the orders of elements of the kernel φ is bounded by the element p^n. Since $\bar{T}/\bar{T}^{\sigma}$ is a totally projective module and the Ulm length of the module is a limit ordinal number, we have $\bar{T}/\bar{T}^{\sigma} = \bigoplus_{\tau<\sigma} A_{\tau}$, where each of the modules A_{τ} has the Ulm length $\leq\tau$ (property (e) from Section 30). First, we assume that the confinality of σ is equal to ω. We can choose a sequence $\tau_0 < \tau_1 < \tau_2 < \cdots$ of ordinal numbers with $\sigma = \sup\{\tau_i \mid i \geq 0\}$ and a sequence $C_0, C_1, C_2, \ldots$ of cyclic submodules in T such that the set of the orders of the submodules is not bounded, $C_i \subseteq T^{\tau_i}$, and $\varphi(C_i) \subseteq \bigoplus_{\tau<\tau_{i+1}} A_{\tau}$ for all i (we choose C_i and then we choose τ_{i+1}). We note that subscripts of nonzero components of elements in $\varphi(C_i)$ are contained in the interval $[\tau_i, \tau_{i+1})$. Let P be the product of the modules C_i for all $i \geq 0$ and let C be the sum of C_i. We write elements of P as formal sums $\sum c_i$, $c_i \in C_i$. For every $\tau < \sigma$, we can define a homomorphism $f_{\tau} : P \to T/T^{\tau}$ as follows. The components c_i of the sum $\sum c_i$ are contained in T^{τ}, beginning with some k. We assume that f_{τ} maps from this sum into the residue class $(c_1 + \cdots + c_{k-1}) + T^{\tau}$. The homomorphisms f_{τ} agree with π_{ρ}^{τ}. Therefore, they induce the homomorphism $T/T^{\tau} \to \bar{T}/\bar{T}^{\tau}$ for every $\tau < \sigma$. The induced homomorphism $L \to \bar{L}$ can be similarly obtained. Let $\psi\colon P \to \bar{L}$ be the composition of these mappings. We note that $\psi C \subseteq \bar{T}/\bar{T}^{\sigma}$. We assume that $a = \sum c_i \in P$ and $\psi a \in \bar{T}/\bar{T}^{\sigma}$. Considering the decomposition for $\bar{T}/\bar{T}^{\sigma}$, we obtain that there exists a subscript j such that the subscripts of nonzero components of the element ψa are contained in the interval $[0, \tau_j)$. Since the subscript sets of nonzero components of elements of $\varphi(C_i)$ and $\varphi(C_j)$ are disjoint for $i \neq j$, we obtain that $\varphi(c_i) = 0$ for $i \geq j$. Therefore $p^n c_i = 0$ for $i \geq j$. Consequently, $p^n a \in C$. We have proved that the kernel of the mapping $P/C \to \bar{L}/(\bar{T}/\bar{T}^{\sigma})$ induced by ψ is bounded. This mapping factors through $L/(T/T^{\sigma})$ by the definition of ψ. The rank of the maximal divisible torsion submodule of the module P/C is not less than $2^{\aleph_0}$. Therefore, the same holds for $L/(T/T^{\sigma})$.

Now we assume that the confinality of σ exceeds ω. It is sufficient to prove that $L/(T/T^{\sigma})$ is a bounded module. Let $a \in L$ and let $\{t_{\tau} + T^{\tau} \mid \tau < \sigma\}$ be the set of all components of the element a, where $t_{\tau} - t_{\rho} \in T^{\tau}$ for $\tau < \rho$. Since $\varphi(t_{\tau}) - \varphi(t_{\rho}) \in \bar{T}^{\tau}/\bar{T}^{\sigma}$, we have that $\varphi(t_{\tau})$ and $\varphi(t_{\rho})$ have equal components in A_{ν} for $\nu \leq \tau$. It follows from the assumption on the confinality of σ that there exists a finite subset in σ containing the set of all subscripts of components of the element $\varphi(t_{\tau})$ for every $\tau < \sigma$. If τ_0 is the supremum of this set, then

$\varphi(t_\tau - t_{\tau_0}) = 0$ for every τ with $\tau_0 \le \tau < \sigma$. This gives

$$p^n a = p^n t_{\tau_0} + T^\sigma \in T/T^\sigma. \qquad \square$$

It follows from the lemma that the module $p^\mu T^\bullet$ has elements of infinite order for every ordinal number μ with $\mu + \omega \le l(T^\bullet)$ and $l(T) = l(T^\bullet)$ if and only if $l(T) \in [\sigma, \sigma + \omega)$, where σ is an ordinal number of confinality $> \omega$.

Let M be a Warfield module with torsion submodule T such that M/T is a divisible module. We take a nice decomposition basis X of the module M (see Lemma 32.10 about the existence of it). We set $\bar{T} = M/RX$. Then the above results hold for the module T. Before we formulate Proposition 36.3 which is crucial for the subsequent presentation, we recall some properties. Let an element $x \in T$ have an infinite order. In this case, $(Rx)^\bullet = \widehat{R}x$, since the cotorsion hull of a reduced torsion-free module coincides with the completion of the module (Proposition 21.3) and $\widehat{R}x \cong \widehat{R}$. For a submodule A of the module $T^\bullet$, we denote by A_* the submodule in $T^\bullet$ such that $A \subseteq A_*$ and $A_*/A = t(T^\bullet/A)$. It is clear that A_* is a pure submodule in $T^\bullet$.

If A is a submodule of a reduced module M such that M/A does not have submodules of the form $R(p^\infty)$, then we can assume that $A^\bullet$ is a submodule in $M^\bullet$ and $M^\bullet/A^\bullet = (M/A)^\bullet$. Indeed, by assumption, we have $\mathrm{Hom}(R(p^\infty), M/A) = 0$, and the result follows from the exact sequence

$$0 \to \mathrm{Ext}(R(p^\infty), A) \to \mathrm{Ext}(R(p^\infty), M) \to \mathrm{Ext}(R(p^\infty), M/A) \to 0.$$

Proposition 36.3 (Files [74]). *Let T be a reduced primary module and let M and N be two pure Warfield submodules in $T^\bullet$ such that $T \subseteq M \bigcap N$ and $\mathrm{End}(M) = \mathrm{End}(N)$. Then $\widehat{R}M = \widehat{R}N$.*

Proof. If M is a primary module, then M is an $\widehat{R}$-module. Consequently, $\widehat{R} \subseteq \mathrm{End}(M) = \mathrm{End}(N)$. Therefore N is an $\widehat{R}$-module and $N = C(N)$. By Lemma 35.3, $N = T = M$.

We assume that M is not a primary module. It follows from Lemma 36.1 that $\mathrm{Hom}(M, T) = \mathrm{Hom}(N, T)$. Let $\lambda = l(T)$. By using Lemma 32.10, we choose a nice decomposition basis X in M such that M/RX is a totally projective module and $p^\lambda M \subseteq RX$. We set $A = RX$. By the remark stated before the proposition, $A^\bullet$ can be embedded in $T^\bullet$ and

$$(M/A^\bullet) = M^\bullet/A^\bullet = T^\bullet/A^\bullet.$$

First, we prove that $N \subseteq (A^\bullet)_*$. It is sufficient to verify that $y \notin N$ provided $y \in T^\bullet \setminus (A^\bullet)_*$. We note that the order in $T^\bullet/A^\bullet$ of the element $y + A^\bullet$ is infinite.

In the factor module $T^\bullet/A^\bullet$, we take the submodule $G = (R(y + A^\bullet))_*$. Then G is a reduced module and $t(G) = M/A$ has the length λ. Consequently, we have $h_G(p^k y + A^\bullet) < \lambda + \omega$ for all $k \geq 0$, since the order of $y + A^\bullet$ is infinite. We assert that there exists an ordinal number μ with $\mu + \omega \leq l(T^\bullet)$ such that $h_G(p^k y + A^\bullet) \leq \mu + k$ for all $k \geq 0$. First, we assume that $l(T) \neq l(T^\bullet)$; then $l(T^\bullet) = l(T) + \omega$. If $h_G(p^k y + A^\bullet) = \lambda + m \geq \lambda$ for some $k \geq 0$, then we set $\mu = \lambda + m$ and we note that $h_G(p^k y + A^\bullet) \leq \mu + k$ for all $k \geq 0$, since $l(t(G)) = \lambda$. It is clear that $\mu + \omega \leq l(T^\bullet)$. If $l(T) = l(T^\bullet)$, then $\sigma \leq l(T) < \sigma + \omega$ for some ordinal number σ of confinality $> \omega$ (see the remark after the proof of Lemma 36.2). We have $l(t(G)) = l(t(G)^\bullet) = \lambda$, whence $h_G(p^k y + A^\bullet) < \lambda$ for all $k \geq 0$. Let μ be the supremum of the heights $h_G(p^k y + A^\bullet)$ for all $k \geq 0$. Then $\mu + \omega < \sigma \leq l(T^\bullet)$ by the confinality of σ. We have proved the existence of μ.

By the remark after Lemma 36.2, we can choose an element $z \in p^\mu T^\bullet$ of infinite order. Since G is a module of rank 1 with totally projective module $t(G)$, we have that $R(y + A^\bullet)$ is a nice submodule in G and $G/R(y + A^\bullet)$ is a totally projective module (see the paragraph before Corollary 32.9). The mapping $ry + A^\bullet \to rz$ $(r \in R)$ from $R(y + A^\bullet)$ into $T^\bullet$ does not decrease the heights with respect to G and $T^\bullet$. Consequently, the mapping can be extended to a homomorphism $G \to T^\bullet$ (Corollary 32.8). Furthermore, we take the composition of the canonical mapping $M \to M/A \subseteq G$ with this homomorphism and extend the composition to an endomorphism α of the module $T^\bullet$. We obtain $\alpha y = z$ and $\alpha A = 0$. Therefore $\alpha \in \mathrm{Hom}(M, T) = \mathrm{Hom}(N, T)$ and $\alpha y \notin T$. Consequently, $y \notin N$.

We show that $M \subseteq \widehat{R}N$. This implies the inclusion $\widehat{R}M \subseteq \widehat{R}N$ and the converse inclusion follows from the symmetry. It is sufficient to prove that $X \subseteq \widehat{R}N$. We take an element $x \in X$. We have $A = Rx \oplus C$ and $A^\bullet = \widehat{R}x \oplus C^\bullet$. Since X is a nice decomposition basis for M and M/A is a totally projective module, the projection $A \to Rx$ can be extended to an endomorphism π of the module M (Corollary 31.11). Since $\pi M \nsubseteq T$, we have $\pi^\bullet N \nsubseteq T$, say, $\pi^\bullet(b) \notin T$ for some $b \in N$. By the proved properties, $N \subseteq (A^\bullet)_*$ and we can assume that $b \in A^\bullet$. Therefore $b = vrx + c$, where v is an invertible element in $\widehat{R}$, $0 \neq r \in R$, and $c \in C^\bullet$. We have $\pi^\bullet(N) \subseteq N$, since $\pi \in \mathrm{End}(M)$. Consequently, $vrx = \pi^\bullet(vrx + c) = \pi^\bullet(b) \in N$. Therefore $rx = v^{-1}\pi^\bullet(b) \in \widehat{R}N$ and $X \subseteq \widehat{R}N$, since X is pure. □

Theorem 36.4 (Files [74]). *We assume that M is a Warfield module and $N = N_1 \oplus D$, where N_1 is a Warfield module and D is a divisible module. If ψ is some isomorphism from* $\mathrm{End}(M)$ *onto* $\mathrm{End}(N)$, *then ψ is a topological isomorphism and $M \cong N$.*

Proof. If D is a nonzero module, then it contains a summand of the form $R(p^\infty)$

or K. Similar to Theorem 33.6, we can take the inverse isomorphism ψ^{-1} and we obtain that M has a summand which is isomorphic to one of the modules $R(p^\infty)$, K, or $\widehat{R}$. The first two cases are impossible. In the remaining case, we obtain $\widehat{R} \subseteq C(M)$; this is impossible by Lemma 35.3. Thus, $N = N_1$ is a reduced module.

First, we assume that $M/t(M)$ is not a divisible module. We choose a primary module P such that $M \oplus P$ is equal to the direct sum of modules M_i, $i \in I$, of rank 1 (Corollary 32.3 and Theorem 30.3). There exists a subscript i such that $M_i/t(M_i)$ is a reduced module; therefore, the module is isomorphic to R. Consequently, $M/t(M)$ has R as a direct summand. Then M also has a similar summand (see Theorem 34.2). Similar to Theorem 33.5, we can verify that N has a summand R and ψ is induced by an isomorphism from M onto N. In particular, ψ is a topological isomorphism.

We assume that $M/t(M)$ is a divisible module. If M is a primary module, then $\mathrm{End}(M)$ is an $\widehat{R}$-algebra; therefore $\mathrm{End}(N)$ is an $\widehat{R}$-algebra. Therefore N is an $\widehat{R}$-module and $N = C(N)$. It follows from Lemma 35.3 that N is a primary module. This can be also proved with the use of standard methods of work with cyclic summands (for example, see the proof of Theorem 33.6). It remains to refer to Theorem 33.2. We can assume that M is a mixed module. It follows from the proved properties that $N/t(N)$ is a divisible module. We set $T = t(N)$. The Kaplansky method (which is used in the process of the proof of Theorem 33.6 (1)) can be also used in the given situation, since M and N are reduced modules. As a result, we can construct an isomorphism $\varphi\colon t(M) \to T$ such that $\psi(\alpha)$ and $\varphi^{-1}\alpha\varphi$ coincide on T for all $\alpha \in \mathrm{End}(M)$. Let $M' = \varphi^\bullet(M) \subseteq T^\bullet$. Then $\mathrm{End}(M') = \mathrm{End}(N)$. If $\beta \in \mathrm{End}(M')$, then $\beta' = \varphi^\bullet\beta(\varphi^\bullet)^{-1} \in \mathrm{End}(M)$. Consequently, $\psi(\beta') \in \mathrm{End}(N)$. However, $\psi(\beta')$ and $(\varphi^\bullet)^{-1}\beta'\varphi^\bullet = \beta$ coincide on T. Consequently, $\beta^\bullet = \psi(\beta')^\bullet$ and $\beta^\bullet \in \mathrm{End}(N)$. We obtain $\mathrm{End}(M') \subseteq \mathrm{End}(N)$. The converse inclusion can be proved with the use of similar arguments. Proposition 36.3 implies $\widehat{R}M' = \widehat{R}N$. By Proposition 35.2, $M' = N$, whence $M \cong N$.

Finally, why is ψ a topological isomorphism? First, we verify that the identification $\mathrm{End}(M') = \mathrm{End}(N)$ is topological. Let $y \in N$. Since $\widehat{R}M' = \widehat{R}N$, we have $y = r_1x_1 + \cdots + r_nx_n$, $r_i \in \widehat{R}$ and $x_i \in M'$. It is clear that if $\alpha \in \mathrm{End}(M')$ and $\alpha x_i = 0$ for all i, then $\alpha^\bullet(y) = 0$ ($\alpha^\bullet$ is an $\widehat{R}$-module endomorphism). It can be asserted that neighborhoods of zero remain neighborhoods of zero under the identification. This implies a topological nature of the identification. The isomorphism ψ is topological, since it is the composition of the identification with the isomorphism induced by $\varphi^\bullet$. (The induced isomorphism always is a topological isomorphism; see Section 33.) □

Corollary 36.5. *If M is a Warfield module, then every automorphism of the algebra* $\mathrm{End}(M)$ *is a topological automorphism.*

Files [74] has constructed two modules M and N such that the isomorphism ψ from Theorem 36.4 is not induced by any isomorphism from M onto N. The corresponding domain R should be incomplete. In the case of the complete domain R, it follows from Theorem 35.5 that ψ is always induced by an isomorphism from M onto N, where we do not impose any limitations on N. In the example of Files, the algebra $\mathrm{End}(M)$ has an external automorphism. In [75], Files determined when the endomorphism ring of a p-local Warfield group (i.e., a Warfield $\mathbb{Z}_p$-module) has external automorphisms.

All exercises are taken from the papers of Files [73, 76]. A sharp module is defined in remarks at the end of the chapter.

Exercise 1. Let M be a reduced R-module which is not primary. Prove that the following conditions are equivalent:

(1) M is a sharp module;

(2) the factor module of the module $\widehat{R}\otimes M$ with respect to the maximal divisible submodule is an Warfield $\widehat{R}$-module of rank 1;

(3) M is contained in Warfield $\widehat{R}$-module M' of rank 1 and M'/M is a torsion-free module.

Exercise 2. (a) A Warfield module of rank 1 is a sharp module.

(b) A reduced torsion-free R-module M is a sharp module if and only if M is isomorphic to some pure R-submodule in $\widehat{R}$.

(c) A domain R is not complete if and only if there exists a sharp R-module of rank > 1.

Exercise 3. If M is a sharp R-module, then $\mathrm{End}_{\widehat{R}}(\widehat{R}M) = \widehat{R}\,\mathrm{End}_R(M)$.

Exercise 4. Prove that a reduced module M has a stable element (the definition of it is presented below) in the following cases:

(1) the rank of M is equal to 1;

(2) M is a Warfield module and the set of height sequences of elements of some decomposition basis for M contains the least element;

(3) $M = A\oplus B$, where A has a stable element and $B/\mathrm{Hom}(A,B)A$ is a primary module.

Remarks. An additional information about the isomorphism problem for endomorphism rings is contained before exercises in Sections 34–36. We present several remarks. Files [76] introduced the notion of a stable element of the mixed module. The properties of such an element resemble properties of an element from the direct summand which is isomorphic to R. Precisely, an element x of the module M is said to be *stable* if the following conditions hold:

(1) there exists an endomorphism $\rho \in \operatorname{End}(M)$ such that $\rho x = x$ and $\rho M/Rx$ is a primary module;

(2) $M/\operatorname{End}(M)x$ is a primary module.

Files has proved the following theorem. We assume that M and N are reduced modules and $t(M)$ is a totally projective module. If each of the modules has a stable element, then every isomorphism $\operatorname{End}(M) \to \operatorname{End}(N)$ is induced by some isomorphism $M \to N$. Some applications are obtained (e.g., to Warfield modules). The paper of May [226] is interesting. It contains the following isomorphism theorem in the weak sense for a quite large class of mixed modules that are not related to totally projective modules or Warfield modules. Let M be a module over a complete domain such that M/M^1 is a mixed module of countable rank. Then the isomorphism $\operatorname{End}(M) \cong \operatorname{End}(N)$ implies the isomorphism $M \cong N$. The conditions of the theorem hold for the module M of countable rank which contains an element such that all coordinates of the height sequence of it are finite. It has been constructed an isomorphism of endomorphism algebras which is not induced by a module isomorphism, i.e., in the general case, the module M does not satisfy the isomorphism theorem in the strong sense. In the paper of Files [73], a reduced module M is said to be *sharp* if the $\widehat{R}$-hull $\widehat{R}M$ of M is a Warfield module of rank 1. It is proved that for a sharp module M, all automorphisms of the algebra $\operatorname{End}(M)$ are internal automorphisms.

There is a large bibliography on isomorphisms of endomorphism rings of modules over various rings (see reviews of Mikhalev [235] and Markov–Mikhalev–Skornyakov–Tuganbaev [219]). The papers of Wolfson [320, 321] and Franzsen–Schultz [89] are devoted to endomorphism rings of modules which are close to free. In the last paper, authors analyze the situation, where an isomorphism $\operatorname{End}(M) \to \operatorname{End}(N)$ is not induced by any semilinear isomorphism $M \to N$, but there exists a semilinear isomorphism $M \to N$. It is discussed the interrelation between this situation and the existence of internal or external automorphisms of the ring $\operatorname{End}(M)$.

There exist studies related to isomorphisms of endomorphism groups and endomorphism semigroups of modules (in this connection, see the mentioned reviews); we mean the additive group and the multiplicative semigroup of the endomorphism ring, respectively.

Let S_1 and S_2 be two rings. An additive isomorphism $f\colon S_1 \to S_2$ is called an *anti-isomorphism* if $f(xy) = f(y)f(x)$ for all $x, y \in S_1$. An anti-isomorphism of a ring onto itself is called an *antiautomorphism*. Wolfson [322] considered the question on conditions under which an anti-isomorphism of endomorphism rings of two locally free modules is induced by a semilinear anti-isomorphism of these modules. For example, if endomorphism rings of two reduced torsion-free modules M and N over a complete domain are anti-isomorphic, then M and N are free finitely generated modules.

The isomorphism problem can be studied for automorphism groups of modules (as well as for other algebraic structures). The following question is a fundamental and very difficult problem: when are modules isomorphic if their automorphism groups are isomorphic? Rickart [261] has obtained the positive answer for vector spaces over division rings of characteristic $\neq 2$; Leptin [203] and Liebert [213] have obtained the positive answer for Abelian p-groups ($p \neq 2$). Corner and Goldsmith [50] have proved that if M and N are reduced p-adic torsion-free modules with $\mathrm{Aut}(M) \cong \mathrm{Aut}(N)$ ($p \neq 2$), then $M \cong N$.

Another direction of the study of the isomorphism problem for endomorphism rings is indicated by papers of Hausen–Johnson [137], Hausen–Praeger–Schultz [138], and Schultz [276]. These papers contain the information about Abelian p-groups G and H that are isomorphic to the radicals $J(\mathrm{End}(G))$ and $J(\mathrm{End}(H))$ (as rings without unity). For example, whether G and H are isomorphic and whether there exists an isomorphism between G and H which induces the given isomorphism of radicals? If I is an arbitrary ideal of the endomorphism ring of some module M and I is contained in the radical $J(\mathrm{End}(M))$, then $1 + I$ is a normal subgroup of the group $\mathrm{Aut}(M)$ (Problem 6 is connected to this subgroup). Schultz [277] used this property and the results of [138, 276] to extend the Leptin–Liebert theorem [203, 213] to the case $p = 2$.

Problem 19. Find classes of modules M such that the isomorphism $J(\mathrm{End}(M)) \cong J(\mathrm{End}(N))$ implies the isomorphism $M \cong N$ (the existence of an isomorphism $M \cong N$ that induces the original isomorphism of radicals).

Problem 20. For which modules M and N, the isomorphism $\mathrm{Aut}(M) \cong \mathrm{Aut}(N)$ implies the isomorphism $M \cong N$?

In these two problems, we call special attention to Warfield modules and torsion-free modules over complete domains. The following problem has a more general nature.

Problem 21. For various modules M and N, study all isomorphisms between $\mathrm{End}(M)$ and $\mathrm{End}(N)$, $J(\mathrm{End}(M))$ and $J(\mathrm{End}(N))$, $\mathrm{Aut}(M)$ and $\mathrm{Aut}(N)$, respectively.

Problem 22. Characterize modules M with the following property. For every module N, the existence of a (resp., topological) isomorphism from $\mathrm{End}(M)$ onto $\mathrm{End}(N)$ implies the existence of an isomorphism from M onto N (resp., implies the existence of an isomorphism from M onto N which induces the original isomorphism of endomorphism rings).

Chapter 8

Modules with many endomorphisms or automorphisms

In Chapter 8, we consider the following topics:

- transitive and fully transitive modules (Section 37);
- transitivity over torsion and transitivity mod torsion (Section 38);
- the equivalence of transitivity and full transitivity (Section 39).

Let x and y be two nonzero elements of some module M. In contrast to the case of vector spaces, it is not necessarily possible to map x onto y by some endomorphism or an automorphism of the module M. One of the obvious obstructions is that either the height sequences $U(x)$ and $U(y)$ can be incomparable or $U(x) > U(y)$. Even if we remove this obstruction, then the required endomorphism or automorphism does not necessarily exist. In the chapter, we consider modules which have sufficiently many (in a certain sense) endomorphisms or automorphisms. Considerable attention is given to mixed completely decomposable modules.

37 Transitive and fully transitive modules

We define transitive and fully transitive modules and present some relatively elementary results on such modules. We recall some notations. Let M be a module and let $x \in M$. Then $o(x)$ and $h(x)$ are the order and the height of the element x in M, respectively. We assume that the height means the "generalized height", i.e., the height in the sense of Section 28. The height sequence $U(x)$ of the element x in M is the sequence consisting of ordinal numbers and the symbol ∞ $(h(x), h(px), h(p^2x), \dots)$. The following partial order can be defined on the set of height sequences:

$$U(x) \leq U(y) \Longleftrightarrow h(p^i x) \leq h(p^i y)$$

for $i = 0, 1, 2, \dots$. By definition, the largest lower bound $\inf(U(x), U(y))$ is the sequence $(\sigma_0, \sigma_1, \sigma_2, \dots)$, where $\sigma_i = \min(h(p^i x), h(p^i y))$.

Let D be the maximal divisible submodule of the module M and let $D \neq 0$. We earlier assumed that $h(x) = \infty$ for every $x \in D$ and $\sigma < \infty$ for every ordinal number σ. We introduce a new symbol ∞^+ which is needed for a correct formulation of the main definitions and the reduction of the study to the case of reduced modules. We set $h(0) = \infty^+$ and $\infty < \infty^+$. The symbol ∞^+ is used only for distinguishing the height of the zero element of the module M and the height of a nonzero element in D; if M is a reduced module, then it is not necessary to use the symbol ∞^+.

Definition 37.1. Let M be a module. We say that

(1) M is a *fully transitive* module if for all $x, y \in M$ with $U(x) \leq U(y)$, there exists an endomorphism of the module M which maps x onto y;

(2) M is a *transitive* module if for all $x, y \in M$ with $U(x) = U(y)$, there exists an automorphism of the module M which maps x onto y.

The restrictions to the height sequences in (1) and (2) are necessary in the sense that for every endomorphism (automorphism) α of the module M, we have $U(x) \leq U(\alpha x)$ $(U(x) = U(\alpha x))$, $x \in M$.

In the process of verifying transitivity or full transitivity, we can assume (if it is convenient) that x and y are nonzero elements. We frequently deal with the following situation (or the situation close to the following situation). Let $M = A \oplus N$ and let $x, y \in A$. If it has been proved that there exists an endomorphism (resp., an automorphism) α of the module A with $\alpha x = y$, then we assume that α is an endomorphism (resp., an automorphism) of the module M; in addition, we assume that α coincides on N with the identity mapping. We also note that the inequality $U(x) \leq U(y)$ always implies the inequality $o(x) \geq o(y)$.

Proposition 37.2. *Any divisible module is transitive and fully transitive.*

Proof. First, we assume that D is a divisible torsion-free module. The following more strong assertion holds. For any two nonzero elements x and y in D, there exists an automorphism of the module D which maps x onto y. If the rank of D is 1, then $D = K$. The endomorphisms of the module K coincide with multiplications by elements of K (Example 12.3). Consequently, all nonzero endomorphisms of the module K are automorphisms. Now it remains to use the relation $y = (yx^{-1})x$. An arbitrary divisible torsion-free module D is a K-space. Consequently, $Kx \oplus G = D = Ky \oplus H$, where the modules G and H are isomorphic to each other. It was just mentioned that there exists an isomorphism $\beta : Kx \to Ky$ with $\beta x = y$. Let γ be some isomorphism from G onto H. Then (β, γ) is an automorphism of the module D which maps x onto y.

Let D be a divisible primary module. We have that the height sequence $U(x)$ of the element $x \in D$ of order p^k has the form $(\infty, \dots, \infty, \infty^+, \dots)$, where the symbol ∞^+ begins with the position $k + 1$. We take any two elements $x, y \in D$ with

$U(x) \leq U(y)$ (resp., $U(x) = U(y)$). Let $D = R(p^\infty)$ and let $c_1, c_2, \dots, c_n, \dots$ be the generator system for D considered in Section 4. Since $U(x) \leq U(y)$, we have $s \leq t$, where $p^s = o(y)$ and $p^t = o(x)$. Consequently, $x, y \in Rc_t$, $x = uc_t$, and $y = p^k vc_t$, where $k \geq 0$ and the elements u and v of R are invertible in the ring R. The multiplication of the module $R(p^\infty)$ by the element $p^k vu^{-1}$ maps x onto y. If $U(x) = U(y)$, then $k = 0$ and we have an automorphism of the module $R(p^\infty)$. If D is an arbitrary divisible primary module, then it follows from the proof of Theorem 6.3 that there exist decompositions

$$P_1 \oplus G = D = P_2 \oplus H,$$

where $x \in P_1$, $y \in P_2$, $P_1, P_2 \cong R(p^\infty)$, and $G \cong H$. By the above, there exists a homomorphism $P_1 \to P_2$ which maps x onto y; this gives an endomorphism of the module D which maps x onto y. If height sequences coincide, we obtain the required automorphism.

Now we assume that D is a divisible mixed module and $D = D_t \oplus D_0$, where D_t is a primary module and D_0 is a torsion-free module. Let $x, y \in D$ and let $U(x) \leq U(y)$ $(U(x) = U(y))$. We have $x = x_1 + x_2$ and $y = y_1 + y_2$, where $x_1, y_1 \in D_t$ and $x_2, y_2 \in D_0$. We consider two possible cases.

Case (a): $x_2 = 0$. If $y_2 \neq 0$, then $U(y_2) = (\infty, \infty, \dots)$ and $U(x_1) \leq U(y) = U(y_2)$, which is impossible. Consequently, $y_2 = 0$ and $x, y \in D_t$. There exists an endomorphism (an automorphism) α of the module D_t and D with $\alpha x = y$.

Case (b): $x_2 \neq 0$. There exists a submodule G such that $D = D_t \oplus G$ and $x \in G$ (the remark after of Theorem 6.1). If $y_2 \neq 0$, then we similarly have $D = D_t \oplus H$ and $y \in H$. Then our argument is similar to the argument used in (a). The relation $y_2 = 0$ is possible only if $U(x) < U(y)$. Since $Rx \cong R$, there exists a homomorphism $\varphi\colon Rx \to D$ with $\varphi x = y$. It can be extended to an endomorphism of the module D, since D is injective. □

Let A be a direct summand of the module M and let $x \in A$. The height sequences of the element x with respect to A and M are equal to each other. Consequently, we usually write $U(x)$ instead of $U_A(x)$ or $U_M(x)$.

Lemma 37.3. *A direct summand A of a fully transitive module M is a fully transitive module.*

Proof. We denote by π the projection $M \to A$. Let $x, y \in A$ and let $U(x) \leq U(y)$. Consequently, $\alpha x = y$ for some $\alpha \in \mathrm{End}(M)$. Then $(\pi\alpha_{|A})x = y$ and $\pi\alpha_{|A} \in \mathrm{End}(A)$. □

Proposition 37.4. *Let M be a module and let D be the maximal divisible submodule in M. The module M is transitive (resp., fully transitive) if and only if the factor module M/D is transitive (resp., fully transitive).*

Proof. We assume that M is a transitive module. We have $M = A \oplus D$ for some A. Since $M/D \cong A$, we need to prove that A is transitive. If $x, y \in A$ and $U(x) = U(y)$, then we take an automorphism α of the module M with $\alpha x = y$. We represent α by the matrix

$$\begin{pmatrix} \beta & \varphi \\ 0 & \gamma \end{pmatrix}$$

with respect to a given decomposition of the module M (as in Proposition 2.4), where β (resp., γ) is an endomorphism of the module A (resp., D), φ is a homomorphism from A into D. It is easy to verify that β is an automorphism of the module A and $\beta x = (\alpha\pi)x = \pi(\alpha x) = \pi y = y$, where π is the projection from M onto A.

Conversely, let A be a transitive module, $x, y \in M$, and let $U(x) = U(y)$. We have $x = x_1 + x_2$ and $y = y_1 + y_2$, where $x_1, y_1 \in A$ and $x_2, y_2 \in D$. We assume that $x_1 = 0$. Then $y_1 = 0$. Otherwise, $U(x_2) = U(y) = U(y_1)$, which is a contradiction. Thus, $x, y \in D$. By Proposition 37.2, there exists an automorphism of the module M which maps x onto y. If $x_1 \neq 0$, then we also have $y_1 \neq 0$. We have decompositions $B \oplus D = M = C \oplus D$, where $x \in B$ and $y \in C$. It follows from $B \cong A \cong C$ that there exists an isomorphism $\alpha\colon B \to C$ with $\alpha x = y$. Assuming that α coincides on D with the identity mapping, we obtain the required automorphism.

We pass to full transitivity. It follows from Lemma 37.3 that it remains to prove that full transitivity of the module A implies full transitivity of the module M. The proof is similar to the proof in the case of transitive modules. □

Till the end of the chapter, we assume that M is a reduced module. We return to our agreement that $h(0) = \infty$. We present examples of transitive and fully transitive modules in each of the three main classes of modules: primary modules, torsion-free modules, and mixed modules.

First, we consider cyclic primary modules. If $x \in R(p^m)$ and $o(x) = p^k$, then the exponent k is denoted by $e(x)$. We have the obvious relation $e(x) + h(x) = m$. Now we assume that $A = Ra$ and $B = Rb$, where the first module is isomorphic to $R(p^m)$ and the second module is isomorphic to $R(p^n)$. Let $x \in A$, $y \in B$, and let $U(x) \leq U(y)$. We verify the existence of a homomorphism $\varphi\colon A \to B$ with $\varphi x = y$. We have $h(x) \leq h(y)$ and $e(y) \leq e(x)$. We have $x = p^s ua$ and $y = p^t vb$, where $s \leq t$ and u, v are invertible elements in R. The relations $e(x) + s = m$ and $e(y) + t = n$ hold. Furthermore, we have

$$e(p^{t-s}u^{-1}vb) = n - t + s = e(y) + t - t + s = e(y) + s \leq e(x) + s = m.$$

Thus, we obtain

$$o(p^{t-s}u^{-1}vb) \leq o(a).$$

We have obtained that the mapping $ra \to rp^{t-s}u^{-1}vb\,(r \in R)$ is a homomorphism $A \to B$ which maps x onto y. For $U(x) = U(y)$, we have $h(x) = h(y)$ and $e(x) = e(y)$. Therefore $o(a) = o(b)$ and $A \cong B$. The same mapping already is an isomorphism.

Proposition 37.5. *A primary module M without elements of infinite height is transitive and fully transitive.*

Proof. Let $x, y \in M$ and let $U(x) = U(y)$. It follows from Corollary 7.6 that we can assume that M is the direct sum of a finite number of cyclic modules. If x and y have order p, then we have

$$Ra \oplus G = M = Rb \oplus H,$$

where $x \in Ra$ and $y \in Rb$ (Corollary 7.5). It follows from the argument presented before the proposition that there exists an automorphism of the module M which maps x onto y. Let $o(x) = p^k$, $k \geq 2$. Since $U(px) = U(py)$, it inductively follows that there exists an automorphism α with $\alpha(px) = py$. If $\alpha x = y'$, then $U(y') = U(y)$. It is sufficient to prove the existence of an automorphism which maps y' onto y. We have $y = y' + z$, where $pz = 0$ and $h(y) \leq h(z)$. There exists a decomposition $M = Ra_1 \oplus \cdots \oplus Ra_s$ with $z \in Ra_s$. We have $y' = y_1 + \cdots + y_s$ with $y_i \in Ra_i$. We need to prove the existence of an automorphism which maps y' onto $y' + z$. We assume that $y_s = 0$. We choose a subscript i with $h(y_i) \leq h(z)$. For simplicity, we assume that $i = 1$. Let $y_1 = p^m u a_1$ and $z = p^n v a_s$, where $n \geq m$ and u, v are invertible elements of the ring R. We have the decomposition

$$Ra_1 \oplus Ra_s = R(a_1 + p^{n-m}u^{-1}va_s) \oplus Ra_s.$$

There exists a mapping α such that α maps a_1 onto $a_1 + p^{n-m}u^{-1}va_s$ and α coincides with the identity mapping on the summands $Ra_2, \dots, Ra_s$. It is clear that α is an automorphism of the module M and $\alpha x = y$. Let $y_s \neq 0$. If $y_s + z = 0$, then, as above, we can construct an automorphism β with $\beta(y' + z) = y'$. Then β^{-1} is the required automorphism. If $y_s + z \neq 0$, then $h(y_s) = h(y_s + z)$. We denote by α the mapping which coincides with the identity mapping on $Ra_1, \dots, Ra_{s-1}$ and maps y_s onto $y_s + z$.

We prove full transitivity. Let $x, y \in M$, $U(x) \leq U(y)$, and let $M = Ra_1 \oplus \cdots \oplus Ra_s$. It is sufficient to consider the case, where y is contained in one of the summands Ra_i. For example, let $y \in Ra_1$. We have $x = x_1 + x'$, where $x_1 \in Ra_1$ and $x' \in Ra_2 \oplus \cdots \oplus Ra_s$. If $h(x_1) \leq h(y)$, then there exists an endomorphism α of the module M with properties $\alpha x_1 = y$ and $\alpha x' = 0$. Otherwise, $U(x) = U(x')$, $U(x') \leq U(y)$, and $U(y + x') = U(x')$. Let γ be an automorphism of the module M with $\gamma x = y + x'$. The composition of γ and the projection $M \to Ra_1$ gives the required endomorphism. □

Hill [143] has proved that a totally projective module is transitive and fully transitive. Griffith [125] found other classes of transitive fully transitive primary modules. In works [231] of Megibben and [37] of Carroll–Goldsmith, primary modules without any transitivity property are constructed. Kaplansky [166] has proved that if 2 is an invertible element in R, then transitivity always implies full transitivity in the primary case. Corner [47] has proved the independence of these two transitivity notions for Abelian 2-groups. Papers of Megibben [233] and Goldsmith [117] also contain results on independence.

For a nonzero element x of some torsion-free module, the relation $h(px) = h(x) + 1$ holds. The height $h(x)$ contains the complete information about the height sequence $U(x)$. For example,

$$U(x) \leq U(y) \Longleftrightarrow h(x) \leq h(y).$$

Proposition 37.6. *Every torsion-free module M over a complete discrete valuation domain R is transitive and fully transitive.*

Proof. In fact, the full transitivity was proved in the beginning of Section 17. Let $x, y \in M$ and let $h(x) = h(y)$. If the elements x and y are linearly dependent, then $M = A \oplus G$, where x and y are contained in the module A which is isomorphic to R. There exists an automorphism of the module A which maps x onto y (this follows from the arguments from Section 17). We assume that x and y are linearly independent elements. Then the pure submodule generated by x and y is equal to $A \oplus B$, where A and B are isomorphic to R. By Corollary 11.5, $M = A \oplus B \oplus H$. We have

$$A_1 \oplus B_1 = A \oplus B = A_2 \oplus B_2,$$

where $x \in A_1$ and $y \in A_2$ (see the beginning of Section 17). Now it is easy to construct an automorphism which maps x onto y. □

A transitive torsion-free module M over an arbitrary (not necessarily complete) domain is fully transitive. Indeed, if $x, y \in M$ and $h(x) \leq h(y)$, then $y = p^k z$, where $h(z) = h(x)$. Let α be an automorphism which maps x onto z. Then $(p^k \alpha)x = y$. Exercises 6–8 contain more serious results on transitivity for torsion-free modules.

We pass to mixed modules. There is an obvious necessary condition of transitivity (full transitivity) for mixed modules. Precisely, the torsion submodule $t(M)$ of a transitive (resp., fully transitive) of the mixed module M should be transitive (resp., fully transitive). This condition is certainly far from sufficient. This can be easily verified by the use of the functor G from Theorem 22.2. Let X be a torsion-free module which is not fully transitive (and there are many such modules). Since $G(X)^1 \cong X$, the mixed module $G(X)$ can be neither fully transitive nor transitive. In addition, $t(G(M))$ is the direct sum of cyclic modules.

Let M be a mixed module such that $t(M)$ does not have elements of infinite height. It follows from Corollary 7.6 that for all elements $a_1, \dots, a_n \in t(M)$, there exists a decomposition $t(M) = A \oplus N$, where A is the direct sum of a finite number of cyclic modules and $a_1, \dots, a_n \in A$. By Theorem 7.2, the module A is a direct summand of the module M. We will frequently use this property.

Theorem 37.7 (Files [77]). *Let M be a mixed module of rank 1. We assume that the torsion submodule $t(M)$ does not have elements of infinite height. Then M is a transitive module and fully transitive module.*

Proof. If $M/t(M) \cong R$, then $M = A \oplus t(M)$, where $A \cong R$. We leave the check of the both transitivities to the reader (see Exercise 1).

Let $M/t(M) = K$. First, we prove the full transitivity. Let $x, y \in M$ and let $U(x) \leq U(y)$. We have $sy = rx + c$, where $s, r \in R$ and $c \in t(M)$. Using the inequality $h(x) \leq h(y)$, we can restrict ourself to the case $s = 1$. Thus, we have $y = rx + c$. We note that $U(x) \leq U(c)$. The module multiplication by r maps x onto rx. Therefore, the proof can be reduced to the case, where $y \in t(M)$. If $x \in t(M)$, then it was mentioned before the theorem, x and y are contained in a direct summand of the module M which is a direct sum of cyclic primary modules. In this case, we can use Proposition 37.5. Therefore, we assume that x is not contained in $t(M)$ and y is contained in $t(M)$. We have $M = N \oplus A$, where $y \in A$ and A is the direct sum of a finite number of cyclic primary modules; in particular, $p^n A = 0$ for some n. According to the decomposition, we have $x = x_0 + a$. Since the module $M/t(M)$ is divisible, $x_0 = p^n x_1 + b$, $x_1 \in N$, $b \in t(N)$. We state that $U(b + a) \leq U(y)$. Since

$$h(p^i(b + a)) = \min(h(p^i b), h(p^i a)),$$

it is sufficient to verify that $h(p^i b) \leq h(p^i y)$ if $h(p^i a) > h(p^i y)$ and $p^i y \neq 0$. In this case,

$$\min(h(p^{i+n} x_1 + p^i b), h(p^i a)) = h(p^i x) \leq h(p^i y) < h(p^i a).$$

It follows from $h(p^i y) < n$ that

$$h(p^{i+n} x_1 + p^i b) = h(p^i b) \quad \text{and} \quad h(p^i b) = h(p^i x) \leq h(p^i y),$$

which is required. Now we have $N = N_0 \oplus A_0$, where A_0 is the direct sum of a finite number of cyclic modules and $b \in A_0$. An element $b + a$ can be mapped onto y by some homomorphism φ from $A_0 \oplus A$ into A. By assuming that φ annihilates N_0, we extend φ to an endomorphism of the module M. Then

$$\varphi x = p^n \varphi(x_1) + \varphi(b + a) = y,$$

since $p^n A = 0$. Consequently, M is a fully transitive module.

We verify that M is transitive. We assume that $U(x) = U(y)$, $x, y \in M$. Then either $x, y \in t(M)$ or the orders of the elements x and y are infinite. If $x, y \in t(M)$, then we again can use Proposition 37.5.

Let $x, y \notin t(M)$. Then $y = rx + c$, where $r \in R$ and $c \in t(M)$. In addition, the element r is invertible, since $U(x) = U(y)$. We use the induction on the order of the element c. For $c = 0$, the module multiplication by r is an automorphism which maps x onto y. We assume that $c \neq 0$. We have $U(px) = U(py)$, $py = rpx + pc$, and $o(pc) < o(c)$. Therefore, we inductively obtain an automorphism α with $\alpha(px) = py$. We set $x' = \alpha x$. Now it is sufficient to find an automorphism of the module M which maps x' onto y. Since $px' = py$, we have $y = x' + a$ with $pa = 0$. Let $M = N \oplus A$, where A is the direct sum of a finite number of cyclic modules and $a \in A$. According to the decomposition, we have $x' = x_1 + a_1$ and $y = x_1 + (a_1 + a)$. If $h(x') = h(x_1)$, then $U(x_1) \leq U(a)$ and there exists a $\varphi \in \mathrm{Hom}(N, A)$ with $\varphi(x_1) = a$, since M is a fully transitive module. In this case, the automorphism

$$\begin{pmatrix} 1 & \varphi \\ 1 & 1 \end{pmatrix}$$

of the module $N \oplus A$ which maps x' to y. To complete the proof, we can assume that $h(x') = h(a_1) < h(x_1)$. Since

$$h(a_1) = h(x') = h(y) = \min(h(x_1), h(a_1 + a)) < h(x_1),$$

we have $h(a_1 + a) = h(a_1)$. Consequently, $U(a_1) = U(a_1 + a)$. We choose an automorphism γ of the module A which maps from a_1 to $a_1 + a$. Then the automorphism $(1, \gamma)$ of the module $M = N \oplus A$ maps x' onto y, which is required. $\square$

For a mixed module of rank 1 with totally projective torsion submodule, Files [77] has obtained a result which is similar to Theorem 37.7. Theorem 37.7 does not hold even for modules of rank 2. Files [77] has constructed a p-adic module M such that $M/t(M)$ is a divisible module of rank 2, $t(M)$ does not have elements of infinite height, but M is neither transitive, nor fully transitive. As for other torsion submodules $t(M)$ in Theorem 37.7 and the necessary condition indicated before the theorem, we give the following remark. The same paper [77] contains the construction of a module M of rank 1 such that $t(M)$ has one of the transitivity properties, but M is neither transitive, nor fully transitive.

It seems that topics related to transitivity (full transitivity) are interesting for Warfield modules. Files [77] has proved that a Warfield module of rank ≤ 2 is transitive and fully transitive. For larger ranks, this is not true. Hill and Ullery [151] have proved that every Warfield module satisfies some weak transitivity.

They considered four different classes of Warfield modules of rank 3 which are not transitive.

Exercise 1. If P is a primary module, then the module $R \oplus P$ is transitive (resp., fully transitive) if and only if the module P is transitive (resp., fully transitive).

Exercise 2. Let $M = A \oplus P$, where A and P are fully transitive torsion-free module and a primary module, respectively. Then M is a fully transitive module.

Exercise 3. Study the situation with transitivity for the module $A \oplus P$, where A is a transitive torsion-free module and P is a transitive primary module.

Exercise 4. Let $M = P \oplus N$, where P is a primary module without elements of infinite height and N is some transitive (resp., fully transitive) module. Is the module M transitive (resp., fully transitive)?

Exercise 5 (Files [77]). Any Warfield module of rank 1 is transitive and fully transitive.

Exercise 6. Let M be a fully transitive torsion-free module and let T be the center of the ring $\mathrm{End}(M)$. Then T is a discrete valuation domain and every T-submodule in M of finite or countable rank is free.

Exercise 7. For a torsion-free module M of finite rank, the following conditions are equivalent:

(1) M is a transitive module;

(2) M is a fully transitive module;

(3) the center T of the ring $\mathrm{End}(M)$ is a discrete valuation domain, $\mathrm{End}_R(T) = \mathrm{End}_T(T)$, and M is a free T-module.

Exercise 8. Let M and N be two fully transitive torsion-free modules. Then every topological isomorphism $\mathrm{End}(M) \to \mathrm{End}(N)$ is induced by some isomorphism $M \to N$.

38 Transitivity over torsion and transitivity mod torsion

In the remaining sections, the studies are mainly directed to mixed modules. In this section, we introduce two weak notions of transitivity which are specially adapted to the use in mixed modules. Some presented results are of general type and other results are related to the class of completely decomposable modules; we study these modules quite deeply.

Definition 38.1. Let M be a module. We say that

(1) M is *transitive over torsion* if for all $x \in M$ and $y \in t(M)$ with $U(x) = U(y)$, there exists an automorphism of the module M which maps from x to y;

(2) M is *transitive mod torsion* if for all $x, y \in M$ with $U(x) = U(y)$, there exists an automorphism α with $\alpha x - y \in t(M)$.

Modules which are fully transitive over torsion or mod torsion are defined similarly. In (1) and (2), it should be written $U(x) \leq U(y)$ and "an endomorphism" instead of $U(x) = U(y)$ and "an automorphism", respectively.

We note that M is transitive mod torsion (resp., fully transitive mod torsion) if and only if $p^k(\alpha x) = p^k y$ for some automorphism (resp., endomorphism) α of the module M and some $k \geq 0$ in all cases, where $U(x) = U(y)$ (resp., $U(x) \leq U(y)$).

The proof of the following simple result is similar to the proof of Lemma 37.3.

Lemma 38.2. *If M is a fully transitive over torsion* (*resp., mod torsion*) *module, then every direct summand of the module M is fully transitive over torsion* (*resp., mod torsion*).

Proposition 38.3. *A module M is fully transitive if and only if M is fully transitive over torsion and fully transitive mod torsion.*

Proof. We assume that the module M is fully transitive over torsion and mod torsion. If $x, y \in M$ and $U(x) \leq U(y)$, then there exist an endomorphism α of the module M and the element $c \in t(M)$ such that $\alpha x = y + c$ (since M is fully transitive mod torsion). Consequently, $U(x) \leq U(\alpha x - y) = U(c)$. Using the property of full transitivity over torsion, we can find an endomorphism β with $\beta x = c$. Then $(\alpha - \beta)x = y$. The converse implication is trivial. □

The analogue of Proposition 38.3 for transitivity does not hold (see Exercise 2). We have only the following result.

Lemma 38.4. *A module M is transitive if and only if M is transitive mod torsion and the following condition holds: in every case, where elements $x \in M$ and*

$y \in M[p]$ satisfy the relation $h(x) = h(x+y)$, there exists an automorphism of the module M which maps x onto $x+y$.

Proof. We assume that M is transitive mod torsion and satisfies the indicated condition. Let z and y be two elements of M which have the same height sequence. Then $y = \alpha z + c$ for some automorphism α and the element c of finite order. We set $x = \alpha z$. It is sufficient to prove that x can be mapped onto $x+c$ by some automorphism. For $c = 0$, we take the identity mapping as such an automorphism. By applying the induction on the order of the element c, we obtain $\beta(px) = p(x+c)$, where β is some automorphism. We note that

$$U(\beta x) = U(x) = U(x+c) = U(\beta x + (x + c - \beta x))$$

and $p(x + c - \beta x) = 0$. By assumption, there exists an automorphism γ which maps βx onto $\beta x + (x+c-\beta x) = x+c$. Then $(\beta\gamma)x = x+c$, which is required. The converse assertion is obvious. □

We pass to the main result of the section.

Theorem 38.5 (Files [78]). *A completely decomposable module is fully transitive mod torsion if and only if the module is transitive mod torsion.*

Proof. First, we assume that a completely decomposable module M is not fully transitive mod torsion. This means that there exist elements $x, y \in M$ such that $U(x) \leq U(y)$ and for all m, it is impossible to map $p^m x$ onto $p^m y$ by any automorphism of the module M. Clearly, we can assume that we have the decomposition $M = M_1 \oplus M_2$, where M_1 has rank 1 and $y \in M_1$. It is obvious that the order of y is infinite. Let $x = x_1 + x_2$, where $x_1 \in M_1$ and $x_2 \in M_2$. We assume that the coordinates of sequences $U(x)$ and $U(x_1)$ coincide at an infinite number of positions, i.e., $h(p^k x) = h(p^k x_1)$ for an infinite set of integers k. It follows from $U(x) \leq U(y)$ that $h(p^k x_1) \leq h(p^k y)$ for an infinite set of integers k. It is easy to show that $up^m y = vp^n x_1$ for some invertible elements $u, v \in R$ and nonnegative integers m and n with $m \leq n$. The composition of the projection from M onto M_1 with the multiplication of M_1 by the element $vu^{-1}p^{n-m}$ maps $p^m x$ onto $p^m y$. This contradiction proves that we can replace x and y by $p^k x$ and $p^k y$ for sufficiently large integer k to assume that $h(p^i x_1) > h(p^i x)$ for all $i \geq 0$. However, then $U(x) = U(x - x_1)$. Therefore

$$U(x) \leq U(x+y) = \inf(U(x_1 + y), U(x - x_1)) \leq U(x - x_1) = U(x).$$

We have that x and $x+y$ have the same height sequences. There does not exist an automorphism α which maps $p^m x$ onto $p^m(x+y)$; otherwise, $(\alpha - 1)(p^m x) = p^m y$, which is impossible. Therefore M is not transitive mod torsion.

Now we assume that M is fully transitive mod torsion. We can assume that $M = M_1 \oplus \cdots \oplus M_n$, where all M_i have rank 1. We take elements x and y of infinite order with equal height sequences. We have

$$x = x_1 + \cdots + x_n, \quad y = y_1 + \cdots + y_n, \quad x_i, y_i \in M, \quad i = 1, \dots, n.$$

We need to obtain an automorphism of the module M which maps $p^m x$ onto $p^m y$ for some $m \geq 0$. It is clear that for any two automorphisms α, β and an integer $k \geq 0$, the elements x and y can be replaced by $\alpha(p^k x)$ and $\beta(p^k y)$, respectively. We repeat such a procedure until it will be obvious that x can be mapped onto y by some automorphism of the module M.

First, we assume that there exists a positive integer m such that $h(p^k x_1) > h(p^k x)$ for all $k \geq m$. Then $h(p^k x) = h(p^k(x - x_1))$ for all such integers k. We can enlarge m (if necessary) to assume that $\varphi(p^m(x - x_1)) = p^m x_1$ for some $\varphi \colon M_2 \oplus \cdots \oplus M_n \to M_1$. According to the decomposition

$$M = M_1 \oplus (M_2 \oplus \cdots \oplus M_n),$$

the matrix

$$\begin{pmatrix} 1 & 0 \\ -\varphi & 1 \end{pmatrix}$$

represents an automorphism α of the module M which maps $p^m x$ onto $p^m(x - x_1) \in M_2 \oplus \cdots \oplus M_n$. Consequently, we can replace x and y by $\alpha(p^m x)$ and $p^m y$, respectively; therefore, we can assume that for $x_1 \neq 0$, the coordinates in $U(x)$ and $U(x_1)$ coincide at an infinite number of positions. We repeat this procedure for each of the remaining components $x_2, \dots, x_n$ (it is important that this procedure does not damage the results of our previous work), and then we make the same actions with each of the components $y_1, \dots, y_n$ of the element y. As a result, we obtain the situation, where for every i, the coordinates in the sequences $U(x)$ and $U(x_i)$ (resp., $U(y)$ and $U(y_i)$) coincide at an infinite number of positions if $x_i \neq 0$ (resp., $y_i \neq 0$). We assert that we can preserve this situation and additionally ensure the property that $x_i = 0 \iff y_i = 0$, $1 \leq i \leq n$. For example, assume that $y_1 \neq 0$, but $x_1 = 0$. Then the order of y_1 is infinite. We denote by π_i the projection from M onto M_i. Since $U(x) = U(y) \leq U(y_1)$ and M is fully transitive mod torsion, we can assume that $y_1 = \varphi x$ for some $\varphi \in \operatorname{End}(M)$. We have the relation

$$y_1 = (\varphi\pi_1)x_2 + \cdots + (\varphi\pi_1)x_n.$$

Since the rank of M_1 is equal to 1, the coordinates of sequences $U(y_1)$ and $U(y)$ coincide at an infinite number of positions and

$$U(y) = U(x) \leq U((\varphi\pi_1)x_i), \quad 2 \leq i \leq n,$$

we have that $U(y_1)$ coincides with at least one sequence $U((\varphi\pi_1)x_j)$, $j \neq 1$, beginning with some position. Corresponding to the decomposition

$$M = M_1 \oplus M_j \oplus \left(\bigoplus_{i \neq 1,j} M_i \right)$$

the matrix

$$\begin{pmatrix} 1 & 0 & 0 \\ \varphi\pi_1 & 1 & 0 \\ 0 & 0 & 1 \end{pmatrix}$$

represents an automorphism β which maps x onto $((\varphi\pi_1)x_j, x_2, \dots, x_n)$. We note that the order of the element $x_1' = (\varphi\pi_1)x_j$ is infinite and that the coordinates of sequences $U(x_1')$ and $U(x)$ coincide at an infinite number of positions. We can replace x by βx; therefore, we can assume that $x_1 \neq 0$ provided $y_1 \neq 0$. We continue similar replacements with respect to all remaining nonzero components of the element y; then we continue similar replacements with respect to every nonzero component of the element x. Thus, we obtain the situation described above. Now it is quite easy to find an integer m and an automorphism γ with $\gamma(p^m x) = p^m y$. First, we note that if $x_i \neq 0$, then $h(p^k x_i) \leq h(p^k y_i)$ for an infinite set of integers k (since $U(x) = U(y) \leq U(y_i)$ and the coordinates in $U(x_i)$ and $U(x)$ coincide at an infinite number of positions). Similarly, $h(p^k y_i) \leq h(p^k x_i)$ for an infinite set of integers k. Since the rank M_i is equal to 1 and x_i, y_i are elements of infinite order, it follows from easy arguments that $p^{m_i} y_i = p^{m_i} u_i x_i$ for some integers $m_i \geq 0$ and invertible elements $u_i \in R$. For every j, let γ coincide on M_j with the multiplication by u_j if $x_j \neq 0$ and let γ coincide on M_j with the identity mapping if x_j (equivalently, y_j) is equal to zero. Let m be the maximal among the integers m_j with $x_j \neq 0$. Then $\gamma(p^m x) = p^m y$, and the proof is completed. □

The following variants of full transitivity are quite useful (especially in the topics related to full transitivity of direct sums). Let M and N be two modules. An ordered pair (M, N) is said to be *fully transitive* if for all $x \in M$ and $y \in N$ with $U(x) \leq U(y)$, there exists a homomorphism $\varphi \colon M \to N$ with $\varphi x = y$. The pair (M, N) is said to be *fully transitive over torsion* if the indicated homomorphism exists for all elements $x \in M$ and $y \in t(N)$ with $U(x) \leq U(y)$. In the previous section, it was proved in fact that every pair of cyclic primary modules is full transitive (see also Exercise 1).

Proposition 38.6. (1) *A pair (M, N) is fully transitive over torsion if and only if it is fully transitive over $N[p]$ (the meaning of this is explained in the proof).*

(2) *A direct sum $M = \bigoplus_{i \in I} M_i$ is fully transitive over torsion if and only if the pair (M_i, M_j) is fully transitive over torsion for all $i, j \in I$.*

Proof. (1) We assume that the pair (M, N) is fully transitive over $N[p]$, i.e., for all $x \in M$ and $y \in N[p]$, we have $\varphi x = y$ for some homomorphism $\varphi\colon M \to N$ provided $U(x) \le U(y)$. Now let $U(x) \le U(y)$, where $x \in M$ and $y \in t(N)$. Using the induction on the order of the element y, we can assume that $\varphi(px) = py$ for some $\varphi\colon M \to N$. Then $y' = y - \varphi x \in N[p]$ and $U(x) \le U(y')$. Consequently, $\psi x = y'$, where $\psi\colon M \to N$. Then $(\varphi + \psi)x = y$.

(2) Let $x \in M$, $y \in M[p]$, and let $U(x) \le U(y)$. It is sufficient to consider the case, where $M = M_1 \oplus \cdots \oplus M_n$. We have $x = x_1 + \cdots + x_n$ and $y = y_1 + \cdots + y_n$, where $x_i, y_i \in M_i$. We have $h(x) = h(x_k)$ for some k. Consequently, $U(x_k) \le U(y) \le U(y_i)$ $(1 \le i \le n)$, since $py = 0$. By assumption $\varphi_i(x_k) = y_i$ for some homomorphisms $\varphi_i : M_k \to M_i$. Then $(\pi_k(\varphi_1 + \cdots + \varphi_n))x = y$, where π_k is the projection from M onto M_k. By (1), the module M is fully transitive over torsion. The proof of the converse assertion is similar to the proof of Lemma 37.3; see also Lemma 38.2. □

Corollary 38.7. *If M is fully transitive over torsion, then the module $\bigoplus_s M$ is fully transitive over torsion for every cardinal number s.*

Corollary 38.8 (Files [78]). *A completely decomposable module is fully transitive over torsion if the torsion submodule of the module does not have elements of infinite height or it is a Warfield module.*

Proof. It follows from Proposition 38.6 that it is sufficient to prove that the pair (M, N) is fully transitive over $N[p]$ provided the rank of M is equal to 1 and either $t(M)$ and $t(N)$ do not have elements of infinite height or M is a Warfield module. We assume that $x \in M$, $y \in N[p]$, and $U(x) \le U(y)$. We consider the first case. First, we assume that the module M is split; let $M = Ra \oplus t(M)$. According to the decomposition, we have $x = x_1 + x_2$. Using the projection from M onto Ra or $t(M)$, we can assume that $x \in Ra$ or $x \in t(M)$. In the first case, it is easy to construct a homomorphism $\varphi\colon Ra \to Ry$ with $\varphi x = y$. If $x \in t(M)$, then the result follows from Proposition 37.5. For the nonsplit module M, the factor module $M/t(M)$ is a divisible module (it is isomorphic to K). We denote by n the height of the element y. We have $x = p^{n+1}a + c$, where $a \in M$ and $c \in t(M)$. We embed the element c in a finitely generated direct summand P of the module M. Then $U(c) \le U(y)$, since c and x have the same height. Consequently, $y = \varphi c$ for some $\varphi\colon P \to Rz$, where $p^n z = y$. The composition of the projection from M onto P with φ maps x onto y.

As for the second possibility, a more general result is true. Precisely, the pair (M, N) is fully transitive. By Corollary 32.8 (b), Rx is a nice submodule in M and M/Rx is a totally projective module. Consequently, mappings $Rx \to N$, which do not decrease heights, can be extended to homomorphisms from M into N (by the same corollary). □

The last corollary is not true for arbitrary completely decomposable modules. For a primary module P, transitivity (resp., full transitivity) over torsion of the module $R \oplus P$ is equivalent to the ordinary transitivity (resp., full transitivity) of it. For some modules P, this leads us to easy examples of completely decomposable modules, which are transitive, but are not fully transitive, and conversely; for example, see Exercise 1 from Section 37.

Exercise 1. Verify that (M, N) and (A, P) are fully transitive pairs, where M, N, and P are primary modules such that M and N do not have elements of infinite height and A is a torsion-free module.

Exercise 2 (Files [77, 78]). There exists a mixed nontransitive module of rank 1 which is transitive over torsion and transitive mod torsion.

Exercise 3. For a pure submodule M of the R-module $\widehat{R}$, the following conditions are equivalent:

(1) M is a transitive module;

(2) M is a fully transitive module;

(3) there exists a pure R-subalgebra T in $\widehat{R}$ such that the R-modules M and T are isomorphic to each other and T is a discrete valuation domain.

A module M is said to be *quasi-pure injective* if every homomorphism $A \to M$, where A is a pure submodule of the module M, can be extended to an endomorphism of the module M.

Exercise 4 (Dobrusin [54]). A reduced module M is quasi-pure injective if and only if either M is pure-injective or $M \cong A \oplus \bigoplus_n S$, where A is a pure fully invariant submodule of an adjusted pure-injective module, S is a pure R-subalgebra in $\widehat{R}$, and S is a discrete valuation domain.

Exercise 5. Is a quasi-pure injective module transitive or fully transitive?

39 Equivalence of transitivity and full transitivity

We return to original notions of transitivity and use our previous work for the proof of the results on interrelations between two transitivity properties for some large classes of mixed modules (Theorem 38.5 is also related to this topic). In particular, we give attention to transitivity and full transitivity of direct sums of copies of a single module.

Theorem 39.1 (Files [78]). *Let M be a completely decomposable module. If the torsion submodule $t(M)$ does not have elements of infinite height or M is a Warfield module, then the following conditions are equivalent*:

(1) *M is a fully transitive module*;

(2) *M is a transitive module*;

(3) *M is fully transitive mod torsion*;

(4) *M is transitive mod torsion.*

Proof. Naturally, we have (1) $\Longrightarrow$ (3) and (2) $\Longrightarrow$ (4). The equivalence of conditions (3) and (4) follows from Theorem 38.5. In addition, the implication (3) $\Longrightarrow$ (1) follows from Corollary 38.8 and Proposition 38.3. It remains to verify the implication (4) $\Longrightarrow$ (2). We use Lemma 38.4. Till the end of the proof, we assume that $x \in M$, $y \in M[p]$, and $h(x) = h(x+y)$. We need to obtain an automorphism of the module M which maps x onto $x+y$.

Assume that $t(M)$ does not contain elements of infinite height. An element y can be embedded in a finitely generated direct summand C of the module M; let $M = N \oplus C$. In this decomposition, we have $x = z + c$. If $h(z) \leq h(c)$, then $U(z) \leq U(y)$. Therefore $y = \varphi z$ for some $\varphi\colon N \to C$ by Corollary 38.8. The automorphism

$$\begin{pmatrix} 1 & \varphi \\ 0 & 1 \end{pmatrix}$$

maps x onto $x+y$. Consequently, we can assume that $h(z) > h(c) = h(x)$; therefore $U(c+y) = U(c)$. There exists an automorphism γ of the module C which maps c onto $c+y$ (Proposition 37.5). Then the automorphism $(1,\gamma)$ of the module M maps x onto $x+y$, which is required.

We assume that M is a Warfield module. Since M is a completely decomposable module, there exists a decomposition $M = A \oplus M'$ such that M' is a completely decomposable module, A is a module of rank 1, $x = x_1 + x_2$, $y = y_1 + y_2$, $h(x) = h(x_1)$, $x_1, y_1 \in A$, and $x_2, y_2 \in M'$. We note that $U(x_1) \leq U(y_2)$. It follows from the proof of Corollary 38.8 that the pair (A, M') is fully transitive. Consequently, we have $y_2 = \varphi(x_1)$ for some $\varphi\colon A \to M'$. We replace x by the image $x_1 + x_2 + y_2$ of x under the automorphism

$$\begin{pmatrix} 1 & \varphi \\ 0 & 1 \end{pmatrix}.$$

Thus, we reduce the original problem to the situation, where $y_2 = 0$. In this situation, if $h(x_1) = h(x_1 + y_1)$, then $x_1 + y_1 = \gamma(x_1)$ for some $\gamma \in \operatorname{Aut}(A)$ (A is transitive by Exercise 5 from Section 37). Then the automorphism $(\gamma, 1)$ maps x onto $x + y_1$. Consequently, we can assume that $h(x_1 + y_1) > h(x_1) =$

$h(x)$. Then also $h(x) = h(x_2)$, since $h(x + y_1) = h(x)$. Now we can choose a summand A' of rank 1 of the module M' such that with respect to the new decomposition $M = A' \oplus M''$, the element y_1 is contained in M'' and the component of the element x in A' has the height sequence $\leq U(y_1)$. Furthermore, we can repeat the argument from the beginning of the paragraph to obtain an automorphism of the module $A' \oplus M''$ which maps x onto $x + y_1$. □

We pass to direct sums of fully transitive modules. We already considered them in Proposition 38.6 and Corollary 38.7.

Proposition 39.2 (Files [78]). *Let M be a completely decomposable module. If M is fully transitive mod torsion, then the module $\bigoplus_s M$ is fully transitive mod torsion for every cardinal number s.*

Proof. It is sufficient to consider the case, where the rank of the module M is finite. Considering Lemma 38.2, we have that it is sufficient to prove the full transitivity of the module $M_1 \oplus M_2$, where M_1 and M_2 are two copies of a completely decomposable module of finite rank which is fully transitive mod torsion. We denote by $\tau : M_2 \to M_1$ the identity mapping. We assume that elements $x, z \in M_1$ and $y \in M_2$ satisfy $U(x+y) \leq U(z)$. We choose some decomposition $M_1 = M_2 = A_1 \oplus \cdots \oplus A_n$, where the ranks of all summands are equal to 1. According to the decomposition, we have $x = x_1 + \cdots + x_n$ and $y = y_1 + \cdots + y_n$. We replace the elements x, y, and z by the elements $p^m x$, $p^m y$ and $p^m z$ for sufficiently large integer m. Thus, we obtain a partition $\{1, \ldots, n\} = I \cup J$ such that $h(p^k x_i) > h(p^k y_i)$ for all $k \geq 0$ provided $i \in I$, and $U(x_i) \leq U(y_i)$ provided $i \in J$. Let π be the projection $M_2 \to \bigoplus_{i \in I} A_i \subseteq M_2$. It follows from properties of the given partition that

$$U(x + (\pi\tau)y) = \inf(U(x), U(y)) = U(x + y).$$

Since M_1 is fully transitive mod torsion, there exist $\varphi \in \mathrm{End}(M_1)$ and $m \geq 0$ such that $\varphi(p^m(x + (\pi\tau)y)) = p^m z$. The endomorphism

$$\begin{pmatrix} \varphi & 0 \\ \pi\tau\varphi & 0 \end{pmatrix}$$

of the module $M_1 \oplus M_2$ maps $p^m(x + y)$ onto $p^m z$. Now we pass to the general case, where $z = z_1 + z_2 \in M$, $z_i \in M_i$, $i = 1, 2$, and $U(x + y) \leq U(z)$. Since $U(z) \leq U(z_1), U(z_2)$, it follows from the proved properties that there exist $\alpha_1, \alpha_2 \in \mathrm{End}(M)$ and $m \geq 0$ with $\alpha_i(p^m(x + y)) = p^m z_i$, $i = 1, 2$. Then

$$(\alpha_1 + \alpha_2)(p^m(x + y)) = p^m z.$$

□

The following result follows from assertions 38.3, 38.7, and 39.2.

Corollary 39.3. *If M is a completely decomposable fully transitive module, then the module $\bigoplus_s M$ is fully transitive for every cardinal number s.*

There exist interesting interrelations between properties of full transitivity of a module M and transitivity of the sum $\bigoplus_s M$, $s \geq 2$. In particular, the module $M \oplus M$, called the *square* of the module M, can be considered as $\bigoplus_s M$. Other interrelations between these two objects are presented in Exercise 7 of Section 14 and Problem 4 (b).

Lemma 39.4. *Let M be an arbitrary module. If $\bigoplus_s M$ ($s \geq 2$) is a transitive module, then M is a fully transitive module.*

Proof. Let $x, y \in M$ with $U(x) \leq U(y)$. The elements $(x, 0, 0, \dots)$ and $(y, x, 0, \dots)$ of the module $\bigoplus_s M$ have the same height sequences. Since this module is transitive, there exists an automorphism α which maps $(x, 0, 0, \dots)$ onto $(y, x, 0, \dots)$. The composition of the embedding M in the first summand M of the module $\bigoplus_s M$, the automorphism α, and the projection onto M gives the required endomorphism of the module M which maps x onto y. □

We have that transitivity of the square of a module implies full transitivity of this module. For a completely decomposable module, the converse assertion holds.

Theorem 39.5 (Files [78]). *A completely decomposable module is fully transitive if and only if the square of the module is transitive.*

Proof. By Lemma 39.4, we need to only verify transitivity of the module $M_1 \oplus M_2$ provided M_1 and M_2 are two copies of some fully transitive completely decomposable module. Let $\tau : M_1 \to M_2$ be the identity mapping. We note that $M_1 \oplus M_2$ is fully transitive by Corollary 39.3; therefore $M_1 \oplus M_2$ is transitive mod torsion by Theorem 38.5. We verify that the direct sum satisfies the condition of Lemma 38.4. We assume that $x, y \in M_1 \oplus M_2$, $h(x) = h(x + y)$ and $py = 0$. According to the decomposition, we have $x = x_1 + x_2$ and $y = y_1 + y_2$. We can assume that $h(x) = h(x_1)$. Consequently, $y_2 = \varphi(x_1)$ for some $\varphi\colon M_1 \to M_2$, since the direct sum is fully transitive and $U(x_1) \leq U(y_2)$. We replace x by the image $x_1 + x_2 + y_2$ of x under the automorphism

$$\begin{pmatrix} 1 & \varphi \\ 0 & 1 \end{pmatrix}.$$

This allows us to assume that $y = y_1 \in M_1$ in the original problem. If $h(x_2) \leq h(y_1)$, then $y_1 = \psi(x_2)$, where $\psi\colon M_2 \to M_1$ and the automorphism

$$\begin{pmatrix} 1 & 0 \\ \psi & 1 \end{pmatrix}$$

maps from x to $x + y_1$. Otherwise, we have

$$h(x_2) > h(x) = h(x_1).$$

We can verify that

$$U(x_1 + y_1) \leq U(\tau y_1),$$

since $h(x_1 + y_1) = h(x_1) = h(x) \leq h(y_1)$. In addition,

$$U(\tau x_1 + x_2) \leq U(y_1),$$

since $h(\tau x_1 + x_2) = h(x_1) = h(x) \leq h(y_1)$. By assumption, there exist homomorphisms $\varphi\colon M_1 \to M_2$ and $\psi\colon M_2 \to M_1$ such that $\varphi(x_1 + y_1) = \tau(y_1)$ and $\psi(\tau x_1 + x_2) = y_1$. It can be verified that the composition

$$\begin{pmatrix} 1 & \tau \\ 0 & 1 \end{pmatrix} \begin{pmatrix} 1 & \varphi \\ \psi & 1 + \psi\varphi \end{pmatrix} \begin{pmatrix} 1 & -\tau \\ 0 & 1 \end{pmatrix}$$

of automorphisms of the module $M_1 \oplus M_2$ maps x onto $x + y_1$. This proves the condition from Lemma 38.4. □

Files [78] notes that Theorem 39.1 holds for every Warfield module. In this direction, Hennecke and Strüngmann [139] have proved that if 2 is an invertible element in R and M is a transitive module, then M is fully transitive provided M is a completely decomposable module or the rank of the maximal divisible submodule of the module $M/t(M)$ does not exceed 1. The first theorem, which is similar to Theorem 39.5, was first proved for primary modules, by Files and Goldsmith [80]. In [139], quite general results in this topic are contained. Let M be either a completely decomposable module, or a Warfield module, or a module such that the maximal divisible submodule of the module $M/t(M)$ has the rank at most 1. Then the following conditions are equivalent:

(1) M is a fully transitive module;

(2) $\bigoplus_s M$ is a fully transitive module for some cardinal number $s > 1$;

(3) $\bigoplus_s M$ is a fully transitive module for all cardinal numbers $s > 1$;

(4) $\bigoplus_s M$ is a transitive module for some $s > 1$;

(5) $\bigoplus_s M$ is a transitive module for all $s > 1$.

This theorem can be applied to primary modules and torsion-free modules. It is not known whether for every module M, the module M is fully transitive if and only if $M \oplus M$ is transitive. It is not also known whether the square $M \oplus M$ is fully transitive if and only if $M \oplus M$ is transitive.

A more general problem is the search for transitivity conditions for direct sums and products of modules. In the corresponding studies, the notion of a fully transitive pair is intensively used. In a more general form of a "fully transitive system of modules", the notion appeared in papers of Grinshpon and Misyakov (see [129, 132]); Files and Goldsmith [80] define this notion later and independently. These topics and other aspects of the theory of fully transitive Abelian groups and modules are presented in [127, 128, 129, 130, 132, 237]. A review (with proofs) of many results of these papers is presented in [131].

Exercise 1 (Files [78]). Prove that Theorem 39.1 is true for all Warfield modules.

Exercise 2. For a completely decomposable module M, the following conditions are equivalent:

(1) M is transitive mod torsion;

(2) $\bigoplus_s M$ is transitive mod torsion for every (equivalently, for some) cardinal number s.

Exercise 3. (a) Let $M = A \oplus B$ and let $\varphi\colon A \to B$ and $\psi\colon B \to A$ be two homomorphisms. Verify that the matrix

$$\begin{pmatrix} 1 & \varphi \\ \psi & 1+\psi\varphi \end{pmatrix}$$

is an invertible matrix.

(b) If $M = A \oplus A$ and $\alpha, \beta \in \mathrm{End}(A)$, then the matrix

$$\begin{pmatrix} \alpha & 1+\alpha\beta \\ 1 & \beta \end{pmatrix}$$

is invertible.

Exercise 4 (see [80, 129, 132, 237], Proposition 38.6 (2)). For a set of primary modules $\{A_i\}_{i\in I}$, the following conditions are equivalent:

(1) $\bigoplus_{i\in I} A_i$ is a fully transitive module;

(2) $\prod_{i\in I} A_i$ is a fully transitive module;

(3) the pair (A_i, A_j) is fully transitive for all $i, j \in I$.

Exercise 5 ([78, 129, 131]). Let $\{A_i\}_{i\in I}$ ($|I| \geq 2$) be a set of torsion-free modules. The following conditions are equivalent:

(1) $\bigoplus_{i\in I} A_i$ is a transitive (resp., fully transitive) module;

(2) $\prod_{i\in I} A_i$ is a transitive (resp., fully transitive) module;

(3) the pair (A_i, A_j) is a fully transitive pair for all $i, j \in I$.

Exercise 6. Let M be a transitive torsion-free module. Every element of the product $\prod_s M$ can be embedded in a direct summand such that the summand is isomorphic to M and the complement summand is a transitive module.

Remarks. A vector space over a division ring can be called "n-transitive" for every n in the well known sense. The notions of a transitive module and a fully transitive module were introduced by Kaplansky. The first study in Kaplansky's book [166] and subsequent papers related to transitivity dealt with Abelian p-groups (the corresponding bibliography is presented in Section 37). Furthermore, transitive and fully transitive torsion-free Abelian groups are systematically studies in papers of professors and postgraduates of the algebra department of Tomsk State University (Russia), see [39, 40, 41, 42, 54, 128, 129, 131, 173, 174, 179, 180]. The main results of papers [39, 40, 41, 42] are also represented in [43]. Chapter 7 of Krylov–Mikhalev–Tuganbaev's book [183] contains a quite complete presentation of the theory of transitive and fully transitive torsion-free Abelian groups. Mixed fully transitive groups are studied in the papers cited before exercises in Section 39.

In the book of Göbel–Trlifaj [109], uniquely transitive modules are defined. Uniquely transitive modules are close to transitive modules and are used in the study of E(S)-algebras which will be considered below.

Sometimes, the study of transitivity properties allows us to better understand the structure of concrete modules. In any case, the study is related to some common topic: interrelations between endomorphisms and automorphisms of algebraic structures. The study of characteristic and fully invariant submodules is related to the same topic. Kaplansky [166] presented the description of fully invariant submodules of primary fully transitive modules. In papers [79] and [201], fully invariant submodules of Warfield modules and balanced projective modules are characterized. Fully invariant subgroups and their lattices are studied in [127, 128, 129, 130] (see also [131]). The problem of additive generating of the endomorphism ring by the automorphism group is also related to the same topic; the problem was discussed in Remarks to Chapter 3.

Problem 23. For which classes of modules, conditions (1)–(3) of Exercises 4 and 5 in Section 39 are equivalent?

Problem 24. For which classes of modules, the theorem of Hennecke and Strüngmann holds (the theorem is formulated after the proof of Theorem 39.5)? In particular, determine interrelations between the following properties of an arbitrary module M:

(1) M is a fully transitive module;

(2) $M \oplus M$ is a fully transitive module;

(3) $M \oplus M$ is a transitive module.

Problem 25. For which modules, the transitivity property is equivalent to the full transitivity property?

Problem 26. Describe transitive modules of rank 1 and Warfield modules.

Problem 27. Study characteristic submodules and the lattices of characteristic submodules in various classes of modules.

A module M over a ring S is called a *multiplication* module if every submodule of M is equal to IM for some ideal I of the ring S.

Problem 28. Study modules which are multiplication modules over their endomorphism rings or modules which are multiplication modules over the centers of their endomorphism rings.

The module property formulated in Exercise 7 of Section 4 and Exercise 4 of Section 6 can be generalized as follows. Let S and R be arbitrary rings and let $e\colon S \to R$ be a central ring homomorphism (i.e., the image of e is contained in the center of the ring R). The left R-module A is considered as a left S-module and a right S-module, where $sa = e(s)a$ and $as = sa$ for $s \in S$ and $a \in A$. An R-module A is called a *module with unique module multiplication* (with respect to e) or *$UM(e)$-module* if the additive group A does not have another module multiplication $\circ$ such that $sa = s \circ a$ for all $s \in S$ and $a \in A$ (cf. the definition of a UA-module before Problem 10). There always exists a homomorphism $e\colon \mathbb{Z} \to R$ such that $e(k) = k \cdot 1$, $k \in \mathbb{Z}$. An $UM(e)$-module has a unique R-module structure.

Problem 29. For a given central homomorphism $e\colon S \to R$, describe all $UM(e)$-modules.

Problem 30. Let $A_1, \ldots, A_n$ be $UM(e)$-modules. Find conditions implying that $A_1 \oplus \cdots \oplus A_n$ is a $UM(e)$-module. In particular, find such conditions if $A_1 = \ldots = A_n$.

The necessary condition is the absence of nonzero derivations

$$\delta\colon R \to \mathrm{Hom}_S(A_i, A_j) \qquad (1 \leq i, j \leq n)$$

such that $\delta(e(S)) = 0$. An additive homomorphism δ is called a *derivation* if $\delta(xy) = x\delta(y) + \delta(x)y$ for all $x, y \in R$ ($\mathrm{Hom}_S(A_i, A_j)$ is an R-R-bimodule). The problem can be formulated in terms of certain homomorphisms from R into the matrix ring with entries in $\mathrm{Hom}_S(A_i, A_j)$.

Again, let $e\colon S \to R$ be a central homomorphism. An R-module A is called a $T(e)$-*module* if the canonical mapping $R \otimes_S A \to R \otimes_R A$ with $x \otimes_S a \to x \otimes_R a$ is an isomorphism, and an R-module A is called an $E(e)$-*module* if $\mathrm{Hom}_R(R, A) = \mathrm{Hom}_S(R, A)$. Applying these definitions to the R-module R, we obtain the notions of a T-*ring* and an E-*ring* (see Section 4 and Exercises 4,5,6,8 in Section 4). For $S = \mathbb{Z}$, E-modules and E-rings are studied in [183, Section 6]. There are various interrelations between the defined three classes of modules. For example, $T(e)$-modules and $E(e)$-modules are $UM(e)$-modules. Every R-module is a $UM(e)$-module if and only if R is a $T(e)$-ring.

Problem 31. Study properties of $T(e)$-modules, $E(e)$-modules, $UM(e)$-modules, and the interrelations between them.

We note that Chapter 13 of the book of Göbel–Trlifaj [109] contains interesting information about E(S)-algebras. An E(S)-algebra R is an $E(e)$-ring R, where S is a commutative ring, e is an embedding, and R is naturally considered as an S-algebra.

In connection to Exercise 1 in § 27, we note that a ring S is said to be *left steady* if the class of all small left S-modules coincides with the class of all finitely generated S-modules (e.g., see [64], [328]). (Small modules are also called *dually slender* modules.) There are some more related notions which are some finiteness conditions. A module A is said to be *self-small* if the image of any homomorphism $A \to \oplus_{i \in I} A_i$, where every module A_i is isomorphic to A and I is an arbitrary subscript set, is contained in a finite sum of some summands A_i (see [183, §31], [71]). One more condition is that the endomorphism ring of the module A is discrete in the finite topology (this topology is considered in § 16). If the finite topology is discrete, then the module is self-small. For countably generated modules, the converse is true (see [183, Corollaries 31.4, 31.5]).

It is natural to present the following definition (e.g., see [159]). Let $\mathcal{X}$ be some class of S-modules. An S-module A is said to be *small with respect to the class* $\mathcal{X}$ or $\mathcal{X}$-*small* if for every subscript set I and any modules $B_i \in \mathcal{X}$, the image of every homomorphism $A \to \oplus_{i \in I} B_i$ is contained in a finite sum of some B_i. Therefore, a module A is self-small if and only if A is small with respect to the class of all isomorphic copies of A; a module A is small if and only if A is small with respect to the class of all modules (or the class of all injective modules).

In the study of $\mathcal{X}$-small modules, we can use the condition that the homomorphism group is discrete in the finite topology (this topology is considered in § 34,

[183, §26], [159]).

It seems to be natural to consider interrelations between the following classes of modules: the class of finitely generated modules, the class of small (or $\mathcal{X}$-small) modules, the class of self-small modules, the class of modules with discrete endomorphism rings (or with discrete homomorphism groups).

Problem 32. (a) Characterize self-small modules over complete discrete valuation domains.

(b) Over discrete valuation domains, study modules which are small with respect to various classes of modules (e.g., study modules which are small with respect to the class of reduced modules).

(c) For a discrete valuation domain S, describe S-modules which are small or self-small modules over their endomorphism rings.

In the remaining problems, we consider modules over arbitrary rings.

Problem 33. Study interrelations between the following classes of modules: the class of finitely generated modules, the class of small (resp., $\mathcal{X}$-small or self-small) modules, the class of modules with discrete endomorphism rings (or homomorphism groups). Find conditions under which all these classes or some these classes coincide? (For example, this is true for projective modules.)

Problem 34. (a) For which rings S, all the above classes coincide?

(b) For which rings S, some pairs of the classes in Problem 33 coincide? For example, describe rings S such that finitely generated S-modules coincide with small S-modules or small (resp., $\mathcal{X}$-small) S-modules coincide with S-modules with discrete endomorphism rings (resp., homomorphism groups)?

References

[1] F. W. Anderson and K. R. Fuller, *Rings and Categories of Modules,* Springer-Verlag, New York (1974).

[2] V. I. Arnautov, S. T. Glavatsky, and A. V. Mikhalev, *Introduction to the Theory of Topological Rings and Modules,* Marcel Dekker, New York (1995).

[3] D. Arnold, "A duality for torsion-free modules of finite rank over a discrete valuation ring," *Proc. London Math. Soc.,* **24**, 204–216 (1972).

[4] D. Arnold, "A duality for quotient divisible Abelian groups of finite rank," *Pacific J. Math.,* **42**, 11–15 (1972).

[5] D. Arnold, "Exterior powers and torsion free modules over discrete valuation rings," *Trans. Amer. Math. Soc.*, **170**, 471–481 (1972).

[6] D. Arnold, "Finite rank torsion-free Abelian groups and rings," *Lect. Notes Math.,* **931**, 1–191 (1982).

[7] D. Arnold, "A finite global Azumaya Theorem in additive categories," *Proc. Amer. Math. Soc.,* **91**, No. 1, 25–30 (1984).

[8] D. Arnold, *Abelian Groups and Representations of Finite Partially Ordered Sets*, Springer-Verlag, New York (2000).

[9] D. Arnold, "Direct sums of local torsion-free Abelian groups," *Proc. Amer. Math. Soc.,* **130**, No. 6, 1611–1617 (2001).

[10] D. Arnold and M. Dugas, "Indecomposable modules over Nagata valuation domains," *Proc. Amer. Math. Soc.,* **122**, 689–696 (1994).

[11] D. Arnold and M. Dugas, "Representation type of posets and finite rank Butler groups," *Colloq. Math.,* **74**, 299–320 (1997).

[12] D. Arnold and M. Dugas, "Co-purely indecomposable modules over discrete valuation rings," *J. Pure and Appl. Algebra,* **161**, 1–12 (2001).

[13] D. Arnold, M. Dugas, and K. M. Rangaswamy, "Finite rank Butler groups and torsion-free modules over a discrete valuation ring," *Proc. Amer. Math. Soc.,* **129**, 325–335 (2001).

[14] D. Arnold, M. Dugas, and K. M. Rangaswamy, "Torsion-free modules of finite rank over a discrete valuation ring," *J. Algebra*, **272**, No. 2, 456–469 (2004).

[15] D. Arnold, R. Hunter, and F. Richman, "Global Azumaya theorems in additive categories," *J. Pure Appl. Algebra,* **16**, 223–242 (1980).

[16] D. Arnold and K. M. Rangaswamy, "Mixed modules of finite torsion-free rank over a discrete valuation domain," *Forum Math.*, **18**, No. 1, 85–98 (2006).

[17] D. Arnold, K. M. Rangaswamy, and F. Richman, "Pi-balanced torsion-free modules over a discrete valuation domain," *J. Algebra*, **295**, No. 1, 269–288 (2006).

[18] D. Arnold and F. Richman, "Subgroups of finite direct sums of valuated cyclic groups," *J. Algebra*, **114**, No. 1, 1–15 (1988).

[19] J. Arnold, "Power series rings over discrete valuation rings," *Pacific J. Math.*, **93**, No. 1, 31–33 (1981).

[20] E. Artin, *Geometric Algebra*, Interscience Publishers, New York (1957).

[21] P. Astuti and H. K. Wimmer, "Stacked submodules of torsion modules over discrete valuation domains," *Bull. Austral. Math. Soc.*, **68**, No. 3, 439–447 (2003).

[22] M. Atiyah, "On the Krull–Schmidt theorem with applications to sheaves," *Bull. Soc. Math. Fr.,* **84**, 307–317 (1956).

[23] M. Auslander, I. Reiten, and S. O. Smalø, *Representation Theory of Artin Algebras*, Cambridge University Press, Cambridge (1995).

[24] G. Azumaya, "Corrections and supplementaries to my paper concerning Krull–Remak–Schmidt's theorem," *Nagoya Math. J.*, **1**, 117–124 (1950).

[25] R. Baer, "Abelian groups that are direct summands of every containing Abelian group," *Bull. Amer. Math. Soc.,* **46**, 800–806 (1940).

[26] R. Baer, "A unified theory of projective spaces and finite Abelian groups," *Trans. Amer. Math. Soc.,* **52**, 283–343 (1942).

[27] R. Baer, "Automorphism rings of primary Abelian operator groups," *Ann. Math.,* **44**, 192–227 (1943).

[28] R. Baer, *Linear Algebra and Projective Geometry,* Academic Press, New York (1952).

[29] C. Mo Bang, "A classification of modules over complete discrete valuation rings," *Bull. Amer. Math. Soc.,* **76**, 381–383 (1970).

[30] C. Mo Bang, "Countable generated modules over complete discrete valuation rings," *J. Algebra,* **14**, 552–560 (1970).

[31] C. Mo Bang, "Direct sums of countable generated modules over complete discrete valuation rings," *Proc. Amer. Math. Soc.,* **28**, No. 2, 381–388 (1971).

[32] H. Bass, *Algebraic K-Theory*, Benjamin, New York (1968).

[33] R. A. Beaumont and R. S. Pierce, "Torsion-free rings," *Ill. J. Math.,* **5**, 61–98 (1961).

[34] R. A. Beaumont and R. S. Pierce, "Some invariant of p-groups," *Michigan Math. J.*, **11**, 138–149 (1964).

[35] N. Bourbaki, *Algébre Commutative*, Hermann, Paris (1961, 1964, 1965).

[36] S. Breaz, "On a class of mixed groups with semi-local Walk-endomorphism ring," *Comm. Algebra,* **30**, No. 9, 4473–4485 (2002).

[37] D. Carroll and B. Goldsmith, "On transitive and fully transitive Abelian p-groups," *Proc. Roy. Irish Acad.,* **96A**, No. 1, 33–41 (1996).

[38] H. Cartan and S. Eilenberg, *Homological Algebra,* Princeton University Press, Princeton (1956).

[39] A. R. Chekhlov, "On decomposable fully transitive torsion-free Abelian groups," *Sib. Mat. Zh.,* **42**, No. 3, 714-719 (2001).

[40] A. R. Chekhlov, "On one class of endotransitive groups," *Mat. Zametki,* **69**, No. 6, 944-949 (2001).

[41] A. R. Chekhlov, "Fully transitive torsion-free Abelian groups of finite p-rank," *Algebra Logika,* **40**, No. 6, 698-715 (2001).

[42] A. R. Chekhlov, "Abelian groups with many endomorphisms," Doctoral dissertation [in Russian], Tomsk (2003).

[43] A. R. Chekhlov and P. A. Krylov, "On 17 and 43 problems of L. Fuchs," *J. Math. Sci.*, **143**, No. 5, 3517-3602 (2007).

[44] A. L. S. Corner, "Every countable reduced torsion-free ring is an endomorphism ring," *Proc. London Math. Soc.,* **13**, No. 3, 687–710 (1963).

[45] A. L. S. Corner, "Endomorphism rings of torsion-free Abelian groups," In: *Proc. Internat. Conf. Theory of Groups,* Canberra (1965), 59–60.

[46] A. L. S. Corner, "On endomorphism rings of primary Abelian groups," *Quart. J. Math. Oxford,* **20**, No. 79, 277–296 (1969).

[47] A. L. S. Corner, "The independence of Kaplansky's notions of transitivity and full transitivity," *Quart. J. Math. Oxford,* **27**, 15–20 (1976).

[48] A. L. S. Corner, "Fully rigid systems of modules," *Rend. Sem. Mat. Univ. Padova,* **82**, 55–66 (1989).

[49] A. L. S. Corner and R. Göbel, "Prescribing endomorphism algebras, a unified treatment," *Proc. London Math. Soc.,* **50**, No. 3, 447–479 (1985).

[50] A. L. S. Corner and B. Goldsmith, "Isomorphic automorphism groups of torsion-free p-adic modules," In: *Abelian Groups, Theory of Modules, and Topology (Padua, 1997),* Dekker, New York (1998), 125–130.

[51] P. Crawley and A. W. Hales, "The structure of torsion Abelian groups given by presentations," *Bull. Amer. Math. Soc.,* **74**, 954–956 (1968).

[52] P. Crawley and A. W. Hales, "The structure of Abelian p-groups given by certain presentations," *J. Algebra,* **12**, 10–23 (1969).

[53] P. Crawley and B. Jónsson, "Refinements for infinite direct decompositions of algebraic systems," *Pac. J. Math.,* **14**, 797–855 (1964).

[54] Yu. B. Dobrusin, "Abelian groups close to algebraically compact," Candidate dissertation [in Russian], Tomsk (1982).

[55] M. Dugas, "On the existence of large mixed modules," *Lect. Notes Math.,* **1006**, 412–424 (1983).

[56] M. Dugas, "On the Jacobson radical of some endomorphism rings," *Proc. Amer. Math. Soc.,* **102**, No. 4, 823–826 (1988).

[57] M. Dugas and R. Göbel, "Every cotorsion-free ring is an endomorphism ring," *Proc. London Math. Soc.,* **45**, 319–336 (1982).

[58] M. Dugas and R. Göbel, "Every cotorsion-free algebra is an endomorphism algebra," *Math. Z.,* **181**, 451–470 (1982).

[59] M. Dugas and R. Göbel, "On endomorphism rings of primary Abelian groups," *Math. Ann.,* **261**, 359–385 (1982).

[60] M. Dugas and R. Göbel, "Endomorphism algebras of torsion modules, II," *Lect. Notes Math.,* **1006**, 400–411 (1983).

[61] M. Dugas and R. Göbel, "Torsion-free Abelian groups with prescribed finitely topologized endomorphism rings," *Proc. Amer. Math. Soc.,* **90**, No. 4, 519–527 (1984).

[62] M. Dugas, R. Göbel, and B. Goldsmith, "Representation of algebras over a complete discrete valuation ring," *Quart. J. Math.,* (2), **35**, 131–146 (1984).

[63] P. C. Eklof, *Set-Theoretic Methods in Homological Algebra and Abelian Groups,* Presses Univ. Montréal, Montréal (1980).

[64] P. C. Eklof, K. R. Goodearl, and J. Trlifaj, "Dually slender modules and steady rings," *Forum. Math.*, **9**, No. 1, 61-74 (1997).

[65] P. C. Eklof and A. H. Mekler, *Almost Free Modules. Set-Theoretic Methods,* North-Holland Publishing Co., Amsterdam (1990).

[66] A. P. Eraskina, "λ-purity of modules over a discrete valuation ring," *Sib. Mat. Zh.*, **14**, 208–212 (1973).

[67] A. Facchini, *Module Theory: Endomorphism Rings and Direct Sum Decompositions in Some Classes of Modules,* Birkhäuser, Basel–Boston–Berlin (1998).

[68] A. Facchini and P. Zanardo, "Discrete valuation domains and rank of their maximal extensions," *Rend. Sem. Mat. Univ. Padova,* **75**, 143–156 (1986).

[69] C. Faith, *Algebra: Rings, Modules, and Categories,* I, Springer, Berlin (1973).

[70] C. Faith, *Algebra II, Ring Theory,* Springer, Berlin (1976).

[71] T. Faticoni, "Categories of modules over endomorphism rings," *Mem. Amer. Math. Soc.*, **103**, No. 492, 1–140 (1993).

[72] Th. G. Faticoni, "Direct sums and refinement," *Comm. Algebra,* **27**, No. 1, 451–464 (1999).

[73] S. T. Files, "Mixed modules over incomplete discrete valuation rings," *Comm. Algebra,* **21**, No. 11, 4103–4113 (1993).

[74] S. T. Files, "Endomorphisms of local Warfield groups," *Contemp. Math.,* **171**, 99–107 (1994).

[75] S. T. Files, "Outer automorphisms of endomorphism rings of Warfield groups," *Arch. Math,* **65**, 15–22 (1995).

[76] S. T. Files, "Endomorphism algebras of modules with distinguished torsion-free elements," *J. Algebra,* **178**, 264–276 (1995).

[77] S. T. Files "On transitive mixed Abelian groups," *Lect. Notes Pure Appl. Math.,* **132**, 243–251 (1996).

[78] S. T. Files, "Transitivity and full transitivity for nontorsion modules," *J. Algebra,* **197**, 468–478 (1997).

[79] S. T. Files, "The fully invariant subgroups of local Warfield groups," *Proc. Amer. Math. Soc.,* **125**, No. 12, 3515–3518 (1997).

[80] S. Files and B. Goldsmith, "Transitive and fully transitive groups," *Proc. Amer. Math. Soc.,* **126**, No. 6, 1605–1610 (1998).

[81] A. A. Fomin, "Duality in some classes of Abelian torsion-free groups of finite rank," *Sib. Mat. Zh.,* **27**, No. 4, 117–127 (1986).

[82] A. A. Fomin, "Invariants and duality in some classes of Abelian torsion-free groups of finite rank," *Algebra Logika,* **26**, No. 1, 63–83 (1987).

[83] A. A. Fomin, "The category of quasi-homomorphisms of torsion-free Abelian groups of finite rank,"*Contemp. Math.,* Part 1, **131**, 91–111 (1992).

[84] A. A. Fomin, "Finitely presented modules over the ring of universal numbers," *Contemp. Math.,* **171**, 109–120 (1994).

[85] A. Fomin and W. Wickless, "Categories of mixed and torsion-free Abelian groups," In: *Abelian Groups and Modules,* Kluwer, Boston (1995), 185–192.

[86] A. Fomin and W. Wickless, "Quotient divisible Abelian groups," *Proc. Amer. Math. Soc.,* **126**, No. 1, 45–52 (1998).

[87] B. Franzen and R. Gobel, "Prescribing endomorphism algebras, the cotorsion-free case," *Rend. Sem. Mat. Univ. Padova*, **80**, 215–241 (1989).

[88] B. Franzen and B. Goldsmith, "On endomorphism algebras of mixed modules," *J. London Math. Soc.,* **31**, No. 3, 468–472 (1985).

[89] W. N. Franzsen and P. Schultz, "The endomorphism ring of locally free module," *J. Austral. Math. Soc.* (Series A), **35**, 308–326 (1983).

[90] L. Fuchs, "On the structure of Abelian p-groups," *Acta Math. Acad. Sci. Hungar.,* **4**, 267–288 (1953).

[91] L. Fuchs, "Notes on Abelian groups." I, *Ann. Univ. Sci. Budapest,* **2**, 5–23 (1959); II, *Acta Math. Acad. Sci. Hungar.,* **11**, 117–125 (1960).

[92] L. Fuchs, *Infinite Abelian Groups. I,* Academic Press, New York–London (1970).

[93] L. Fuchs, *Infinite Abelian Groups. II,* Academic Press, New York–London (1973).

[94] L. Fuchs, *Abelian p-Groups and Mixed Groups,* University of Montreal Press, Montreal (1980).

[95] L. Fuchs and L. Salce, *Modules over valuation domains,* Lect. Notes Pure Appl. Math., **97**, Marcel Dekker, New York (1985).

[96] L. Fuchs and L. Salce, *Modules over Non-noetherian Domains. Mathematical Surveys and Monographs,* Amer. Math. Soc., **84**, Providence, RI (2001).

[97] P. Gabriel, "Des catégories abéliennes," *Bull. Soc. Math. Fr.,* **90**, 323–448 (1962).

[98] R. Göbel, "Endomorphism rings of Abelian groups," *Lect. Notes Math.,* **1006**, 340–353 (1983).

[99] R. Göbel, "Modules with distinguished submodules and their endomorphism algebras," In: *Abelian Groups (Curacao, 1991),* Dekker, New York (1993), 55–64.

[100] R. Göbel and B. Goldsmith, "Essentially indecomposable modules which are almost free," *Quart. J. Math. Oxford* (2), **39**, 213–222 (1988).

[101] R. Göbel and B. Goldsmith, "Mixed modules in L," *Rocky Mountain J. Math.,* **19**, 1043–1058 (1989).

[102] R. Göbel and B. Goldsmith, "On almost-free modules over complete discrete valuation rings," *Rend. Sem. Mat. Univ. Padova,* **86**, 75–87 (1991).

[103] R. Göbel and B. Goldsmith, "On separable torsion-free modules of countable density character," *J. Algebra,* **144**, 79–87 (1991).

[104] R. Göbel and B. Goldsmith, "Cotorsion-free algebras as endomorphism algebras in L — the discrete and topological cases," *Comment. Math. Univ.* **34**, No. 1, 1–9 (1993).

[105] R. Göbel and W. May, "The construction of mixed modules from torsion modules," *Arch. Math.,* **48**, 476–490 (1987).

[106] R. Göbel and W. May, "Independence in completions and endomorphism algebras," *Forum Mathematicum,* **1**, 215–226 (1989).

[107] R. Göbel and A. Opdenhövel, "Every endomorphism of a local Warfield module of finite torsion-free rank is the sum of two automorphisms," *J. Algebra,* **233**, 758–771 (2000).

[108] R. Göbel and A. Paras, "Splitting off free summand of torsion-free modules over complete dvrs," *Glasgow Math. J.,* **44**, 349–351 (2002).

[109] R. Göbel and J. Trlifaj, *Approximations and Endomorphism Algebras of Modules,* de Gruyter Expositions in Mathematics, 41, Walter de Gruyter, Berlin (2006).

[110] R. Göbel and B. Wald, "Separable torsion-free modules of small type," *Houston J. Math.,* **16**, 271–287 (1990).

[111] H. P. Goeters, "The structure of Ext for torsion-free modules of finite rank over Dedekind domains," *Comm. Algebra*, **31**, No. 7, 3251–3263 (2003).

[112] B. Goldsmith, "Endomorphism rings of torsion-free modules over a complete discrete valuation ring," *J. London Math. Soc.,* **2**, No. 3, 464–471 (1978).

[113] B. Goldsmith, "Essentially-rigid families of Abelian p-groups," *J. London Math. Soc.,* (2), **18**, 70–74 (1978).

[114] B. Goldsmith, "Essentially indecomposable modules over a complete discrete valuation ring," *Rend. Sem. Mat. Univ. Padova,* **70**, 21–29 (1983).

[115] B. Goldsmith, "An essentially semi-rigid class of modules," *J. London Math. Soc.,* (2), **29**, No. 3, 415–417 (1984).

[116] B. Goldsmith, "On endomorphisms and automorphisms of some torsion-free modules," In: Abelian Group Theory, Gordon and Breach, New York, 417–423 (1987).

[117] B. Goldsmith, "On endomorphism rings of non-separable Abelian p-groups," *J. Algebra,* **127**, 73–79 (1989).

[118] B. Goldsmith and W. May, "The Krull–Schmidt problem for modules over valuation domains," *J. Pure Appl. Algebra,* **140**, No. 1, 57–63 (1999).

[119] B. Goldsmith, S. Pabst, and A. Scott, "Unit sum numbers of rings and modules," *Quart. J. Math.*, **49**, 331–344 (1998).

[120] B. Goldsmith and P. Zanardo, "On the analogue of Corner's finite rank theorem for modules over valuation domains," *Arch. Math.,* **60**, No. 1, 20–24 (1993).

[121] B. Goldsmith and P. Zanardo, "The Walker endomorphism algebra of a mixed module," *Proc. Roy. Irish Acad.,* Sect. A, **93**, No. 1, 131–136 (1993).

[122] B. Goldsmith and P. Zanardo, "Endomorphism rings and automorphism groups of separable torsion-free modules over valuation domains," In: *Abelian Groups, Theory of Modules, and Topology (Padua, 1997)*, Dekker, New York (1998), 249–260.

[123] K. R. Goodearl, *Ring Theory,* Marcel Dekker, New York–Basel (1976).

[124] G. Grätzer, *General Lattice Theory*, Academie-Verlag, Berlin (1978).

[125] P. Griffith, "Transitive and fully transitive primary Abelian groups," *Pacific J. Math.,* **25**, 249–254 (1968).

[126] P. A. Griffith, *Infinite Abelian Group Theory,* The University of Chicago Press, Chicago and London (1970).

[127] S. Ya. Grinshpon, "Fully invariant subgroups of separable Abelian groups," *Fundam. Prikl. Mat.,* **4**, No. 4, 1281–1307 (1998).

[128] S. Ya. Grinshpon, "Fully invariant subgroups of Abelian torsion-free groups and their lattices," *Fundam. Prikl. Mat.,* **6**, No. 3, 747–759 (2000).

[129] S. Ya. Grinshpon, "Fully invariant subgroups of Abelian groups and full transitivity," Doctoral dissertation [in Russian], Tomsk (2000).

[130] S. Ya. Grinshpon, "Fully invariant subgroups of Abelian groups and full transitivity," *Fundam. Prikl. Mat.,* **8**, No. 2, 407–473 (2002).

[131] S. Ya. Grinshpon and P. A. Krylov, "Fully invariant subgroups, full transitivity, and homomorphism groups of Abelian groups," *J. Math. Sci.*, **128**, No. 3, 2894–2997 (2005).

[132] S. Ya. Grinshpon and V. M. Misyakov, "On fully transitive Abelian groups," In: *Abelian Groups and Modules* [in Russian], No. 6, Tomsk. gos. univ., Tomsk (1986), 12–27.

[133] M. Harada, *Factor Categories with Applications to Direct Decomposition of Modules,* Marcel Dekker, New York (1983).

[134] D. K. Harrison, "Infinite Abelian groups and homological methods," *Ann. Math.,* **69**, No. 2, 366–391 (1959).

[135] D. K. Harrison, "On the structure of Ext," In: *Topics in Abelian Groups (Proc. Sympos., New Mexico State Univ., 1962)*, Scott, Foresman and Co., Chicago, Ill. (1963), 195–209.

[136] J. Hausen, "Modules with the summand intersection property," *Comm. Algebra*, **17** (1989), No. 1, 135–148.

[137] J. Hausen and J. A. Johnson, "Determining Abelian p-groups by the Jacobson radical of their endomorphism rings," *J. Algebra,* **174**, 217–224 (1995).

[138] J. Hausen, C. E. Praeger, and P. Schultz, "Most Abelian p-groups are determined by the Jacobson radical of their endomorphism rings," *Math. Z.,* **216**, 431–436 (1994).

[139] G. Hennecke and L. Strüngmann, "Transitivity and full transitivity for p-local modules," *Arch. Math.,* **74**, 321–329 (2000).

[140] I. N. Herstein, *Noncommutative Rings,* John Wiley and Sons, Inc., New York (1968).

[141] P. Hill, "Sums of countable primary groups," *Proc. Amer. Math. Soc.,* **17**, 1469–1470 (1966).

[142] P. Hill, "Ulm's theorem for totally projective groups," *Notices Amer. Math. Soc.,* **14**, 940 (1967).

[143] P. Hill, "On transitive and fully transitive primary groups," *Proc. Amer. Math. Soc.,* **22**, 414–417 (1969).

[144] P. Hill, "Endomorphism rings generated by units," *Trans. Amer. Math. Soc.,* **141**, 99–105 (1969). Errata: *Trans. Amer. Math. Soc.*, **157**, 511 (1971).

[145] P. Hill, "The classification problem," In: *Abelian Groups and Modules, Udine, 1984*, Springer, Vienna (1984), 1–16.

[146] P. Hill, M. Lane, and C. Megibben, "On the structure of p-local groups," *J. Algebra,* **143**, 29–45 (1991).

[147] P. Hill and C. Megibben, "On the theory and classification of Abelian p-groups," *Math. Z.,* **190**, 17–38 (1985).

[148] P. Hill and C. Megibben, "Axiom 3 modules," *Trans. Amer. Math. Soc.,* **295**, No. 2, 715–734 (1986).

[149] P. Hill and C. Megibben, "Torsion-free groups," *Trans. Amer. Math. Soc.,* **295**, 735–751 (1986).

[150] P. Hill and C. Megibben, "Mixed groups," *Trans. Amer. Math. Soc.,* **334**, 121–142 (1992).

[151] P. Hill and W. Ullery, "The transitivity of local Warfied groups," *J. Algebra,* **208**, 643–661 (1998).

[152] P. Hill and W. Ullery, "Isotype subgroups of local Warfield groups," *Comm. Algebra,* **29**, No. 5, 1899–1907 (2001).

[153] P. Hill and J. K. West, "Subgroup transitivity in Abelian groups," *Proc. Amer. Math. Soc.,* **126**, No. 5, 1293–1303 (1998).

[154] R. Hunter, "Balanced subgroups of Abelian groups," *Trans. Amer. Math. Soc,* **215**, 81–98 (1976).

[155] R. Hunter and F. Richman, "Global Warfield groups," *Trans. Amer. Math. Soc.,* **266**, No. 1, 555–572 (1981).

[156] R. Hunter, F. Richman, and E. Walker, "Warfield modules," *Lect. Notes Math.,* **616**, 87–123 (1977).

[157] R. Hunter, F. Richman, and E. Walker, "Existence theorems for Warfield groups," *Trans. Amer. Math. Soc.,* **235**, 345–362 (1978).

[158] R. Hunter and E. Walker, "S-groups revisited," *Proc. Amer. Math. Soc.,* **82**, 13–18 (1981).

[159] A. V. Ivanov, "A problem on Abelian groups," *Mat. sb*, **105**, No. 4, 525-542 (1978).

[160] N. Jacobson, *Structure of Rings,* Amer. Math. Soc., Providence, RI (1968).

[161] R. Jarisch, O. Mutzbauer, and E. Toubassi, "Calculating indicators in a class of mixed modules," *Lect. Notes Pure Appl. Math.,* **182**, 291–301 (1996).

[162] R. Jarisch, O. Mutzbauer, and E. Toubassi, "Characterizing a class of simply presented modules by relation arrays," *Arch. Math.,* **71**, 349–357 (1998).

[163] R. Jarisch, O. Mutzbauer, and E. Toubassi, "Characterizing a class of Warfield modules by relation arrays," *Rocky Mountain J. Math.*, **30**, No. 4, 1293–1314 (2000).

[164] R. Jarisch, O. Mutzbauer, and E. Toubassi, "Characterizing a class of Warfield modules by Ulm submodules and Ulm factors," *Arch. Math.*, **76**, 326–336 (2001).

[165] T. Jech, *Set Theory,* Academic Press, New York (1978).

[166] I. Kaplansky, *Infinite Abelian Groups,* University of Michigan Press, Ann Arbor, Michigan (1954, 1969).

[167] I. Kaplansky, "Projective modules," *Ann. Math.,* **68**, 372–377 (1958).

[168] I. Kaplansky, "Modules over Dedekind rings and valuation rings," *Trans. Amer. Math. Soc.,* **72**, 327–340 (1952).

[169] I. Kaplansky and G. Mackey, "A generalization of Ulm's theorem," *Summa Brasil. Math.,* **2**, 195–202 (1951).

[170] F. Kasch, *Modules and Rings,* Academic Press, London–New York (1982).

[171] P. Keef, "An equivalence for categories of modules over a complete discrete valuation domain," *Rocky Mountain J. Math.*, **27**, No. 3, 843–860 (1997).

[172] G. Kolettis, Jr., "Direct sums of countable groups," *Duke Math. J.,* **27**, 111–125 (1960).

[173] P. A. Krylov, "Strongly homogeneous torsion-free Abelian groups," *Sib. Mat. Zh.*, **24**, No. 2, 77–84 (1983).

[174] P. A. Krylov, "Fully transitive torsion-free Abelian groups," *Algebra Logika*, **29**, No. 5, 549–560 (1990).

[175] P. A. Krylov, "Mixed Abelian groups as modules over their endomorphism rings," *Fundam. Prikl. Mat.*, **6**, No. 3, 793–812 (2000).

[176] P. A. Krylov, "Affine groups of modules and their automorphisms," *Algebra Logika,* **40**, No. 1, 60–82 (2001).

[177] P. A. Krylov, "Hereditary endomorphism rings of mixed Abelian groups," *Sib. Math. Zh.,* **43**, No. 1, 83–91 (2002).

[178] P. A. Krylov, "The Jacobson radical of the endomorphism ring of an Abelian group," *Algebra Logika,* **43**, No. 1, 60–76 (2004).

[179] P. A. Krylov, "Endomorphism rings and structural theory of Abelian groups," Doctoral dissertation [in Russian], Tomsk (1991).

[180] P. A. Krylov and A. R. Chekhlov, "Torsion-free Abelian groups with a large number of endomorphisms," *Tr. Mat. Inst. Steklova,* Suppl. 2, 156–168 (2001).

[181] P. A. Krylov and E. D. Klassen, "The center of endomorphism ring of a split mixed Abelian groups," *Sib. Mat. Zh.*, **40**, No. 5, 1074–1085 (1999).

[182] P. A. Krylov, A. V. Mikhalev, and A. A. Tuganbaev, "Endomorphism rings of Abelian groups," *J. Math. Sci.,* **110**, No. 3, 2683–2745 (2002).

[183] P. A. Krylov, A. V. Mikhalev, and A. A. Tuganbaev, *Endomorphism Rings of Abelian Groups,* Kluwer Academic Publishers, Dordrecht–Boston–London (2003).

[184] P. A. Krylov and E. G. Pakhomova, "Abelian groups and regular modules," *Math. Notes*, **69**, No. 3, 364–372 (2001).

[185] P. A. Krylov and A. A. Tuganbaev, "Modules over discrete valuation domains, I," *J. Math. Sci.*, **145**, No. 4, 4997–5117 (2007).

[186] P. A. Krylov and A. A. Tuganbaev, "Modules over discrete valuation domains, II," *J. Math. Sci.*

[187] L. Ya. Kulikov, "On the theory of Abelian groups of arbitrary cardinality," *Mat. Sb.*, **9**, 165–182 (1941).

[188] L. Ya. Kulikov, "On the theory of Abelian groups of arbitrary cardinality," *Mat. Sb.*, **16**, 129–162 (1945).

[189] L. Ya. Kulikov, "Generalized primary groups, I," *Trudy Moscov. Mat. Obshch.*, **1**, 247–326 (1952).

[190] L. Ya. Kulikov, "Generalized primary groups, II," *Trudy Moscov. Mat. Obshch.*, **2**, 85–167 (1953).

[191] A. G. Kurosh, "Primitive torsionsfreie abelsche gruppen vom endlichen range," *Ann. Math.*, **38**, 175–203 (1937).

[192] A. G. Kurosh, *The Theory of Groups,* Chelsea Publishing Co., New York (1960).

[193] E. L. Lady, "On classifying torsion free modules over discrete valuation rings," *Lecture Notes Math.*, **616**, 168–172 (1977).

[194] E. L. Lady, "Splitting fields for torsion-free modules over discrete valuation rings. I," *J. Algebra,* **49**, 261–275 (1977).

[195] E. L. Lady, "Splitting fields for torsion-free modules over discrete valuation rings. II," *J. Algebra,* **66**, 281–306 (1980).

[196] E. L. Lady, "Splitting fields for torsion-free modules over discrete valuation rings. III," *J. Algebra,* **66**, 307-320 (1980).

[197] T. Y. Lam, *A First Cource in Noncommutative Rings,* Springer-Verlag, New York (1991).

[198] J. Lambek, *Lecture on Rings and Modules,* Blaisdell Publishing Co., Waltham (1968).

[199] M. Lane, "A new charterization for p-local balanced projective groups," *Proc. Amer. Math. Soc.*, **96**, 379–386 (1986).

[200] M. Lane, "Isotype submodules of p-local balanced projective groups," *Trans. Amer. Math. Soc.*, **301**, No. 1, 313–325 (1987).

[201] M. Lane, "Fully invariant submodules of p-local balanced projective groups," *Rocky Mountain J. Math.*, **18**, No. 4, 833–841 (1988).

[202] M. Lane and C. Megibben, "Balance projectives and Axiom 3," *J. Algebra,* **111**, 457–474 (1987).

[203] H. Leptin, "Abelsche p-gruppen und ihre automorphismengruppen," *Math. Z.,* **73**, 235–253 (1960).

[204] L. Levy, "Torsion-free and divisible modules over non-integral domains," *Canadian J. Math.,* **15**, No. 1, 132–151 (1963).

[205] W. Liebert, "Endomorphism rings of Abelian p-groups," In: *Études sur les Groupes Abéliens (Symposium, Montpellier, 1967),* Springer, Berlin (1968), 239–258.

[206] W. Liebert, "Characterization of the endomorphism rings of divisible torsion modules and reduced complete torsion-free modules over complete discrete valuation rings," *Pacific J. Math.,* **37**, No. 1, 141–170 (1971).

[207] W. Liebert, "Endomorphism rings of reduced torsion-free modules over complete discrete valuation rings," *Trans. Amer. Math. Soc.,* **169**, 347–363 (1972).

[208] W. Liebert, "Endomorphism rings of reduced complete torsion-free module over complete discrete valuation rings," *Proc. Amer. Math. Soc.,* **36**, No. 1, 375–378 (1972).

[209] W. Liebert, "The Jacobson radical of some endomorphism rings," *J. Reine Angew. Math.,* **262/263**, 166–170 (1973).

[210] W. Liebert, "On-sided ideals in the endomorphism rings of reduced complete torsion-free modules and divisible torsion modules over complete discrete valuation rings," *Symposia Mathematica,* **13**, 273–298 (1974).

[211] W. Liebert, "Endomorphism rings of free modules over principal ideal domains," *Duke Math. J.,* **41**, 323–328 (1974).

[212] W. Liebert, Endomorphism rings of Abelian p-groups," *Lect. Notes Math.,* **1006**, 384–399 (1983).

[213] W. Liebert, "Isomorphic automorphism groups of primary Abelian groups," In: *Abelian Group Theory (Oberwolfach, 1985),* Gordon and Breach, New York (1987), 9–31.

[214] P. Loth, *Classification of Abelian Groups and Pontryagin Duality,* Gordon and Breach, Amsterdam (1998).

[215] P. Loth, "Characterizations of Warfield groups," *J. Algebra,* **204**, 32–41 (1998).

[216] S. MacLane, *Homology,* Academic Press, New York (1963).

[217] S. MacLane, *Categories for the Working Mathematician,* Springer, Berlin (1971, 1998).

[218] A. Mader, *Almost Completely Decomposable Groups,* Gordon and Breach, Amsterdam (1998).

[219] V. T. Markov, A. V. Mikhalev, L. A. Skornyakov, and A. A. Tuganbaev, "Endomorphism rings of modules and lattice of submodules," In: *Progress in Science and Technology, Contemporary Problems of Mathematics, Algebra. Topology. Geometry* [in Russian], Vol. 21, All-Union Institute for Scientific and Technical Information (VINITI) Akad. Nauk SSSR (1983), 183–254.

[220] E. Matlis,"Cotorsion modules," *Mem. Amer. Math. Soc.,* **49**, (1964).

[221] E. Matlis, "The decomposability of torsion-free modules of finite rank," *Trans. Amer. Math. Soc.,* **134**, No. 2, 315–324 (1968).

[222] W. May, "Endomorphism rings of mixed Abelian groups," *Contemp. Math.,* **87**, 61–74 (1989).

[223] W. May, "Isomorphism of endomorphism algebras over complete discrete valuation rings," *Math. Z.,* **204**, 485–499 (1990).

[224] W. May, "Endomorphism algebras of not necessarily cotorsion-free modules," In: *Abelian Groups and Noncommutative Rings*, Amer. Math. Soc., Providence, RI (1992), 257–264.

[225] W. May, "Endomorphisms over incomplete discrete valuation domains," *Contem. Math.,* **171**, 277–285 (1994).

[226] W. May, "The theorem of Baer and Kaplansky for mixed modules," *J. Algebra,* **177**, 255–263 (1995).

[227] W. May, "The use of the finite topology on endomorphism rings," *J. Pure Appl. Algebra,* **163**, 107–117 (2001).

[228] W. May, "Torsion-free modules with nearly unique endomorphism algebras," *Commun. Algebra,* **31**, No. 3, 1271–1278 (2003).

[229] W. May and E. Toubassi, "Endomorphisms of rank one mixed modules over discrete valuation rings," *Pacific J. Math.,* **108**, No. 1, 155–163 (1983).

[230] W. May and P. Zanardo, " Modules over domains large in a complete discrete valuation ring," *Rocky Mountain J. Math.,* **30**, No. 4, 1421–1436 (2000).

[231] C. Megibben, "Large subgroups and small homomorphisms," *Michigan Math. J.,* **13**, 153–160 (1966).

[232] C. Megibben, "Modules over an incomplete discrete valuation ring," *Proc. Amer. Math. Soc.,* **19**, 450–452 (1968).

[233] C. Megibben, "A nontransitive, fully transitive primary group," *J. Algebra,* **13**, 571–574 (1969).

[234] B. van der Merwe, "Unique addition modules," *Comm. Algebra*, **27**, No. 9, 4103–4115 (1999).

[235] A. V. Mikhalev, "Isomorphisms and anti-isomorphisms of endomorphism rings of modules," In: *First International Tainan–Moscow Algebra Workshop, Tainan, 1994,* de Gruyter, Berlin (1996), 69–122.

[236] A. P. Mishina and L. A. Skornyakov, *Abelian Groups and Modules*, Amer. Math. Soc., Providence, RI (1976).

[237] V. M. Misyakov, "Full transitivity of Abelian groups," Candidate dissertation [in Russian], Tomsk (1992).

[238] S. H. Mohamed and B. J. Muller, *Continuous and Discrete Modules,* Cambridge Univ. Press, Cambridge (1990).

[239] G. Monk, "One sided ideals in the endomorphism ring of an Abelian p-group," *Acta Math. Acad. Sci.,* **19**, 171–185 (1968).

[240] J. H. Moore, "A characterization of Warfield groups," *Proc. Amer. Math. Soc.,* **87**, 617–620 (1983).

[241] O. Mutzbauer and E. Toubassi, "A splitting criterion for a class of mixed modules," *Rocky Mountain J. Math.,* **24**, 1533–1543 (1994).

[242] O. Mutzbauer and E. Toubassi, "Extending a splitting criterion on mixed modules," *Contemp. Math.,* **171**, 305–312 (1994).

[243] O. Mutzbauer and E. Toubassi, "Classification of mixed modules," *Acta Math. Hungar.,* **72**, No. 1–2, 153–166 (1996).

[244] M. Nagata, *Local Rings,* Wiley Interscience, New York (1962).

[245] R. J. Nunke, "Modules of extensions over Dedekind rings," *Ill. J. Math.,* **3**, 222–241 (1959).

[246] R. J. Nunke, "Purity and subfunctors of the identity," In: *Topics in Abelian Groups,* Chicago, Illinois (1963), 121–171.

[247] R. J. Nunke, "Homology and direct sums of countable Abelian groups," *Math. Z.,* **101**, 182–212 (1967).

[248] R. S. Pierce, "Homomorphisms of primary Abelian groups," In: *Topics in Abelian Groups,* Chicago, Illinois (1963), 215–310.

[249] R. S. Pierce, "Endomorphism rings of primary Abelian groups," In: *Proc. Colloq. Abelian Groups. Tihany, 1963,* Académiai Kiadó, Budapest (1964), 125–137.

[250] B. I. Plotkin, *Groups of Automorphisms of Algebraic Systems,* Wolters-Noordhoff Publishing, Groningen (1972).

[251] L. Procházka, "A generalization of a Prüfer-Kaplansky theorem," *Lect. Notes Math.,* **1006**, 617–629 (1983).

[252] H. Prüfer, "Untersuchungen über die Zerlegbarkeit der abzählbaren primären Abelschen Gruppen," *Math. Z.,* **17**, 35–61 (1923).

[253] G. E. Puninski and A. A. Tuganbaev, *Rings and Modules* [in Russian], Soyuz, Moscow (1998).

[254] P. Ribenboim, "On the completion of a valuation ring," *Math. Ann.,* **155**, 392–396 (1964).

[255] F. Richman, "The constructive theory of KT-modules," *Pacific J. Math.*, **61**, 621–637 (1975).

[256] F. Richman, "Mixed local groups," *Lect. Notes Math.*, **874**, 374–404 (1981).

[257] F. Richman, "Mixed groups," *Lecture Notes Math.*, **1006**, 445–470 (1983).

[258] F. Richman, "Nice subgroups of mixed local groups," *Commun. Algebra*, **11**, No. 15, 1629–1642 (1983).

[259] F. Richman and E. A. Walker, "Extending Ulm's theorem without group theory," *Proc. Amer. Math. Soc.*, **21**, 194–196 (1969).

[260] F. Richman and E. A. Walker, "Filtered modules over discrete valuation domains," *J. Algebra*, **199**, No. 2, 618–645 (1998).

[261] C. E. Rickart, "Isomorphic groups of linear transformations," *Amer. J. Math.*, **72**, 451–464 (1950).

[262] J. Rotman, "A note on completions of modules," *Proc. Amer. Math. Soc.*, **11**, 356–360 (1960).

[263] J. Rotman, "Mixed modules over valuation rings," *Pacific J. Math.*, **10**, 607–623 (1960).

[264] J. Rotman, "A completion functor on modules and algebras," *J. Algebra*, **9**, 369–387 (1968), Erratum, **28**, 210–213 (1974).

[265] J. Rotman and T. Yen, "Modules over a complete discrete valuation ring," *Trans. Amer. Math. Soc.*, **98**, 242–254 (1961).

[266] B. Roux, "Anneaux non commutatifs de valuation discrète ou finie, scindés, I," *C. R. Acad. Sci. Paris. Sér. I Math.*, **302**, No. 7, 259–262 (1986).

[267] B. Roux, "Anneaux non commutatifs de valuation discrète ou finie, scindés, II," *C. R. Acad. Sci. Paris. Sér. I Math.*, **302**, No. 8, 291–293 (1986).

[268] B. Roux, "Anneaux non commutatifs de valuation discrète, scindés, en caractéristique zéro," *C. R. Acad. Sci. Paris. Sér. I Math.*, **303**, No. 14, 663–666 (1986).

[269] B. Roux, "Hautes σ-dérivations et anneaux de valuation discrète non commutatifs, en caractéristique zéro," *C. R. Acad. Sci. Paris, Sér. I Math.*, **303**, No. 19, 943–946 (1986).

[270] B. Roux, "Anneaux de valuation discrète complets, scindés, non commutatifs, en caractéristique zéro," *Lecture Notes Math.*, **1296**, 276–311 (1987).

[271] L. H. Rowen, *Ring Theory. I*, Academic Press, New York (1988).

[272] L. Salce and P. Zanardo, "Rank-two torsion-free modules over valuation domains," *Rend. Sem. Mat. Univ. Padova*, **80**, 175–201 (1988).

[273] A. D. Sands, "On the radical of the endomorphism ring of a primary Abelian group," In: *Abelian Groups and Modules*, Springer, Vienna (1984), 305–314.

[274] P. Schultz, "Notes on mixed groups. I," In: *Abelian Groups and Modules, (Udine, 1984)*, Springer, Vienna (1984), 265–278.

[275] P. Schultz, "Notes on mixed groups. II," *Rend. Sem. Mat. Univ. Padova,* **75**, 67–76 (1986).

[276] P. Schultz, "When is an Abelian p-group determined by the Jacobson radical of its endomorphism ring," *Contem Math.,* **171**, 385–396 (1994).

[277] P. Schultz, "Automorphisms which determine an Abelian p-group," In: *Abelian Groups, Theory of Modules, and Topology (Padua, 1997)*, Dekker, New York (1998), 373–379.

[278] R. O. Stanton, "An invariant for modules over a discrete valuation ring," *Proc. Amer. Math. Soc.,* **49**, No. 1, 51–54 (1975).

[279] R. O. Stanton, "Decomposition bases and Ulm's theorem," *Lect. Notes Math.,* **616**, 39–56 (1977).

[280] R. O. Stanton, "Relative S-invariants," *Proc. Amer. Math. Soc.,* **65**, 221–224 (1977).

[281] R. O. Stanton, "Decomposition of modules over a discrete valuation ring," *J. Austral. Math. Soc.,* **27**, 284–288 (1979).

[282] R. O. Stanton, "Almost affable Abelian groups," *J. Pure Appl. Alg.,* **15**, 41–52 (1979).

[283] A. E. Stratton, "Mixed modules over an incomplete discrete valuation ring," *Proc. London Math. Soc.,* **21**, 201–218 (1970).

[284] A. E. Stratton, "A note on the splitting problem for modules over an incomplete discrete valuation ring," *Proc. London Math. Soc.* (3), **23**, 237–250 (1971).

[285] A. A. Tuganbaev, "Poorly projective modules," *Sib. Mat. Zh.,* **21**, No. 5, 109–113 (1980).

[286] A. A. Tuganbaev, "Serial rings and modules," *Mat. Zametki*, **48**, No. 2, 99–106 (1990).

[287] A. A. Tuganbaev, *Semidistributive Modules and Rings,* Kluwer Academic Publishers, Dordrecht–Boston–London (1998).

[288] A. A. Tuganbaev, *Distributive Modules and Related Topics,* Gordon and Breach Science Publishers, Amsterdam (1999).

[289] A. A. Tuganbaev, *Rings Close to Regular*, Kluwer Academic Publishers, Dordrecht–Boston–London (2002).

[290] H. Ulm, "Zur Theorie der abzählbar-unendlichen abelschen Gruppen," *Math. Ann.,* **107**, 774–803 (1933).

[291] R. G. Underwood, "Galois theory of modules over a discrete valuation ring," In: *Recent Research on Pure and Applied Algebra*, Nova Sci. Publ., Hauppauge, NY (2003), 23–45.

[292] P. Vámos, "Decomposition problems for modules over valuation domains," *J. London Math. Soc.,* **41**, 10–26 (1990).

[293] R. Vidal, "Anneaux de valuation discrète complets, non nécessairement commutatifs," *C. R. Acad. Sci., Paris,* Sér. A-B, **283** (1976).

[294] R. Vidal, "Un exemple d'anneax de valuation discrète complet, non commutatif quin'est pas un anneau de Cohen," *C. R. Acad. Sci., Paris,* Sér. A-B, **284** (1977).

[295] R. Vidal, "Anneaux de valuation discrète complets non commutatifs," *Trans. Amer. Math. Soc.,* **267**, No. 1, 65–81 (1981).

[296] C. Vinsonhaler and W. Wickless, "Dualities for torsion-free Abelian groups of finite rank," *J. Algebra,* **128**, No. 2, 474–487 (1990).

[297] C. L. Walker, "Local quasi-endomorphism rings of rank one mixed Abelian groups," *Lect. Notes Math.,* **616**, 368–378 (1977).

[298] C. L. Walker and R. B. Warfield, Jr., "Unique decomposition and isomorphic refinement theorems in additive categories," *J. Pure Appl. Algebra,* **7**, 347–359 (1976).

[299] E. A. Walker, "Ulm's theorem for totally projective groups," *Proc. Amer. Math. Soc.,* **37**, 387–392 (1973).

[300] E. A. Walker, "The groups P_β," In : *Symposia Mathematica XIII. Gruppi Abeliani,* Academic Press, London (1974), 245–255.

[301] K. D. Wallace, "On mixed groups of torsion free rank one with totally projective primary components," *J. Algebra,* **17**, 482–488 (1971).

[302] R. B. Warfied, Jr., "Homomorphisms and duality for torsion-free groups," *Math. Z.,* **107**, 189–200 (1968).

[303] R. B. Warfield, Jr., "A Krull–Schmidt theorem for infinite sums of modules," *Proc. Amer. Math. Soc.,* **22**, No. 2, 460–465 (1969).

[304] R. B. Warfield, Jr., "Decompositions of injective modules," *Pacific J. Math.,* **31**, 263–276 (1969).

[305] R. B. Warfield, Jr., "An isomorphic refinement theorem for Abelian groups," *Pacific J. Math.*, **34**, No. 1, 237–255 (1970).

[306] R. B. Warfield, Jr., "Exchange rings and decomposition of modules," *Math. Ann.,* **199**, 31–36 (1972).

[307] R. B. Warfield, Jr., "Classification theorems for p-groups and modules over a discrete valuation ring," *Bull. Amer. Math. Soc.,* **78**, No. 1, 88–92 (1972).

[308] R. B. Warfield, Jr., "Simply presented groups," In: *Proceedings of the Special Semester on Abelian Groups,* University of Arizona, Tucson, 1972.

[309] R. B. Warfield, Jr., "A classification theorem for Abelian p-groups," *Trans. Amer. Math. Soc.,* **210**, 149–168 (1975).

[310] R. B. Warfield, Jr., "Classification theory of Abelian groups," I: Balanced projectives," *Trans. Amer. Math. Soc.,* **222**, 33–63 (1976).

[311] R. B. Warfied, Jr., "The structure of mixed Abelian groups," *Lect. Notes Math.*, **616**, 1–38 (1977).

[312] R. B. Warfield, Jr., "Cancellation of modules an groups and stable range of endomorphism rings," *Pacific J. Math.,* **91**, No. 2, 457–485 (1980).

[313] R. B. Warfield, Jr., "Classification theory of Abelian groups, II: Local theory," *Lect. Notes Math.,* **874**, 322–349 (1981).

[314] B. D. Wick, "A projective characterization for SKT-modules," *Proc. Amer. Math. Soc.,* **80**, 39–43 (1980).

[315] B. D. Wick, "A classification theorem for SKT-modules," *Proc Amer. Math. Soc.,* **80**, 44–46 (1980).

[316] G. V. Wilson, "Modules with the summand intersection property," *Comm. Algebra,* **14**, No. 1, 21–38 (1986).

[317] R. Wisbauer, *Foundations of Module and Ring Theory,* Gordon and Breach, Philadelphia (1991).

[318] K. G. Wolfson, "An ideal-theoretic characterization of the ring of all linear transformation," *Amer. J. Math.,* **75**, 358–386 (1953).

[319] K. G. Wolfson, "Isomorphisms of the endomorphism rings of torsion-free modules," *Proc. Amer. Math. Soc.,* **13**, No. 5, 712–714 (1962).

[320] K. G. Wolfson, "Isomorphisms of the endomorphism ring of a free module over a principal left ideal domain," *Mich. Math. J.,* **9**, 69–75 (1962).

[321] K. G. Wolfson, "Isomorphisms of the endomorphism rings of a class of torsion-free modules," *Proc. Amer. Math. Soc.,* **14**, No. 4, 589–594 (1963).

[322] K. G. Wolfson, "Anti-isomorphisms of endomorphism rings of locally free modules," *Math. Z.,* **202**, 151–159 (1989).

[323] A. V. Yakovlev, "On direct decompositions of p-adic groups," *Algebra i Analiz,* **12**, No. 6, 217–223 (2000).

[324] A. V. Yakovlev, "On direct decompositions of S-local groups," *Algebra i Analiz,* **13**, No. 4, 229–253 (2001).

[325] P. Zanardo, "Kurosch invariants for torsion-free modules over Nagata valuation domains," *J. Pure Appl. Algebra,* **82,** 195–209 (1992).

[326] O. Zariski and P. Samuel, *Commutative Algebra,* I, Van Nostrand, Princeton (1958).

[327] O. Zariski and P. Samuel, *Commutative Algebra,* II, Van Nostrand, Princeton (1960).

[328] J. Žemlička, "Steadiness is tested by a single module," *Abelian groups, rings and modules,* (*Perth* 2000), 301-308, *Contemp. Math.*, **273** (2001).

[329] L. Zippin, "Countable torsion groups," *Ann. Math.,* **36**, 86–99 (1935).

Symbols

$A_1 \oplus \cdots \oplus A_n$	(finite) direct sum of modules $A_1, \ldots, A_n$
$\bigoplus_{i \in I} A_i$	direct sum of modules A_i, $i \in I$
A^s or $\bigoplus_s A$	direct sum of s copies of the module A
$\prod_{i \in I} A_i$	direct product of modules A_i, $i \in I$
$\sum_{i \in I} A_i$	sum of the submodules A_i, $i \in I$
$\overline{A}$	closure of the submodule A in the p-adic topology
$\mathrm{Aut}_R(M)$	automorphism group of the module M
$C(M)$	largest $\widehat{R}$-submodule of the module M
$\mathrm{End}_{\mathcal{E}}(A)$	endomorphism ring of an object A in an additive category $\mathcal{E}$
$\mathrm{End}_R(M)$	endomorphism ring of the R-module M
$\mathrm{End}_{\overline{W}}(M)$	endomorphism ring of the module M in the category Walk
$\mathrm{End}_{\underline{W}}(M)$	endomorphism ring of the module M in the category Warf
$\mathrm{Ext}_R(M, N)$	extension group of the R-module N by the R-module M
$\mathrm{Fin}(M)$	ideal of finite endomorphisms of the module M
$\mathrm{Fin}(M, N)$	group of finite homomorphisms from M into the module N
$f(\sigma, M)$	σth Ulm–Kaplansky invariant of the module M
$f_\sigma(M, A)$	σth Ulm–Kaplansky invariant of M with respect to A
$\varphi\|_A$	restriction of a homomorphism φ to the submodule A
$g(e, M)$	Warfield invariant of the module M
$J(S)$	Jacobson radical of the ring S
$h(e, M)$	Stanton invariant of the module M
$h_M(a)$ or $h(a)$	height of an element a in the module M
$h^*(a)$	generalized height of an element a
$\mathrm{Hom}_{\mathcal{E}}(A, B)$	totality of morphisms in a category $\mathcal{E}$
$\mathrm{Hom}_R(M, N)$	group of homomorphisms from ${}_RM$ into ${}_RN$
$\mathrm{Hom}_{\overline{W}}(M, N)$	morphism group in the category Walk
$\mathrm{Hom}_{\underline{W}}(M, N)$	morphism group in the category Warf
$\mathrm{Im}(\varphi)$	image of a module or ring homomorphism φ
$\mathrm{Ker}(\varphi)$	kernel of a module or ring homomorphism φ
$\mathrm{Ines}(M)$	ideal of inessential endomorphisms of the module M
$\mathrm{Ines}(M, N)$	group of inessential homomorphisms from M into N
$\ell(M)$	length of the module M

$\widehat{M}$	completion of the module M in the p-adic topology
mod-S	category of all right S-modules
M^1	first Ulm submodule of the module M
M^σ	σth Ulm submodule of the module M
M_σ	σth Ulm factor of the module M
$M^\bullet$	cotorsion hull of the module M
$M \otimes_R N$	tensor product of two modules M_R and $_RN$
$M[p^n]$	the submodule $\{m \in M \mid p^n m = 0\}$ of M
$o(a)$	order of an element a
$p^n M$	the submodule $\{p^n m \mid m \in M\}$ of M
$r(M)$	rank of the module M
$r_o(M)$	torsion-free rank of the module M
$r_p(M)$	p-rank of the module M
$\widehat{R}M$	$\widehat{R}$-hull of the module M
R_p	residue division ring R/pR
R_{p^n}	factor ring $R/p^n R$
$R(p^n)$	cyclic R-module $R/p^n R$
$R(p^\infty)$	quasicyclic R-module K/R
RX	the submodule of $_RM$ generated by the subset X
S-mod	category of all left S-modules
$\mathcal{TF}$	category of quasihomomorphisms
$t(M)$	torsion submodule of the module M
$U_M(a)$ or $U(a)$	height sequence of an element a in the module M
Walk	Walker category
Warf	Warfield category

Index

P

Q

R

S